Functional Fillers and Nanoscale Minerals

New Markets / New Horizons

Edited by Jon J. Kellar

Published by

Society for Mining, Metallurgy, and Exploration, Inc.

Society for Mining, Metallurgy, and Exploration, Inc. (SME)
8307 Shaffer Parkway
Littleton, Colorado, USA 80127
(303) 973-9550 / (800) 763-3132
www.smenet.org

SME advances the worldwide mining and minerals community through information exchange and professional development. SME is the world's largest association of mining and minerals professionals.

ISBN-13: 978-0-87335-247-5
ISBN-10: 0-87335-247-5

Library of Congress Cataloging-in-Publication Data

Functional fillers and nanoscale minerals : new markets/new horizons / edited by Jon J. Kellar.
p. cm.
Includes bibliographical references.
ISBN-13: 978-0-87335-247-5
ISBN-10: 0-87335-247-5
1. Nanostructured materials. 2. Nanotechnology. I. Kellar, J. J. (Jon J.)

TA418.9.N35F83 2006
620'.5--dc22 2005057607

Contents

Preface

Mineral additives are widespread in industrial manufacturing processes. Mineral fillers are used to extend raw materials and cut costs. In more recent times, minerals and associated inorganics have frequently been used for their functionality and other mineral-specific qualities.

Another rapidly emerging technological area is that of nanoscale minerals. This emergence of nanoscale minerals parallels the global pursuit of nanotechnology, and the use of nanoscale minerals will undoubtedly play an important role in low-cost, high-performance application of nanotechnology.

The Functional Fillers and Nanoscale Minerals Symposium was first introduced at the 2003 SME Annual Meeting. The success of the first symposium led to the second symposium of the same name at the 2006 SME Annual Meeting. That symposium is represented by the 21 papers in this volume. These papers include contributions by many leading international researchers and, in particular, highlight the emerging markets and applications of functional fillers and nanoscale minerals. Other topics covered include new markets/new horizons, nanocomposites and nanofluids, functional fillers, formation and characterization of nanoparticles/surfaces, and applications of nanomaterials and nanominerals.

The papers from the symposium were written for mineral suppliers, industrial users of mineral fillers, and those concerned with new trends in mineral processing and nanotechnology.

The editor is especially grateful for the hard work of the paper authors, session chairs (Drs. Srinivas Veeramasuneni, Cesar Basilio, Sharad Mathur, S.A. Ravishankar, and Courtney Young), as well as the advice and direction provided by Dr. Brij Moudgil.

PART 1

New Markets/New Horizons

Understanding Interfacial Phenomena in Clay/Polymer Systems

David J. Chaiko*

The delamination and dispersion of organoclays in hydrophobic polymeric systems involves a complex balancing of surface chemical interactions between the basal and edge surfaces of the organoclay and the polymer matrix. The proper design of surfactant coatings used to make organoclays remains a critical challenge. At a minimum, modification of the clay surface is needed to satisfy wetting requirements, but the literature is replete with examples showing that surface wetting alone is insufficient to ensure significant improvements in the physico-chemical properties of the nanocomposite. The thermodynamic factors that affect organoclay swelling and dispersion are discussed and a basis for the design of self-dispersing organoclays for polyolefin systems is presented. A new method for characterizing the charge distribution on the clay surface is presented. Preliminary data indicate that the charge sites within the clay crystal lattice can be highly segregated. This gives rise to an amphiphilic organoclay surface which, without further modification, is impossible to exfoliate in polymers like the olefins. However, when properly designed, organoclays can produce significant increases in gas barrier at concentrations that are consistent with those of commonly used polymer additives like stabilizers and colorants. Finally, the effects of isostatically-applied pressure on organoclay morphology are examined. A newly discovered organoclay phase is described that is colorless and transparent. Photographic evidence suggests that isostatic pressure induces a phase transition from a solid to a nanofluid. This new organoclay phase can be compression molded into colorless, transparent, and flexible films without the use of activators, solvents, or polymer additives.

INTRODUCTION

The compounding of fillers such as glass fibers, carbon fibers, and particulates such as calcium carbonate, metal oxides, and hydroxides into polyolefins requires surface treatment to reduce the surface energy of the filler. Exfoliation and dispersion of nanoscale particles, like the phyllosilicates and carbon nanotubes also require surface treatment to facilitate their homogeneous, nanoscale dispersion within polyolefins.

---

* Argonne National Laboratory, Argonne, Illinois

Common surface treatments include ion exchange reactions with surfactants to increase the hydrophobicity of the particle surface. For example, hydrotalcite, which is an anion exchanger, can be treated with stearic acid to promote dispersion in polymers like polyethylene and PVC. In a similar manner, montmorillonite, which is a cation exchanger, can be treated with high-molecular-weight quaternary amines such as dimethyl dihydrogenated tallow ammonium chloride, dimethyl benzyl hydrogenated tallow ammonium chloride, and methyl benzyl dihydrogenated tallow ammonium chloride to promote the formation of a low-energy surface.

During the last twenty years, a large number of research articles and patents have been published in the field of clay/polymer nanocomposites. The majority of these publications deal with various aspects of the compounding and characterization of systems which are more aptly described as intercalates, wherein the clay filler is neither exfoliated nor homogeneously dispersed at the nanoscale. Even when exfoliation in polyolefins has been achieved, usually with the assistance of dispersing aids such as maleated oligomers, there has still been a failure to produce significant improvements in gas barrier or mechanical properties. This suggests that surface wetting alone is insufficient to ensure significant improvements in the chemical and mechanical properties of the nanocomposite.

Despite a vast volume of nanocomposite literature, there are no systematic studies within this literature which pertain to the colloidal properties of mineral/surfactant systems or attempt to explain how they might affect both the exfoliation process and the physico-chemical properties of the clay/polymer system. For example, the incredible complexity of the clay ion exchange process and the characterization of the colloidal species resulting from these reactions (e.g., phase diagrams) have not received the attention they deserve. This is unfortunate, because without these types of information, we do not have a context to properly understand the data gathered in all other clay/polymer nanocomposite studies.

This paper presents an overview of the various thermodynamic and kinetic factors that can influence both the colloidal properties of organoclays and their ability to disperse and exfoliate in hydrophobic polymers like the polyolefins.

SURFACTANT-MODIFIED SMECTITE CLAYS

The preparation of a clay/polymer nanocomposite begins with the selection of an appropriate clay mineral. Jordan and his colleagues, including Houser, were the first researchers to engage in a systematic effort to prepare clay/polymer composites (Jordan 1949; Jordan, Hook, and Finlayson 1950; Jordan and Williams 1954; Hauser 1955). Their pioneering work was initiated and financed by the National Lead Company in a desire to develop value-added products from their Wyoming bentonite deposits. The approach was based on the earlier work of Smith, during the 1930s, in which organophilic bentonites were prepared by ion exchange with organic salts (Smith 1936).

Bentonite is a smectite clay having a 2:1 layer structure in which a two-dimensional octahedral sheet, containing Al^{3+} as the principle cation, is sandwiched between tetrahedral sheets composed of SiO_4. A permanent negative charge, within either the octahedral or tetrahedral sheet, is generated by isomorphous substitution of the principal cations. For example, in the Wyoming bentonite sold by Southern Clay Products (Cloisite Na^+) the lattice charge arises from the substitution of Fe^{2+} within the octahedral sheet. The base exchange capacity (CEC) of the clay is approximately 100–120 milliequivalents/100 g clay.

TABLE 1 Chemical analysis (wt %) after ion exchange of a poly(ethylene glycol) treated clay with 110 milliequivalents of didecyldimethyl ammonium bromide/100 g clay

Na	Ca	N	Amine	PEG	LOI
0.11	0.05	1.2	27.9	10.6	38.6

The basal charge density can be distributed in one of three possible modes: regular, random, or segregated. To help better characterize the structure of the clay lattice, Lagaly (Lagaly 1994) developed a technique to measure the magnitude and the homogeneity of charge distribution. In his approach, this information is derived from measurements of the increase in interlayer swelling as a function of the number of carbon atoms in n-alkyl amines. With his technique, Lagaly found that montmorillonites with CEC values between 70 and 110 milliequivalents/100 g clay display an average area per base-exchange site of about 77.7 Å^2 (Lagaly 1994). However, it is important to recognize that in natural montmorillonites, homogeneous charge distribution appears to be the exception rather than the norm. Improved analytical techniques for quantifying the degree of interlayer charge segregation are desperately needed, as this property will have a profound impact on the ability to disperse and exfoliate organically-modified clays in polymer systems. Fortunately, it appears that the degree of charge segregation on the basal surface can be derived from competing ion exchange/polymer adsorption reactions, as the following discussion illustrates.

It is known that complete monolayer coverage by poly(ethylene glycol) on Na^+-bentonites occurs at approximately 0.27 g of polymer/g of clay. The intercalate structures that are produced with sodium bentonites display a characteristic basal spacing of 17.7 Å (Aranda and Ruiz-Hitzky 1992; Chaiko 2003b) regardless of the polymer's molecular weight. A complete monolayer is produced at this polymer/clay weight ratio. When the polymer loading is expressed in terms of surface area, 0.27 g of poly(ethylene glycol), in a monolayer configuration, covers approximately 800 m^2 of basal surface area.

Recently published data (Chaiko 2003a) indicate that ion exchange, with quaternary amines like didecyldimethyl ammonium bromide, results in the partial removal of poly(ethylene glycol) polymer from the pretreated clay surface. Furthermore, this loss was in direct proportion to the extent of ion exchange by the quaternary amine. Exchange with didecyldimethyl ammonium bromide resulted in expansion of the basal spacing from 17.7 to 25.9 Å. This increase in basal spacing indicates that the quaternary amine adopts a paraffin type morphology (i.e., hemimicelle structure) and that the alkyl chains from adjacent platelets become fully interdigitated after the organoclay is filtered and dried. During the ion exchange of the polymer-coated clay, the polymer concentration on the basal surface was reduced from complete monolayer coverage of 0.27 g polymer/g clay to partial coverage of 0.173 g polymer/g clay (i.e., 10.6 wt % relative to the weight of organoclay) (see Table 1). Within experimental error, the remaining equivalents of poly(ethylene glycol) agree stoichiometrically in a 1:1 mole ratio with the equivalent cationic charge arising from sodium and calcium ions left on the organoclay (see Table 1).

These data suggest that the primary mode of attachment for the polymer is through ion dipole interactions and that the polymer leaves the surface of the clay in proportion to the loss of sodium and calcium ions. Poly(ethylene glycol) is known to form crown ether structures with various cations including Na^+ and Ca^{2+}, but there is no evidence that it forms such structures with high-molecular-weight ammonium salts

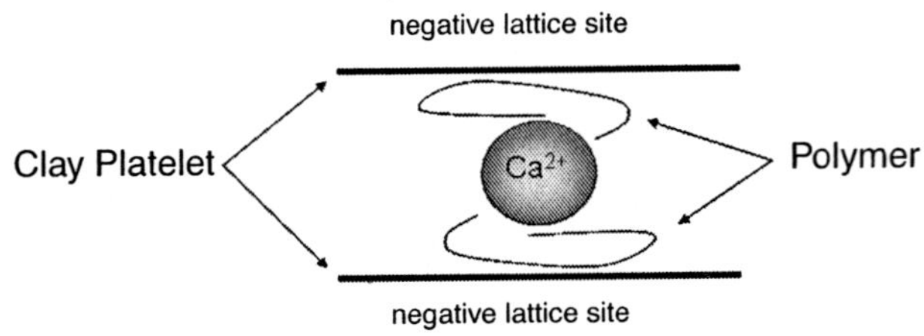

FIGURE 1 Pinning sites generated between adjacent clay platelets by multicharged ions such as calcium

like didecyldimethyl ammonium bromide. Since multivalent ions such as Ca^{2+} can generate pinning sites that lock adjacent clay platelets together (Shi, Lan, and Pinnavaia 1996), the data suggest that two adjacent platelets each contribute one polymer molecule to the calcium complex (see Figure 1). The poly(ethylene glycol) remaining on the clay (i.e., 0.173 g/g clay) would be sufficient to cover approximately 64% of the total basal surface area. This is 447 m^2/g clay.

The direct association of poly(ethylene glycol) with Na^+ and Ca^{2+} ions suggests that these ions are located at what might be regarded as non-exchangeable sites. Furthermore, they appear to be isolated and thereby able to accommodate the presence of the poly(ethylene glycol) without interfering with the packing of the quaternary amine monolayer The association of only one polymer chain per Na^+ ion and the basal spacing of 25.9 Å, both suggest that these particular sites are predominantly located at external surfaces rather than within the organoclay galleries. Isolated sodium ions may be more difficult to exchange with quaternary amines since they offer no opportunity for the stabilizing effect of the chain-chain interactions (e.g., hemimicelle formation) that occur in densely-packed surfactant monolayers.

The implications of charge segregation are indeed profound. For example, extensive charge segregation and hemimicelle formation would give rise to a two-dimensional, amphiphilic structure. The complex colloidal behavior of organoclays can now be viewed in terms of their amphiphilic surface structure and their tendency to form lamellar liquid crystal (LLC) phases. It should be noted that the case for a two-dimensional amphiphilic structure is supported by the fact that fully-exchanged organoclays can spread and form monolayers at air/water interfaces (Slabaugh 1964; Kotov et al. 1994). This spreading behavior was never adequately explained for an organoclay surface that was believed to be homogeneous and hydrophobic.

Using the data in Table 1, we can estimate the total surface area occupied by the polymer and the quaternary amine. This value is approximately 824 m^2/g clay, and it was arrived at by summing the area occupied by 0.172 g polymer (i.e., 447 m^2) and the area covered by the quaternary amine hemimicelle(s)–377 m^2. The area per molecule for the quaternary amine that is used in this calculation is 45 Å^2, which is the value obtained for dimethyl dioctadecylammonium chloride monolayers spread at the air/water interface (Slabaugh 1964). The calculated area of 824 m^2/g clay agrees remarkably well with the commonly accepted value of approximately 800 m^2/g bentonite. More refined estimates of hydrophilic/lipophilic domain sizes should be made possible by conducting a series of measurements with varying polymer molecular weights. As the polymer molecular weight is increased and the area per polymer molecule begins to exceed the area surrounding the isolated charge sites, we should begin to see deviations from the 1:1 cation charge/polymer stoichiometry.

Given the occurrence of a segregated charge distribution, we can expect fully exchanged, hydrophobic organoclays to be difficult to dewater. We can also expect an amphiphilic structure to make it difficult, if not impossible, to disperse and exfoliate organoclays in nonpolar solvents and hydrophobic polymers. The case for a two-dimensional amphiphilic surface is indeed supported by the difficulty encountered in drying organoclays to less than several weight percent moisture, and the fact that solids concentrations above about 25 wt % are rarely reached in organoclay filter cakes. Furthermore, the exfoliation of organoclays in nonpolar solvents, like toluene, xylene, and aliphatic hydrocarbons, requires the simultaneous use of polar activators, heat, and high-shear mixing (Jordan, Hook, and Finlayson 1950).

PHASE BEHAVIOR IN THE ORGANOCLAY/OIL SYSTEM

In general terms, it should be possible to predict the phase behavior of organoclay/oil and organoclay/polyolefin systems from the corresponding oil/water/surfactant phase diagrams. A schematic phase diagram is shown in Figure 2. While the micelle phases are not accessible in the organoclay system, the structural characteristics of lamellar phases in the clay/surfactant and water/surfactant systems are virtually identical. The LLC phase is isolated within the center of the ternary diagram (see Figure 2) and connected by tie lines to the oil phase. The modest size of the LLC domain indicates that these types of phases have a limited ability to solubilize organics via intercalation. Even in highly swollen LLC phases, where the hydrocarbon chains of the surfactant adopt a fully extended conformation, the terminal methyl groups of the surfactant chains always remain in contact except in special cases (Ward et al. 1984). This is understood to be due to entropic effects, wherein the surfactant chains exhibit an order parameter that is intermediate between that of a liquid and a solid (Ward et al. 1984). In other words, the LLC structure is not thermodynamically compatible with an isotropic organic phase.

While organoclays will spontaneously swell in the presence of a suitable organic solvent (e.g., C_{12-18} alkanes), they will not spontaneously exfoliate in an excess of the solvent. Thus, it can be expected that the same entropic effects that control the swelling and dispersion of LLC surfactant phases will also present a thermodynamic barrier preventing the unlimited swelling and exfoliation of organoclays in polymer systems. As we shall see, other variables which play an important role in determining phase behavior of organoclays are temperature and pressure. These will be discussed in the next two sections.

Thermal Transitions

The first order melt transition observed in surfactant LLC systems is also detectable in organoclays (Chaiko and Leyva 2005). For extensive solvation of the surfactant hydrocarbon layer to occur, the hydrocarbon layer must be in a non-crystalline state. Therefore, to promote miscibility of the organoclay with a solvent, wax, or polymer, it would seem prudent to match the thermal transitions of the organoclay surface with those of the organic matrix. It is especially important to match the freezing point of the organoclay surface with that of the polymer matrix to prevent phase separation during cooling from melt processing temperatures (Chaiko and Leyva 2005).

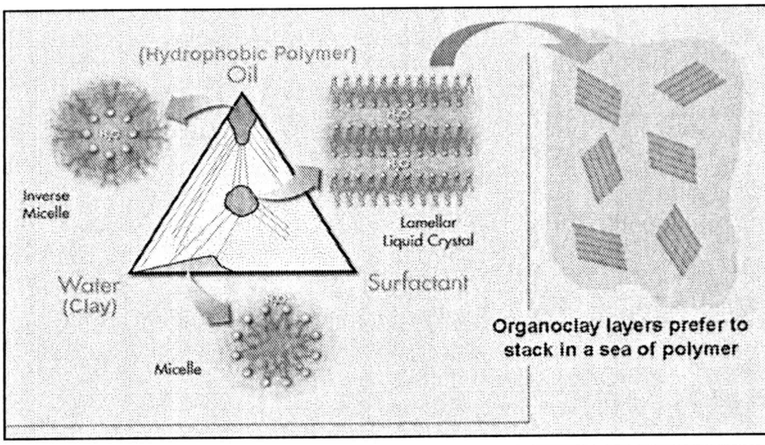

FIGURE 2 Schematic diagram of the surfactant/oil/water system and its analogy with the surfactant/clay/polymer system

Ammonium salts containing single, long-chain hydrocarbon radicals display high melting points that more closely match those of the polyolefins than do surfactants containing two or more long alkyl chains. The effect of chain length(s) on the melting point of the ammonium salts is illustrated in Table 2. The general trend in melting temperature is: trimethyl alkyl ammonium > triethyl alkyl ammonium > dimethyl dialkyl ammonium > tertiary amine. Ammonium salts having melting points above the Hoffman degradation temperature (~150°C) generally decompose at their respective melting points. The majority of the long-chain surfactants in Table 2 are stable to very high temperatures but dispersion and exfoliation at temperatures below their respective melting points would be problematic.

To a certain degree, the nature of the anion affects the melting point of the ammonium salt (see Table 2). When an organoclay prepared by ion exchange with an ammonium salt, the negatively charged clay surface becomes the anion in a newly formed, complex ammonium salt. While the scientific literature is silent on this subject, it appears that the melting points of organoclays are similar to those of the ammonium halide salts. For example, the differential scanning calorimetric measurements of an organoclay prepared from cetyl trimethyl ammonium bromide do not display a melt transition at temperatures below about 200°C (see Figure 3). The data in Figure 3 were collected from a fully-dried organoclay sample.

Pressure-Induced Transitions

Since organoclays are compressible, we may expect pressure to be an important variable which can influence not only the kinetics of swelling and exfoliation but also the stability fields of the phase domains. Surprisingly, there are no reports in the literature pertaining to the study of pressure on the colloidal properties of organoclays. This is surprising, because these types of data are certainly relevant to the organoclay processing conditions encountered during extrusion compounding of clays into polymers.

TABLE 2 Melting points of amines and ammonium salts

Amine	Melting Point, °C
Hexadecyl trimethyl ammonium bromide	>230 (dec)
Octadecyl trimethyl ammonium bromide	250 (dec)
Hexadecyl dimethyl ethyl ammonium chloride	190 (dec)
Triethyl methyl ammonium bromide	307 (dec)
(2-hydroxyethyl) trimethyl ammonium chloride	302 (dec)
Triethyl hexyl ammonium bromide	114–117
Dodecyl amine	30–32
Trimethyl phenyl ammonium bromide	237 (dec)
Trimethyl phenyl ammonium chloride	215 (dec)
Trimethyl phenyl ammonium iodide	227 (dec)

(dec) indicates a decomposition temperature

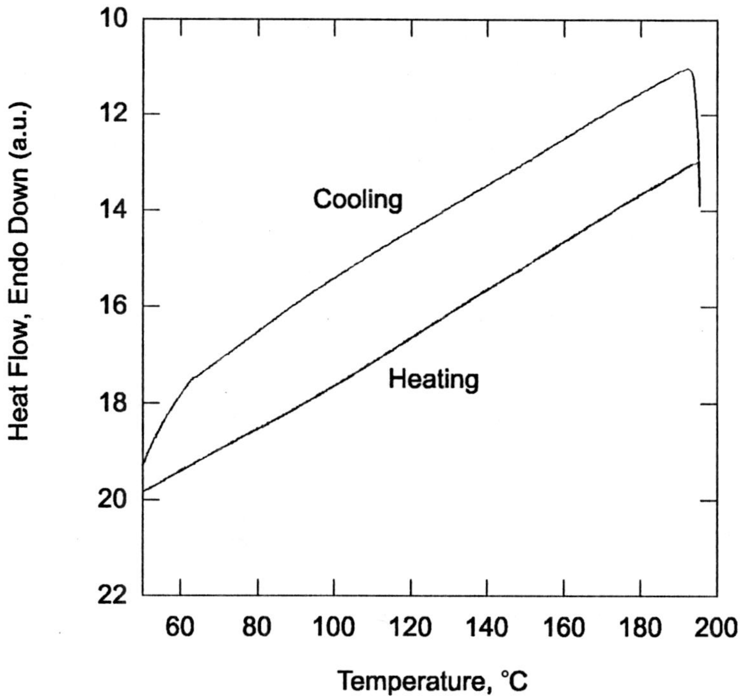

FIGURE 3 DSC measurements of cetyl trimethyl ammonium bentonite. Note the lack of any melt/freeze transitions.

When contemplating the application of pressure we have a choice between uniaxially or isostatically applied force. For example, the high-shear environment of an intermeshing, twin-screw extruder is an example of a uniaxially compressed system combined with dispersive mixing. If one were to compression mold (i.e., uniaxial pressing) a commercial organoclay, as with a hydraulic press, the result would be the production of a brittle, white, opaque film. The white color and the opacity are due to extensive particle-particle interactions and particle agglomerates with length scales greater than the wavelengths of visible light. Conversely, we have found that compression molded films of organoclays containing an appropriate hydrophobic edge treatment are transparent, dark green, and possess a sharp d_{001} reflection (Chaiko 2002). In other words, these edge-activated organoclays display certain physico-chemical properties found in surfactant LLC systems. They are capable of fluid-like flow but only under extremely high yield stresses. And, like surfactant LLC phases, these organoclays become transparent to visible light when the clay platelets are aligned in a single plane.

In contrast to uniaxial compression, isostatic compression of edge-modified organoclays has revealed the existence of an entirely new organoclay phase. While not yet fully characterized, the photographs in Figure 4 illustrate some of the unique physical properties of this new phase. The colorless and transparent particles in Figure 4a, b are approximately 1–2 mm in length and were produced during isostatic compression of the neat organoclay at room temperature. The condensed particles are strong enough to be readily isolated from the organoclay powder by screening. Figure 4c shows the flexible, transparent, and colorless film that was produced by compression molding a collection of individual particles at 130°C. The film displayed in Figure 4c consists solely of an organoclay that was prepared from a yellow-green Wyoming bentonite!

There is also evidence that isostatic compression at room temperature, induces a transition between solid and nanofluid states. It should be noted this sample was slowly compressed over the course of several hours to prevent compressive heating of the organoclay. In Figure 4d, gas bubbles are clearly evident inside the condensed organoclay phase. This can be interpreted as evidence that under pressurized conditions, the organoclay becomes a highly-viscous nanofluid capable of trapping compressed gasses. At room temperatures and pressures, the condensed phase is hard and insoluble in mineral oil, but it readily dissolves to produce a colorless, transparent nanocomposite at temperatures above 100°C.

As opposed to uniaxial compression with shear, isostatic compression offers the possibility of preserving the fractal structure of the organoclay floccs. Processing below the melting temperature of the organoclay may help preserve the fractal structure generated during ion exchange, thereby preserving the large surface areas inherent in these materials. Nevertheless, complete mapping of the phase diagram would require examining the influence of composition, temperature, and pressure.

Isostatic compression may offer a simple and direct route to the preparation of new types of mineral-based nanofluids displaying unique properties. Once in a nanofluid state, it may be possible to admix the organoclay with other components such as polymers, oligomers, and monomers.

DESIGNING FUNCTIONAL ORGANOCLAYS

In this last section, we will examine the potential benefits of nanocomposite systems with respect to gas barrier improvement. In the face of characterization data suggesting that the charge distribution within the clay lattice is highly isolated, we are challenged to develop organoclay chemistries that either account for the physico-chemical properties arising from two-dimensional amphiphilic structures or eliminate them altogether. Another issue that we must address is the limited ability of LLC phases to swell and exfoliate. Longer term, we must learn more about the physico-chemical properties of clay-based nanofluids as they offer a simple route to nanocomposite production.

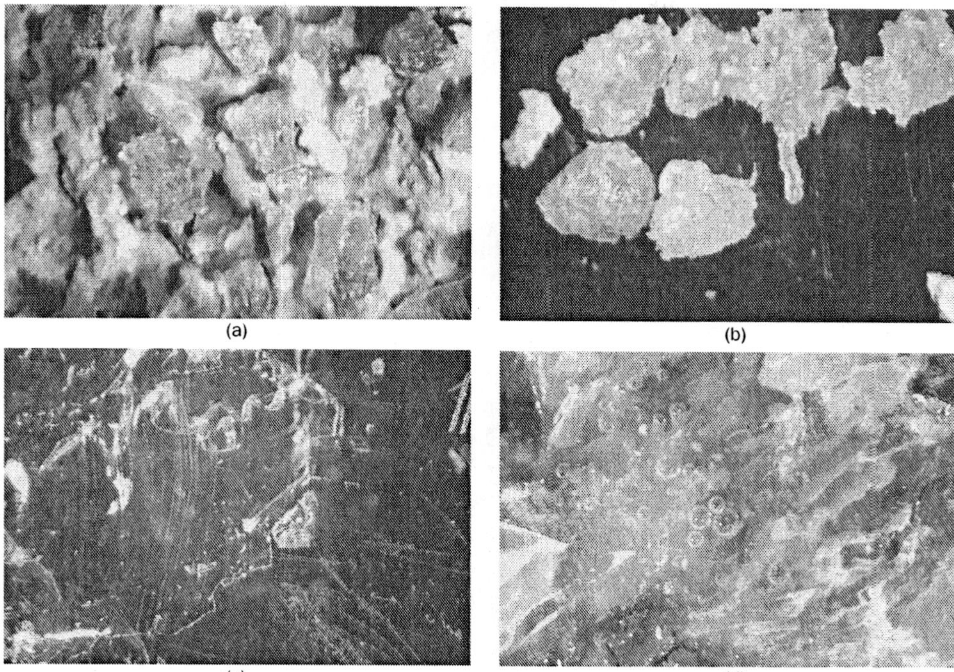

FIGURE 4 Transformations of organoclays produced by isostatic compression: a) colorless clay phase condensed from the neat organoclay phase, b) isolated colorless organoclays phase, particle sizes are 1–2 mm, c) uniaxial compression of the colorless phase at 120°C produces colorless, transparent, and flexible films, d) presence of trapped gas in the condensed clay phase suggests that the organoclay becomes a nanofluid under isostatic compression

To address the problems associated with the incompatibility of LLC phases with isotropic phases in addition to the isolated charge distribution within the clay crystal lattice, a hybrid organoclay incorporating both surfactant and polymer modification was prepared. A Wyoming bentonite (Cloisite Na^{+}) was exchanged with dimethyl dihydrogenated tallow ammonium chloride after hydrophobic treatment of the clay edges with 1-hydroxydodecane-1,1-diphosphonate. However, instead of using poly(ethylene glycol), polyethylene-*block*-poly(ethylene glycol) was used to solvate the isolated charge sites on the clay surface. The copolymer had a polyethylene chain length of approximately 50-carbons and an HLB value of 10 (Sigma-Aldrich). Powder diffraction measurements of the resulting organoclay, using a conventional laboratory diffractometer, revealed the lack of d_{001} reflections. This indicates that the amine chains are fully extended in a lipid bilayer structure. In addition to solvating isolated cation sites, the adsorption of the block copolymer also serves to disrupt the LLC structure and more the freezing temperature of the organoclay surface closer to that of the polyethylene.

When melt compounded into high density polyethylene (HDPE), the organoclay produced a colorless nanocomposite even with clay concentrations of 50 wt% or higher. This lack of color indicates that photonic effects generated by the Fe^{3+}/Fe^{2+} redox couple within the clay crystal lattice (Keller 1953) are somehow dependent on the lamellar spacing of the clay platelets. The highly-ordered phases, which result from extremely high clay loading, appear to be acting as photonic bandgap (PBG) structures (Yablonovitch 1993). All PBG structures are highly ordered and can display unique photonic effects when periodicity appears in one-, two-, or three-dimensions. In PBG structures, photons of certain frequencies are forbidden. If these highly ordered nanocomposites are in fact 1-dimensional PBG structures, we should expect the appearance or disappearance of color to be dependent upon the basal spacing of the organoclay, the intercalate, and nanocomposite phases. In fact, we have observed that highly-ordered organoclay systems can appear either colorless or display a range of colors including green, yellow, brown, and black. From x-ray measurements of these materials, we have discovered that their color depends on both the basal spacing and the length scale over which the one-dimensional ordering occurs. In fact, we have succeeded in producing colorless, transparent intercalates in polypropylene at organoclay concentrations of 40 wt%. In contrast, materials of identical composition, but which were randomly ordered over the length scales of visible light, were dark green and opaque (Chaiko 2005).

In typical clay/polymer nanocomposite formulations, the organoclay is present at concentrations of 1 to 10 wt%. In fact, Nanocor recommends 6 wt% as an optimum concentration in their product literature (Nanocor 2002). To illustrate the potential impact of full exfoliation, the organoclay modified with polyethylene-*block*-poly(ethylene glycol) was diluted with HDPE, by melt compounding, to a final inorganic clay concentration (i.e., ash content) of just 0.3 wt%. Compression molded films were then produced and the water vapor transmission rates (WVTR) at 30°C were measured for the nanocomposite and a reference HDPE film. Transmission rates were calculated from the steady-state portions of the transmission curves and then normalized for film thickness. The resulting permeabilities for the neat polymer and the nanocomposite were 0.065 and 0.02 g·mil/100 in^2·d·atm, respectively. Reducing the WVTR by 70% in HDPE would be considered a significant achievement at any clay loading. It is all the more impressive that this performance was achieved at a clay loading of less than 1 wt%. Water vapor transmission data are not reported on the Nanocor website for polyolefin homopolymers. In fact, a search of the literature failed to find any report of increased water vapor barrier in organoclay/HDPE systems. The fact that this type of performance can be attained with only 3,000 ppm clay is remarkable and fully justifies continued research into the production of clay/polymer nanocomposites.

CONCLUSIONS

There are a number of interesting conclusions that we can draw from the results of the very simple experiments described in this paper. First and foremost, the fact that even though 70 years have passed since the first organobentonites were prepared, there is indeed more to learn about the colloidal properties of clay/surfactant systems. Secondly, it should be possible to use the color of the organoclay crystal lattice along with its intensity as a simple analytical tool to quantify the ratio between the exfoliated and unexfoliated states. As discussed, the characteristic color of the organoclay, while arising from the isomorphous substitution within the crystal lattice, appears to be dependent upon the basal spacing of the organoclay and the long-range ordering of intercalate structures. Observations have shown that the organoclays prepared from

Wyoming bentonites become colorless when the galleries are expanded to the lipid bilayer and exfoliated states.

The electro-optical properties of these nanoscale materials certainly deserve more attention. In addition to applications in polymer nanocomposites, highly dispersible organoclays may also find applications in nonlinear optics, switching devices, liquid crystal displays, and sensors. Without doubt, the newly identified nano-fluid states that are generated by compressing neat organoclays deserve further attention from the nanotechnology community. Their ability to trap and hold gases under high pressure would make them useful in gas storage and sequestration if gas transmission rates within these phases are low.

Regarding the use of nanoclays to enhance the barrier properties of the polyolefins, it is significant that a 70% reduction in the WVTR of HDPE was achieved. It is all the more impressive that this performance was achieved at clay loadings of only 3,000 ppm. This concentration is consistent with those of other polymer additives, such as clarifiers, stabilizers, and dyes. In this regard, highly dispersible mineral phases can truly become high-performance additives. There should be no doubt that significant commercial development of clay-based nanocomposites will occur as the colloidal properties of the minerals and their organically-modified derivatives become better understood.

ACKNOWLEDGMENTS

This work was supported by the DOE's FreedomCAR & Vehicle Technologies Office under contract No. W-31-109-ENG-38 and by the Chemical Engineering Division, Argonne National Laboratory. The alkyl diphosphonate was provided by Dr. Tim Hirzel, at Solutia.

REFERENCES

Aranda, P. and Ruiz-Hitzky, E. 1992. Poly(ethylene oxide)-silicate intercalation compounds. *Chemistry of Materials*. 4:1395.

Bailey, S.W. 1980. Structures of layer silicates. In *Crystal Structures of Clay Minerals and Their Identification*. Edited by G.W. Brindley and G. Brown. London, Mineralogical Society. 1–124.

Chaiko, D.J. 2002. The colloid chemistry of organoclays. In *Affordable Materials Technology—Platform to Global Value and Performance*. Edited by B.M. Rasmussen, L.A. Pilato, and H.S. Kliger. Long Beach CA, Society for the Advancement of Material and Process Engineering. 1064.

Chaiko, D.J. Process for the preparation of organoclays. U.S. Patent No. 6,521,678 (18 February 2003).

Chaiko, D.J. 2003b. New poly(ethylene oxide)–clay composites. *Chemistry of Materials*. 15:1105.

Chaiko, D.J. 2005. Unpublished data.

Chaiko, D.J. and Leyva, A.A. 2005. Thermal transitions and barrier properties of olefinic nanocomposites. *Chemistry of Materials*. 17:13.

Hauser, E.A. 1955. *Silicic Science*. New York, D. Van Nostrand Company.

Jordan, J.W. 1949. Organophilic bentonites. I swelling in organic liquids. *Journal of Physics and Colloid Chemistry*. 53:294.

Jordan, J.W., Hook, B.J., and Finlayson, C.M. 1950. Organophilic bentonites. II organic liquid gels. *Journal of Physics and Colloid Chemistry*. 54:1196.

Jordan, J.W. and Williams, F.J. 1954. Organophilic bentonites. III inherent properties. *Kolloid-Zeitschrift*. 137:40.

Keller, W.D. 1953. Illite and montmorillonite in green sedimentary rocks. *Journal of Sedimentary Petrology*. 23:3.

Kotov, N.A., Meldrum, F.C., Fendler, J.H., Tombácz, E., and Dékány, I. 1994. Spreading of clay organocomplexes on aqueous solutions: construction of langmuir-blodgett clay organocomplex monolayer films. *Langmuir*. 10:3797.

Lagaly, G. 1994. Layer charge determination by alkylammonium ions. In *Layer Charge Characteristics of 2:1 Silicate Clay Minerals*. Edited by A.R. Mermut Boulder, CO: The Clay Minerals Society. 2.

Nanocor. 2002. Polyolefin Nanocomposites, Technical data sheet, P-807. Arlington Heights, IL

Shi, H., Lan, T., and Pinnavaia, T.J. 1996. Interfacial effects on the reinforcement properties of polymer–organoclay nanocomposites. *Chemistry of Materials*. 8:1584.

Slabaugh, W.H. 1964. The spreading of organo-bentonites on water. *Journal of Colloid Science*. 19:165.

Smith, C.R. Compounds of bentonite with organic bases and process of producing same. U.S. Patent No. 2,033,856 (10 March 1936).

Ward, A.J.I., Friberg, S.E., Larsen, D.W., and Rananavare, S.B. 1984. Order parameters of hydrocarbons in a lamellar liquid crystal. *Journal of Physical Chemistry*. 88:826.

Weyl, W.A. 1951. Light absorption as a result of an interaction of two states of valency of the same element. *Journal of Physics and Colloid Chemistry*. 55:507.

Yablonovitch, E. 1993. Photonic band-gap structures. *Journal of the Optical Society of America B*. 10:283.

Characterization Issues in Evaluating Toxicity of Nanoparticles: Size and Shape

Kevin W. Powers*

This paper describes the issues relating to the measurement of nanoparticle size, shape and dispersion when evaluating the toxicity in nanostructured materials. A short summary is provided of the recent findings of the first national nanotoxicity workshop held at the University of Florida in November 2004. Toxicologists often need assistance in understanding and accommodating many of the unique attributes of nanoscale materials as they begin to assess potential health and environmental effects. There is a growing consensus that the complexity of these issues requires a multidisciplinary approach to nanoparticle toxicology that includes biomedical personnel, environmental and physical scientists as well as engineers trained in particle technology.

INTRODUCTION

Nanostructured materials promise to revolutionize many key areas of science and technology. As these materials become more common in industrial applications, there is a need to assess the effects of these materials on human and biological organisms (Royal Society 2004). In an unusual display of unity, environmental groups, government regulatory agencies, and even the commercial sector itself all have recommended caution in promoting the widespread use of free nanoparticulate materials before potential ill effects are addressed (Weiss 2004). As biological and environmental scientists begin to examine the interactions of nanomaterials with living organisms, there is a growing consensus that complete and accurate particle characterization is an essential part of assessing the potential toxicity of nanoparticles in biological systems. Because the key parameters affecting biological activity are at this point largely unknown, characterization of test materials should be as complete as practical (Thomas and Sayre 2005; Oberdorster et al. 2005). Complete characterization includes such measurements as size and

* Particle Engineering Research Center, University of Florida, Gainesville, Florida

size distribution, shape and other morphological features (e.g., crystallinity, porosity, surface roughness), bulk chemistry of the material, solubility, surface area, state of dispersion, surface chemistry and other physico-chemical properties. Exhaustive characterization of test materials will be time consuming, expensive and complex. However, the extent of testing should be tempered with judgment.

In most cases only the properties needed to describe the material or those expected to affect toxicity need to be examined. But how does the researcher know which properties to focus on? How and at what point in the dosimetry–response profile should these properties be measured? Is it sufficient to simply measure the as-received material? Is it possible to measure the desired properties in situ? Can we adequately reconstruct particle properties from post mortem or histological studies? Do we need to address each nanoparticulate system individually or can properties or behavior be deduced by knowledge derived from other studies? These are all important questions and answering them will require continued investment from the scientific, regulatory, and commercial communities.

University of Florida and NIEHS Workshop 2004: Developing Experimental Approaches for the Evaluation of Toxicological Interactions of Nanoscale Materials

In November 2004, an international workshop was held at the University of Florida to discuss issues concerning the evaluation of manufactured nanomaterials for environmental and human toxicity (Roberts et al. 2004). International experts representing the fields of toxicology, the physical sciences and particle technology met to discuss these issues within the context of the following specific breakout topics: a) characterization and dosimetry of nanoscale materials, b) delivery of nanoscale materials to test systems, c) toxicology study protocols appropriate for nanoscale materials, d) detection and quantification of nanoscale materials in test systems/organisms and the environment, e) laboratory safety and disposal issues, and f) specific issues related to uptake and toxicity to the respiratory, skin, and immune systems. The published workshop report is available at the University of Florida website www.nanotoxicology.ufl.edu.

Over the course of the workshop, several central themes emerged from the presentations and discussions:

- It was recommended that that the physical and chemical characterization of nanoscale materials in these studies be much more complete than has been the case in the toxicology literature appearing to date.
- Most participants agreed that "nanotoxicology" need not be a new scientific discipline. While nanoscale materials clearly have unique physical and chemical properties that may lead to unpredictable distribution and effects within biological systems, there was general agreement that the manifestation of biological interactions of nanoscale materials will likely be the same as for any potentially hazardous agent.
- Participants stressed the need to approach nanotoxicology studies from a multidisciplinary approach and recommended that government agencies explore ways to create and promote linkages between toxicologists and experts in materials science, physics, chemistry and other appropriate disciplines.

The workshop participants generally agreed that there are a number of fundamental properties that should be measured for test materials used in nanotoxicity studies. These include size and shape, state of dispersion, bulk physical and chemical properties, surface chemistry and surface area. There are many standard analytical techniques for measuring bulk and surface physical and chemical properties. However quantifying nanoparticle size, shape and agglomerate structure are problematic due to a propensity for these properties to change, particularly with regard to the state of dispersion. The following sections focus on the issues of size, shape and dispersion along with methodologies available to quantify them.

PARTICLE SIZE DISTRIBUTION

Definition of Size

All particles have size and shape. The size of the particulates must first be defined for us to classify them as nano materials. The definition of a nanoscale material has been generally agreed upon to include any material with at least one dimension smaller than 100 nanometers (Roco 2001; Royal Society 2004):

> The size range that holds so much interest is typically from 100nm down to the atomic level (approximately 0.2 nm), because it is in this range (particularly at the lower end) that materials can have different or enhanced properties compared with the same materials at a larger size. The two main reasons for this change in behaviour are an increased relative surface area, and the dominance of quantum effects.
>
> *Nanosciences and Nanotechnologies: Opportunities and Uncertainties,*
> Royal Society Report, 29 July 2004

Particulate systems are only occasionally perfect spheres and even less often monodispersed. Despite this fact, in the particle technology community, particle size is most often defined as the diameter of a sphere which is equivalent in the selected property (e.g., mass), to the particle measured. This "equivalent sphere" model enables one to conveniently plot size distributions of irregularly shaped systems using a single value (diameter) along a single axis. The property most often described is volume, that is, the diameter of a sphere of equal volume to the particle(s) in question, or volume diameter (Allen 2004a). *Note: There are other equivalent spheres such as aerodynamic diameter (diameter of a spherical particle with the equivalent aerodynamic behavior), sedimentation diameter, mobility diameter, etc.*

Size distributions are normally depicted as a log-normal histogram with particle diameter on the abscissa and the ordinate representing the quantity of particles in a given size class. The selection of size classes is arbitrary but most instrument manufacturers use geometrically increasing bins. The quantity axis can depict any property but most often is selected to represent the relative volume of particles in each size class, the relative number of particles in each size class or the calculated surface area of the particles in that size class (See Figure 1).

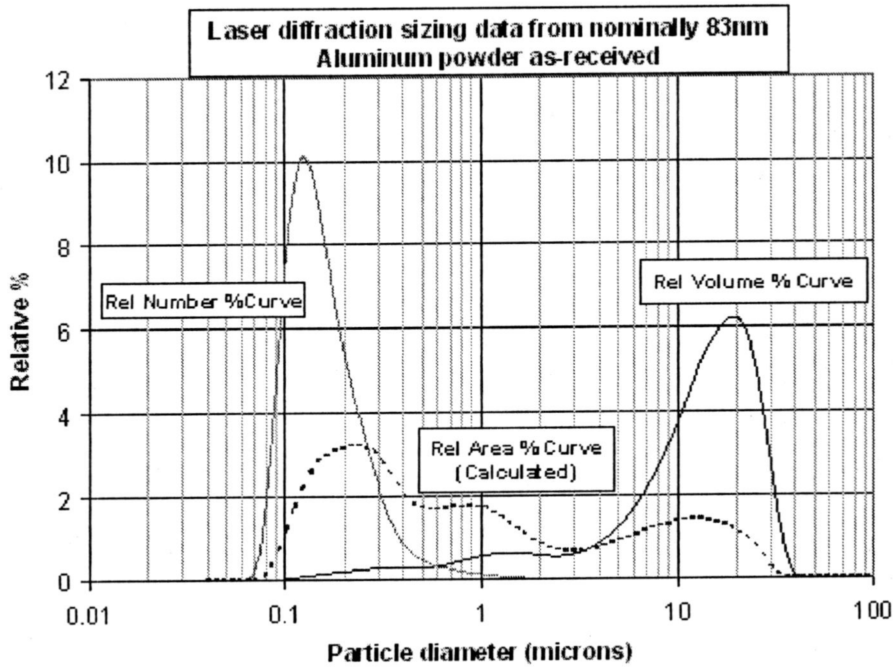

FIGURE 1 Laser diffraction size data for "nanoscale" aluminum powder used for in vitro toxicity experiments. Note the apparent difference in size when depicted as a number distribution versus an area or volume distribution. This is due to the fact that volume scales as the cube of the particle diameter and calculated area as the square. Each curve, if presented by itself, would give an incomplete picture of the particle size distribution/state of agglomeration of the sample. The three curves will overlay only for an ideal spherical, monodisperse, unagglomerated system.

Why Size Is Important to Toxicity

The size can clearly have a dramatic effect on the way in which an organism responds upon exposure to foreign particulate materials. There are actually several aspects to this issue: a) Size can govern where in the body the particulates are deposited (exposure). For example, well established models have been developed for predicting where in the lung particulates of a given aerodynamic diameter are deposited (ICRP 1994; Heyder 2004). b) Size is a factor in the ability of the body to clear foreign particles (clearance mechanisms) (Renwick et al. 2001). The ciliatory system in the lung is designed to clear particulate matter from the upper airways (NRC 1989). Phagocytosis by macrophages and giant cells is a common mechanism by which the body's innate immune system attempts to clear particles on the order of a few microns or less (Lucarelli et al. 2004). c) Size can be a factor in the ultimate fate (and location) of particles that are not cleared (translocation, fibrosis) (Oberdorster et al. 1994; Nemmar et al. 2001). d) Particle size can potentially influence direct mechanisms and extent of toxicity (cytotoxicity, necrosis and mutagenicity). For particulates that strictly satisfy the criteria of nanosize, little research has been conducted regarding differential size effects in the 1–100 nanometer range.

How to Measure Size

There are wide a variety of methods for determining particle size distributions including light scattering, dynamic mobility analysis, time of flight, microscopy, and surface area measurements among many others. ISO committee TC24SC4 ("Particle sizing methods other than sieving") develops and writes standard practices addressing over 10 different physical principles used for the measurement of particle size. At least 6 of these are applicable for size measurement into the sub 100 nanometer range (e.g., ISO 13321, ISO 13320, ISO 15901). To determine the size distribution of a sample, one ideally requires a well-dispersed system and enough particles must be measured to achieve the desired statistical reliability. For a monodisperse system the latter requirement is relatively easy. However, as polydispersity increases it becomes necessary to measure a large number of particles to accurately portray the size distribution (Masuda and Inoya 1971).

For nanoscale particulates, an ensemble method of measurement (such as centrifugal sedimentation, laser or dynamic light scattering) is normally preferred but this may not be practical or the system may be difficult to maintain in a dispersed state. For dry as-received powders, BET surface area can provide an estimated average size (based on a nonporous spherical model) and has the added advantage of providing a direct measurement of specific surface area (SSA) and micro or meso porosity key parameters of interest (ISO 15901; Allen 2004b). Microscopy is one of the most powerful techniques and often relied upon exclusively to provide valuable information regarding size, shape, morphology, etc. (Knapp 1996). However, as mentioned above, the microscopist should ensure that enough particles are examined to provide a statistically valid representation of the full size distribution. This can be very difficult and time consuming. It should also be noted that microscopy and image analysis normally provide only two dimensional images so care must be taken to avoid bias due to orientation effects. For nanomaterials (<100nm), some type of atomic force microscopy or electron microscopy is normally required to capture images with the necessary resolution. Each of these methods is subject to artifacts caused by sample preparation and/or special analysis conditions. For example, high resolution TEM requires thin sample sections or particles of limited diameter in order for the electron beam to penetrate through the sample. It is also conducted under a high vacuum which by definition excludes volatile solvents or solvent systems. Salts or other dissolved species present in the original dispersed sample may crystallize into fine particles and obscure details. Tissue sample preparation, fixation and staining is an art requiring significant skill to avoid artifacts and preserve detail.

Where to Measure the Size

It seems a bit odd to ask the question, "Where shall we measure the size of the particles?" until one considers that with most ensemble techniques it is actually the agglomerate size that is measured–not primary particle size. The state of dispersion can and does change as a function of environment. Additionally, particles may be separated or "size classified" by various means between "the bottle and the cell." For example, a flame pyrolysis method may be used to synthesize nanoparticles for various uses (Kammler 2001). Upon creation, particles begin to coagulate according to their size, concentration, temperature, residence time, pressure and a variety of other conditions. Hard aggregates may be formed through viscous or chemical sintering; fine particles may adhere to surfaces due to static charges or thermal diffusion and large particles may settle out of the gas. The size distribution and dispersion state continues to change through transport, handling and delivery to the final application. For toxicity testing, the size distribution "as dosed" might be quite different than the same system "as-generated" or "as-received." In the animal, physiological clearance mechanisms may reject larger particles and membranes or vasculature may size exclude based on their physical or chemical properties, once again altering the size distribution.

Consequently, there are four points at which we believe particle characterization should be addressed.

- The first location, *perhaps the most obvious*, is the "as-received" powder. This is also the most straightforward but as any particle scientist will attest is often not easy.
- The second, *perhaps the most critical*, is the "as dosed" or "as exposed." Any experimental data will be of limited use if details regarding exposure are undefined.
- The third location, *perhaps the most difficult*, is at the point(s) of interaction with the organism. What is the size, shape and state of agglomeration of the particles or nanostructures as they interact with the cells, tissues or organs? We consider this the most difficult because invasive techniques usually cannot be used without compromising the integrity of the organism and invalidating the test. Noninvasive techniques such as radio or fluorescent labeling of particle surfaces have the potential for changing the surface chemistry and hence the mechanisms of interaction with the organism.
- The fourth and final location, *perhaps the most useful*, is post mortem or histological examination of the cells, tissues and organs exposed to the test materials. Though plagued with issues of artifacts, statistical reliability and interpretation, most of what we learn about the toxic effects of the particles is by examining the effects of the particles on cell and tissue structures.

It has been our experience that all methods of size analysis are subject to certain pitfalls or artifacts. There is no "absolute" method of determining nanoparticle size distributions and multiple techniques should be used wherever possible to develop a more complete understanding of the system. The analyst should be knowledgeable and skilled in the techniques employed. Table 1 shows a number of techniques used to measure particle size in the nanometer range along with advantages, disadvantages and an estimate of their usefulness in the four areas discussed above.

PARTICLE SHAPE

Definition of Particle Shape

The concept of shape seems totally intuitive until one attempts to quantify any but the most geometrically regular shapes. Occasionally particulate systems have readily definable shapes such as spheres, rods, or defined crystal morphologies. For most real systems, shapes are much more variable and many "shape factors" have been devised to attempt to quantify them. Several of the most useful include, sphericity or circularity, aspect ratio, convexity, fractal dimension and others. However the definitions of many shape factors are not well standardized which potentially can lead to confusion (Hentschel and Page 2003). For example circularity is most often defined by the perimeter-area ratio of a two dimensional image. However perimeter is often measured in several different

TABLE 1 Common particle sizing techniques applicable to sub 100 nanometer particle systems

Size Measurement Technique	Nominal Size Range	Advantages	Disadvantages	As-received	As-dosed	In-situ	Post-mortem
Dynamic light scattering[a]	0.7nm–5 microns	Ensemble method also used for zeta potential	Less reliable as distribution broadens	yes	possible	maybe	no
Centrifugal sedimentation[a,b]	10nm–10 microns	Good for broad size distributions		yes	no	no	no
Laser diffraction[b]	40nm–3mm	Broad dynamic range—wet or dry measurements	Assumes spherical particles—shape effects unknown	yes	possible	no	no
Scanning/dynamic mobility analysis[h]	5 nm–2μm	Good for size distributions	Dry low pressure technique—small sample sizes	yes	yes	no	no
Atomic force microscopy[c]	.5 nm–several microns	Good resolution and 3-D imaging	Can only see surface	yes	no	no	yes
Electron microscopy[b,g]	.5 nm–several microns	Good resolution and imaging	Sample preparation and vacuum	yes	no	no	yes
Field flow fractionation[c]	2nm–2 microns	Good resolution of size distributions	Must be used in conjunction with other techniques (e.g., light scattering)	yes	no	no	no
Size exclusion chromatography[d,e]	5nm–2 microns	Good resolution	Slow and requires good calibration	yes	no	no	maybe
Small angle X-ray (or neutron) scattering[i]	1nm–1 micron	Good for solid state embedded systems	Requires high concentrations of particles—cumbersome	yes	no	no	no
Time of flight mass spec[c]	100–60K Daltons	Good for very low molecular size particles and fragments	Expensive—representative sampling difficult	yes	possible	no	no
Acoustic techniques[f]	20nm–10 microns	Good for concentrated systems	Need at least 1wt% of particles	yes	no	no	no

[a](Berne and Pecora 2000); [b](Bootz et al. 2004); [c](Jillavenkatesa and Kelly 2002); [d](Fritz et al. 2000); [e](Bootz et al. 2005); [f](Dukhin et al. 1999); [g](Sjostrom et al. 1995); [h](Burtscher 2005); [i](Glatter, O. and Kratky, O. 1982)

ways such as perimeter of a rectangular box drawn around the particle, perimeter of the ellipse with the same major and minor axes or length of the outside edge of the threshold pixels.

Why Shape Is Important to Toxicity

Particle shape has been implicated in several forms of toxic effects, most relating to the inhalation toxicity of certain inorganic fibers such as asbestos (Berry 2005). The toxicity of quartz has been definitively connected to its crystal structure, other forms of silica having been shown to be much less toxic. Although there are a few examples of high aspect ratio fibers showing increased toxicity, it is not generally proven that all such shapes are dangerous particularly at the nano scale (Hart et al. 1998). Shape, along with size and state of agglomeration, may also affect the disposition and translocation of particles in the organism.

How to Measure Shape

Microscopy is one of the most powerful techniques and is often relied upon exclusively to provide valuable information regarding size, shape, and morphology. For nanomaterials (<100nm), electron microscopy is normally required to capture images with the necessary resolution. These are currently the only techniques that provide reliable information regarding shape at these length scales. However, the microscopist should ensure that enough particles are examined to provide a statistically valid representation of the full size or shape distribution (Masuda and Inoya 1971). This can be very difficult and time consuming and may require the image analysis of literally thousands of individual particles.

It should also be noted that electron microscopy normally provides only two dimensional images so care must be taken to avoid bias due to orientation effects. High resolution microscopy is subject to artifacts caused by sample preparation and/or special analysis conditions. For example, high resolution TEM requires high vacuum and thin sample sections or particles of limited diameter in order for the electron beam to penetrate through the sample. Tissue sample preservation, fixation and staining is an art requiring significant skill to avoid artifacts and preserve detail. Quantification of shape data can be complex and there are numerous ways to express shape information and little standardization of definitions.

STATE OF DISPERSION

Definition of Dispersion

The state of dispersion of a particulate system describes the extent to which particles are agglomerated–that is, held together in groups or clusters by attractive interparticle forces. These forces have a variety of sources the most fundamental of which are the attractive Van der Waals or "Dispersion" forces. The magnitude of these forces are a function of the fundamental atomic properties of the surface atoms and the proximity of these surfaces to each other.

For most ensemble size measurements techniques what is really measured is the size distribution of the agglomerates. Thus, the measured size distribution is highly

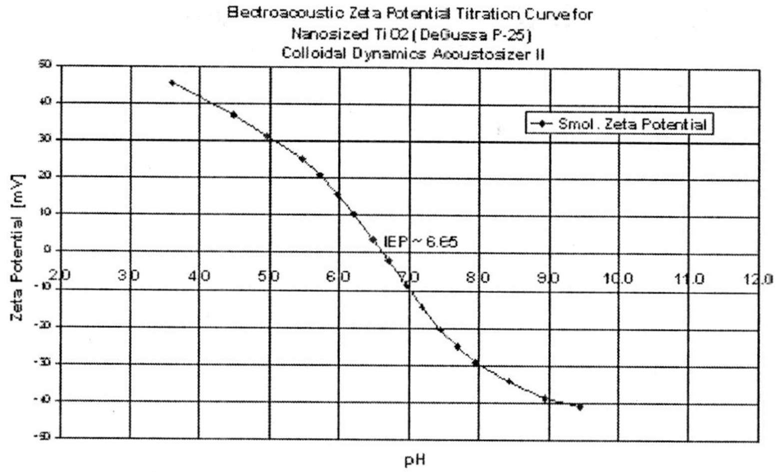

FIGURE 2 Zeta potential titration curve for nanosize TiO_2 powder (DeGussa P-25)

dependent on the state of dispersion of the system. Any ensemble particle size measurement must be interpreted in this context. Due to attractive forces (Van der Waals and others), particles will tend to agglomerate in suspension unless stabilized by surface charge or steric effects. Most aqueous suspensions of hydrophilic polymers or metal oxides will specifically adsorb or desorb hydrogen ions to generate a surface charge. Homogeneous powders that develop a surface charge high enough to overcome interparticle attraction will form more stable dispersions.

The point of zero charge is approximated by measuring the isoelectric point (IEP)—that is, the pH at which the zeta potential is zero. The IEP is the pH at which the dispersion typically has the least native stability in aqueous solutions. The IEP is material dependent but typically values lie in the range from pH 3 (e.g., silica) to pH 9 (e.g., aluminum oxide). Generally the surface charge becomes progressively more positive as pH decreases (acidic solutions) and more negative as pH increases. Figure 2 illustrates a typical zeta potential titration for metal oxide nanopowder (DeGussa P-25 TiO_2) showing an isoelectric point at pH 6.6. Sometimes surfactants are used to change the surface charge or sterically stabilize suspensions through the use of long hydrocarbon chains attached to a polar head. However, not all surfactants are compatible with the physiological environment and many can cause cell lysis and/or other adverse effects.

Particle size affects dispersibility in two ways. For large particles or agglomerates sedimentation becomes an issue. Stable, well dispersed ultrafine and nano systems generally do not suffer from this effect. However, the smaller the particle size, the greater the relative Van der Waals forces per unit mass. This means that, once agglomerated, it becomes progressively more difficult to redisperse nanoscale materials as size decreases. Therefore fine particle dispersions have a greater tendency towards agglomeration unless stabilized by the mechanisms discussed above.

Particles can be de-agglomerated by applying shear (mixing, sonication, grinding, turbulence) but unless the characteristics are present for a stable dispersion (by native surface charge at a given pH, surfactants, or steric stabilization) the system will be prone to re-agglomerate rapidly. Thus particles that are not stabilized by pH or surfactant may agglomerate rapidly when introduced into a different environment such as a highly buffered (high ionic strength) physiological/ biological fluid.

Dry Powder Dispersion

Although most physiological environments are comprised of wet systems, dry aerosol dispersions are associated with exposure by inhalation or deposition on the skin. If the dispersion of fine powders is difficult in solution, it is doubly so as a dry powder. In air, powders cannot be easily stabilized by charge. Though dielectric powders may support a static charge, this serves only to attract the particles to the nearest surface with an opposite (or neutral) polarity. (This is why fine dust can collect or stick on vertical surfaces.) Additives such as surfactants are generally ineffective or counterproductive in the dry state (although there are some flow aids or glidants used for dry dispersion of larger particle sizes). Fine and ultrafine powders (here defined as powders <2.5 microns and .1 micron respectively) that have once been dried generally cannot be fully re-dispersed to their original size distributions without extreme effort.

How to Measure the State of Dispersion

Particle and colloid scientists have been struggling with this issue for decades. The only way to absolutely define the "state of dispersion" is to quantify all agglomerates for size, number of particles, strength of interparticle forces, shape, structure, and distribution of the agglomerates in the system. This is simply not possible.

Most often, an assessment of the state of agglomeration is made by taking two particle size measurements; one of the particles in the agglomerated state and one with the particles dispersed as close to the "primary" particle size distribution as possible. Many have used these and other values to attempt to create some sort of dispersion "index" which represents with one number the state of dispersion (Jillavenkatesa and Kelly 2002). The difficulty in assigning such numbers is that they often do not adequately take into account the effect of polydispersity and the fact that one is never sure how completely the "well dispersed" system represents the primary particle size distribution. Agglomeration rate and strength (or floc strength) studies should be pursued when evidence of agglomerate induced effects appear evident. Experimental protocols for these experiments have been reviewed elsewhere (Jarvis et al. 2005).

Despite these limitations some assessment can usually be made (e.g., a statistical comparison) of dispersed versus agglomerated size distributions. Qualitative observations can also be derived from imaging. The use of multiple size analysis techniques are particularly valuable for this. Figure 3 illustrates an as-received nano size aluminum powder (nominal size 83nm) analyzed by laser diffraction, BET surface area and electron microscopy. What is the true size and state of agglomeration of this powder? By light scattering it has a volume median size of 14.5 microns, certainly not the nano size attributed to it by the manufacturer. By number, the median size is 151 nanometers. And by surface area the size is calculated to be 80 nanometers.

In Figure 4 a series of electron micrographs (TEM and SEM) of the same powder give additional insight into the size and state of agglomeration. Note that large "particles" can be attributed to both polydispersity and agglomeration. By using multiple techniques, it is clear that there is both significant agglomeration and polydispersity in this "nanoparticulate" system. Finally, Figure 5 shows these same particles in an endosome

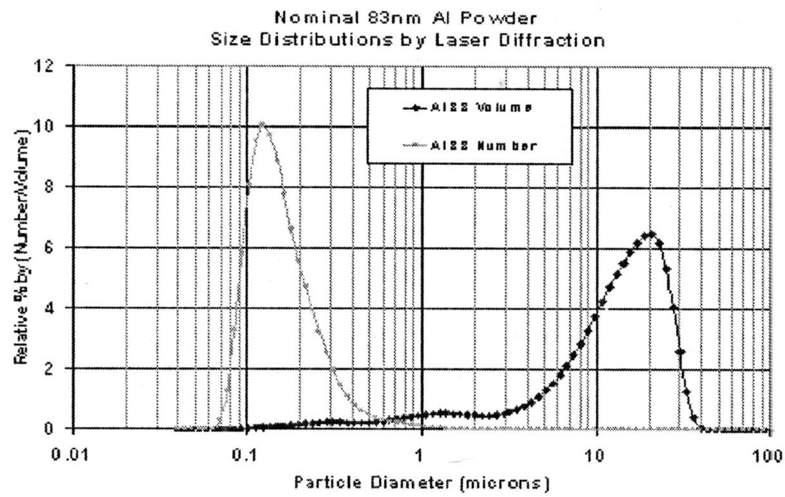

FIGURE 3 As-received size analysis data for a dry agglomerated nano aluminum powder of nominal 83 nanometers size

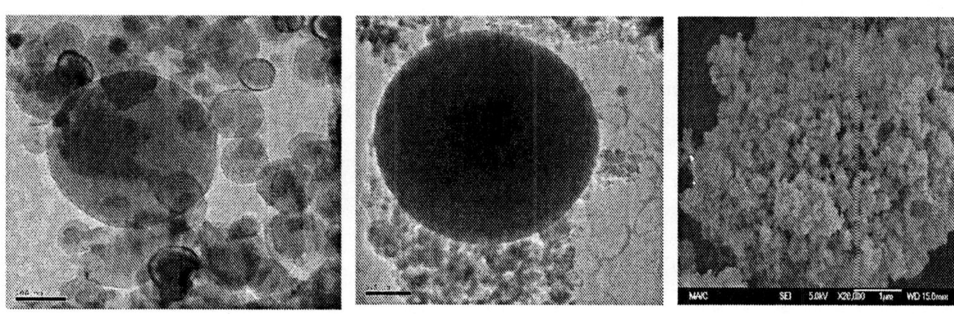

FIGURE 4 Electron microscopy of an "83 nanometer" aluminum powder

inside of a cell after an in vitro toxicity test. Though individual particles can be resolved, it is unclear whether they were transported there as individuals or as agglomerates.

Dispersion and Toxicity

The tendency to agglomerate may at first seem irrelevant to the toxicologist introducing particles to assess cytotoxicity, immune or inflammatory response. However, if such a response is size dependent, the state of agglomeration may be a significant factor. For example, particle agglomerate size has been shown to play a role in the ability of macrophages to phagocytize particles and in the tendency for particle penetration of cell membranes and translocation through the tissues, lymph or circulatory system (Nemmar et al. 2001; Renwick et al. 2001). There is some evidence that very fine particles may escape an immune or inflammatory response observed for larger particles of the same material (Gutwein and Webster 2002). It is difficult to draw firm conclusions regarding

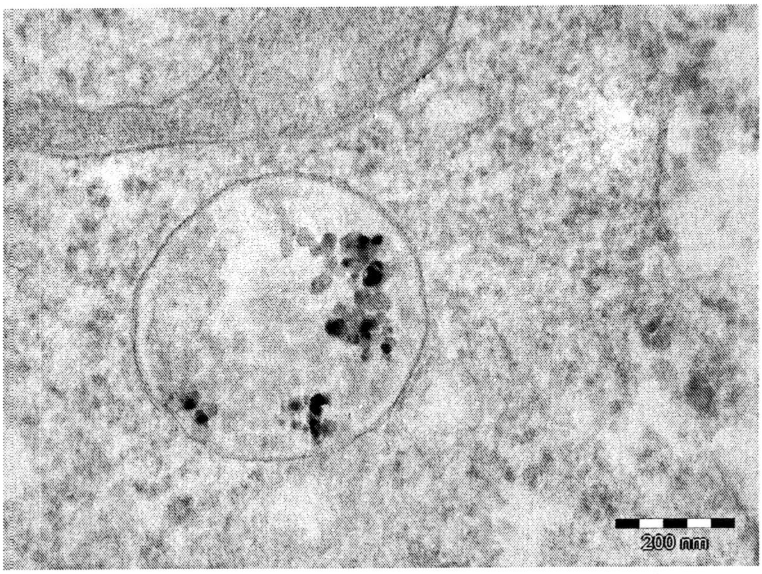

FIGURE 5 Aluminum nano particles inside of an endosome in an A549 cell from an in-vitro toxicity experiment

size effects from existing literature due to the fact that studies are performed with different cell lines, tissues, animal models, different types of ultrafine particles and methods of administration. The dispersion of particles during dosing and in-situ is seldom addressed other than qualitative observations made from post microscopic analysis. Clearly more detailed research is required on the specific effects of size and state of agglomeration.

RECOMMENDED PRACTICES

Basic Particle Characterization

The measurement of particle size and shape is not a new analysis problem. Particle scientists have dealt with the issue for decades and have developed recommended practices which are recorded in literature, texts, national and international standards (ISO, ASTM, USP, ANSI). Many of these practices have been adapted for difficult measurement environments similar to those facing the toxicology community. For basic particle characterization several fundamental principles include:

- Sampling. Due to the large number of particles in a given mass of a nano particle system it is imperative that the particles measured are representative of the bulk material. This is the first and foremost principle of particle characterization. The broader the size distribution, the more significant the errors will be if the sample is not representative.

- Dispersion. Primary particle characteristics should be measured in the most dispersed state achievable. Some estimate of the state of dispersion should be made at the point of interaction with the organism.
- Statistical reliability. Enough particles must be measured to ensure that the desired limits of precision and accuracy are achieved. For broad particle size distributions this may amount to tens of thousands of particles.
- The physical principle of measurement should be selected most appropriate to the application. For example: if measuring airborne deposition of a dry powder into the lungs of an animal subject, the principle of measurement might be a time of flight, cascade impaction or some other technique which measures aerodynamic diameters of the powder.
- The particle size should be measured under conditions as close as possible to the environment at the point application. For example, if in-vitro cell studies are being conducted, the particle size should be measured in cell culture media or at least the same pH and ionic strength conditions.

CONCLUSIONS

A firm understanding of the properties of particulate materials is essential for their characterization and for determining toxicological effects in the physiological environment. This is particularly true when trying to determine the mechanism of toxicity. The issues involved become quite complex. Outside of the body the environmental conditions can be manipulated to promote the maximum dispersion for measuring size and surface chemistry. In the biological environment however, one is restricted to the conditions under which the organism lives. Often properties can only be measured after particles are removed or tissues fixed, potentially introducing artifacts into the measurements. Until better techniques for characterization in these environments are developed, researchers must use the tools available to reconstruct the particle properties in situ and how they interact with cells, tissues and organs. All tests should include basic characterization of size, shape and state of dispersion. To some extent, the objective of the study and the nature of the nanoscale material will determine the specific parameters to be measured.

ACKNOWLEDGMENTS

Support from National Science Foundation Grant No. EEC-9402989 and United States Air Force Contract No. F08830-03-1-0010 is gratefully acknowledged. SEM and TEM microscopy was provided by Ms. Maria Palazuelos, Ms. Kerry Siebein and Dr. Greg Erdos.

BIBLIOGRAPHY

Allen, T. (2004a). *Particle Size Measurement, Vol I: Powder Sampling and Particle Size Measurement*, 5th Ed., Chapman & Hall.

Allen, T. (2004b). *Particle Size Measurement, Vol II: Surface Area and Pore Size Determination*, 5th Ed., Chapman & Hall.

Berne, B., and Pecora, R. (2000). *Dynamic Light Scattering*, Dover Publications, N.Y.

Berry, C. C., Dalby, M. J., McCloy, D., and Affrossman, S. (2005). The fibroblast response to tubes exhibiting internal nanotopography. *Biomaterials* 26, 4985–92.

Bootz, A., Russ, T., Gores, F., Karas, M., and Kreuter, J. (2005). Molecular weights of poly(butyl cyanoacrylate) nanoparticles determined by mass spectrometry and size exclusion chromatography. *European Journal of Pharmaceutics and Biopharmaceutics* 60, 391–399.

Bootz, A., Vogel, V., Schubert, D., and Kreuter, J. (2004). Comparison of scanning electron microscopy, dynamic light scattering and analytical ultracentrifugation for the sizing of poly(butyl cyanoacrylate) nanoparticles. *European Journal of Pharmaceutics and Biopharmaceutics* 57, 369–375.

Burtscher, H. (2005). Physical characterization of particulate emissions from diesel engines: a review. *Journal of Aerosol Science*, 36:896–932.

Chakraborty, B. R., Haranath, D., Chander, H., Hellweg, S., Dambach, S., and Arlinghaus, H. F. (2005). TOF-SIMS and laser-SNMS investigations of dopant distribution in nanophosphors. *Nanotechnology,* 16, 1006–1015.

Choi, H. C., Ahn, H. J., Jung, Y. M., Lee, M. K., Shin, H. J., Kim, S. B., and Sung, Y. E. (2004). Characterization of the structures of size-selected TiO2 nanoparticles using X-ray absorption spectroscopy. *Appl Spectrosc* 58, 598–602.

Donaldson, K., Brown, D., Clouter, A., Duffin, R., MacNee, W., Renwick, L., Tran, L., and Stone, V. (2002). The pulmonary toxicology of ultrafine particles. *J Aerosol Med* 15, 213–20.

Donaldson, K., Li, X. Y., and MacNee, W. (1998). Ultrafine (nanometre) particle mediated lung injury. *J Aerosol Sci* 29, 553–560.

Dukhin, A. S., Shilov, V. N., Ohshima, H., and Goetz, P. J. (1999). Electroacoustic phenomena in concentrated dispersions: New theory and CVI experiment. *Langmuir* 15, 6692–6706.

Fritz, J. S., and Gjerde, D. T. (2000). *Ion Chromatography*, Wiley-VCH.

Glatter, O., and Kratky, O. (1982). *Small-Angle X-ray Scattering*, Academic Press, New York.

Guo, J. (2004). Synchrotron radiation, soft-X-ray spectroscopy and nanomaterials. *International Journal of Nanotechnology* 1, 193–225.

Gutwein, L., and Webster, T., 2002, "Osteoblast and chrondrocyte proliferation in the presence of alumina and titania nanoparticles," *Journal of Nanoparticle Research*, 4, 231–238, 2002.

Hart, G. A., and Hesterberg T. W., 1998, "In vitro toxicity of respirable-size particles of diatomaceous earth, and crystalline silica compared with asbestos and titanium dioxide," *J. Occupational Environmental Medicine*, 40(1), 29–42.

Hentschel, M., and Page, N. (2003). Selection of descriptors for particle shape characterization, *Particle and Particle Systems Characterization,* 20, 25–38.

Heyder. J. (2004). Deposition of Inhaled Particles in the Human Respiratory Tract and Consequences for Regional Targeting in Respiratory Drug Delivery, *The Proceedings of the American Thoracic Society*, 1:315–320.

International Commission on Radiological Protection (ICRP) (1994). Human respiratory tract models for radiological protection. *Ann. ICRP 24.*

ISO 9276, Representation of results of particle size analysis, Parts 1–3, ISO TC24/SC4, International Standards Organization Publication, 2003.

Jarvis, P., Jefferson, B., Gregory, J., and Parsons, S. A. (2005). A review of floc strength and breakage. *Water Res* 39, 3121–3137.

Jillavenkatesa, A., and Kelly, J. (2002) Nanopowder characterization: challenges and future directions. *J Nanopart Research,* 4, 463–468.

Jillavenkatesa, A., Dapkunas, S., and Lum, H. (2001). Particle Size Characterization, *NIST Practice Guide SP 960-1*, National Institute of Standards and Technology, 2001.

Kammler, H., Mädler, L., and Pratsinis, S. (2001). Flame Synthesis of Nanoparticles, *Chemical Engineering and Tech.*, 24,6, 583–596.

Knapp, J., Barber, T., and Lieberman, A., Eds., 1996, *Liquid- and Surface-borne Particle Measurement Handbook*, Marcel Decker Ltd.

Lucarelli, M., Gatti, A., Sararino G., Quattroni, P., Marinelli, L., Monari, E., and Boraschi, D., 2004, Innate defence functions of macrophages can be biased by nano-size ceramic and metallic particles, *Eur. Cytokine Netw.*, 15, 4, 339–346.

Masuda, H., and Inoya, K. (1971). Theoretical study of the scatter of experimental data due to particle size distribution. *Journal of Chemical Engineering, Japan* 4, 60–67.

Nemmar, A., Vanbilloen, H., Hoylaerts, M. F., Hoet, P. H., Verbruggen, A., and Nemery, B. (2001). Passage of intratracheally instilled ultrafine particles from the lung into the systemic circulation in hamster. *Am J Respir Crit Care Med* 164, 1665–8.

NRC 1989, *Biologic markers in Pulmonary Toxicology*, Subcommittee on Pulmonary Toxicology Committee on biologic Markers, National Research Council, National Academy Press, Washington, D.C.

Oberdorster, G., Maynard, A., Donaldson, K., Castranova, V., Fitzpatrick, J., Ausman K., Carter, J., Karn, B., Kreyling, W., Lai, D., Olin, S., Monteiro-Riviere, N., Wartheit, D., and Yang, H. (2005). Principles for characterizing the potential human health effects from exposure to nanomaterials: elements of a screening strategy. *Particle and Fibre Toxicology*, doi: 10.1186/ 1743-8977-2-82:8

Oberdorster, G., Ferin, J., Gelein, R., Soderholm, S. C., and Finkelstein, J. (1992). Role of the Alveolar Macrophage in Lung Injury—Studies with Ultrafine Particles, *Environ Health Persp* 97, 193–199.

Oberdorster, G., Oberdorster, E., and Oberdorster, J. (2005). Nanotoxicology: an emerging discipline evolving from studies of ultrafine particles, *Environ Health Perspect* 113, 823–39.

Oberdorster G., et al. (1994). Correlation between particle size, in vivo particle persistence, and lung injury, *Environmental Health Perspect.* 102, Supp. 5, 173–179.

Renwick, L. C., Donaldson, K., and Clouter, A. (2001). Impairment of alveolar macrophage phagocytosis by ultrafine particles, *Toxicol Appl Pharmacol* 172, 119–27.

Roberts, S., Bucher, J., Masten, S., Moudgil, B., Powers, K., and Walker, N. (2004). Developing Experimental Approaches for the Evaluation of Toxicological Interactions of Nanoscale Materials, Final Workshop Report 3–4 November, 2004, 1–37; University of Florida, Gainesville, FL, USA. www.nanotoxicology.ufl.edu.

Roco, M. C. (2001). International strategy for nanotechnology research and development, *Journal of Nanoparticle Research* 3: 353–360.

Royal Society. (2004). Nanoscience and nanotechnologies: opportunities and uncertainties, RS Policy document 19/04.

Sjostrom, H., Stafstrom, S., Boman, M., and Sundgren, J.-E. (1995). Superhard and elastic carbon nitride thin films having fullerenelike microstructure. American Physical Society.

Thomas, K., and Sayre, P. (2005). Research strategies for safety evaluation of nanomaterials, Part I: evaluating the human health implications of exposure to nanoscale materials, *Toxicol Sci* 87, 316–21.

Weiss, R. (2004). For Science, Nanotech Poses Big Unknowns, Sunday, February 1, *Washington Post*, Washington, D.C.

Current and Emerging Applications of Nanomaterials Involving Nanoscale Fillers and Minerals

Shawn P. Decker*

The "nanomaterials" revolution is impacting the minerals business with clays and filler minerals subject to processing developments, creating commercial grades of nano-size minerals. Minerals currently being developed in this field include clays, talc, titanium dioxide, and calcium carbonate. Nanomaterials are already seeing use in everyday domestic and commercial applications, and consumption of some 100–150,000 metric tons of nanomaterials is expected by 2006. Market applications for nano-minerals include: wood preservatives, marine antifouling, thermoplastics, permanent coatings, environmental catalysts, abrasion-resistant coatings, sunscreens, deodorants, oral care, glass polishing, and semiconductor polishing. This chapter examines existing and emerging nanomaterials applications involving nanoscale fillers and minerals.

INTRODUCTION

What Is Nanotechnology?

Nanotechnology is the creation of functional materials, devices, and systems through control / manipulation of matter on the nanometer (1 to 100+ nm) length scale and the exploitation of novel properties and phenomena developed at that scale. One way to look at this in terms of a general reference is that there are 25,400,000 nanometers in a single inch. In general, nanotechnology can be broadly segmented into two categories. The first is *nano-materials* such as carbon nanotubes, metallic nanoparticles, nano-minerals, and biological nanoscale systems. The second is *nano-electronics* such as

* Center for Accelerated Applications at the Nanoscale, South Dakota School of Mines and Technology, Rapid City, South Dakota

micro- / nano-electromechanical systems (MEMS / NEMS) and nanoscale integrated circuits / memory devices.

What Are Nano-minerals?

Crystalline and amorphous silicates, especially clay minerals, and metal oxides and hydroxides are the most important representatives of nano-minerals. Common characteristics of nano-minerals are that they possess very small grain sizes and, as a consequence, large reactive surface areas. These surfaces are often modifiable and can be imparted with chemicals that can generate catalytic properties.

Common examples of nano-clays include:

- Montmorillonite
- Vermiculite
- Kaolinite
- Mica
- Halloysite
- Talc

Common examples of nano-minerals include:

- Calcium carbonate
- Alumina
- Cerium oxide
- Titanium dioxide
- Zinc oxide
- Zirconium
- Magnesium hydroxide

Often nano-minerals are formed from naturally occurring minerals such as montmorillonite, kaolinite, mica, and halloysite. As such, naturally occurring nano-minerals have importance to a number of earth sciences such as mineralogy, geochemistry, geology, and environmental science. In terms of applications, disciplines such as materials science, polymers, chemistry, chemical and petroleum engineering, construction engineering, pharmacy and food science are relevant.

PRODUCTION OF NANO-MINERALS

There are a wide variety of techniques for producing nanoparticles:

- Condensation from a vapor
- Chemical synthesis
- Solid-state processes such as high energy milling

Nanoparticles that are produced by these methods can then be coated with hydrophilic or hydrophobic substances depending on the desired subsequent application. Production of nano-minerals is rather amenable to both the chemical synthesis and solid-state processing methods. Another convenient way in which to distinguish these two methods are the so-called "bottom-up" versus "top-down" techniques.

Bottom-Up Production Methods

One way to look at the "bottom-up" production methods is as assembling nanostructures on an atom by atom or molecule by molecule basis. This method, in essence, can be thought as starting with a core nucleus of an atom or molecule and continually adding to this foundation and then terminating the synthesis or production at the desired particle size.

Most of the recent developments in nanopowder production have been in the bottom-up approaches and many different methods have been successfully used to grow nanosized crystals and particles. The major issues to overcome with these methods are controlling the particle growth and then preventing the newly formed particles from agglomerating once they are formed. Subsequent issues are ensuring that the reactions are complete, leaving no unwanted chemicals (e.g., no residual zinc metal in the case of ZnO). While there are a number of more advanced techniques for bottom-up preparation, some common examples are precipitation techniques for silica, calcium carbonate, alumina hydrate and magnesium hydroxide.

Top-Down Production Methods

"Top-down" production methods are often thought of as an extension of traditional methods of producing large quantities of fine and nano-crystaline powder. Generally these involve some form of high energy wet milling with the addition of milling aids and long milling times. Unfortunately, many of these techniques suffer from difficulties in ensuring all the particles are milled properly that often result in product fractions that tend to have a long "tail" representing the un-milled product. Also, the longer milling times tend to result in more milling impurities resultant from the impact of the product with the internal chambers of the mill. Removal of these impurities and grinding aids can cause subsequent problems with the end-product.

Most ultra-fine minerals and fillers are made by using top-down production methods. Conventional grinding technology and can rarely get below 1 micron. Another disadvantage of these techniques is that 50 to 70% of all the energy used in mineral processing is for comminution. However, the efficiency is typically on the order of 1%, with most of the energy going into heat and wear on the equipment. Consequently, there are considerable needs and opportunities to develop new "top-down" techniques. One example is "vibro-energy" milling as in Micro Grinding Systems machines can reduce mica down to 1 micron.

APPLICATIONS OF NANO-MINERALS

Nano-minerals and nano-clays are used to create a number of useful products. Areas that produce the greatest revenue for nanoparticle minerals and clays are chemical-mechanical polishing, magnetic recording tapes, sunscreens, automotive catalyst supports, bio-labeling, electro-conductive coatings and optical fibers.

Established "nano" applications include:

- Nanocomposites
- Sunscreens and cosmetics
- Abrasion and corrosion resistant coatings
- Anti-fouling marine paints and coatings

- Protective and glare-reducing coatings for eyeglasses and windshields
- Pharmaceutical tablets
- Wood preservatives
- Stain-free clothing
- Dental-bonding agents
- Bactericidal ceramic tile coatings
- Automobile catalytic converters
- Self-cleaning windows
- Environmental catalysts

Emerging "nano" applications include:

- Cooling wafers to replace compressors in cars, refrigerators, air conditioners and multiple other devices utilizing no chemicals or moving parts
- Sensors for airborne chemicals or other toxins
- Photo-voltaic (solar cells), fuel cells and portable power to provide inexpensive, clean energy
- Supercapacitors for high performance electronics
- Air and liquid filtration devices
- New high-performance materials

Already nano-minerals and subsequent processed materials are starting to play key roles in infrastructure development and maintenance, agriculture, ceramics industries, and environmental mitigation. The need for scientific research on fine-grained nano-minerals is unmistakable because they are starting to play an ever-expanding role in our society.

While there are numerous current and potential applications of nano-minerals, so-called polymer nanocomposites are by far the largest market application.

NANOCOMPOSITES

In general, nanocomposites are materials with a nanoscale structure that improve the macroscopic properties of products. Typically, nanocomposites are clay, polymer or carbon, or a combination of these materials with nanoparticle building blocks.[1] The definition of nanocomposite material has broadened significantly to encompass a large variety of systems such as one-dimensional, two-dimensional, three-dimensional and amorphous materials, made of distinctly dissimilar components and mixed at the nanometer scale.

Classification of Nanocomposites

Nanocomposites, materials with nanoscale separation of phases, can generally be divided into two types: inorganic / organic composites and multilayer structures. The general class of inorganic/organic nanocomposite materials is a rapidly growing area of research. Significant effort is focused on the ability to obtain control of the nanoscale structures via innovative synthetic approaches. The properties of nanocomposite materials depend not only on the properties of their individual parents, but also on their morphology and interfacial characteristics.[2]

Inorganic/Organic Nanocomposites. The term inorganic/organic nanocomposite is generally used to describe a system in which small inorganic (typically clay-based) particles are dispersed within an organic polymer matrix at a scale on the order of nanometers. The inorganic components can be three-dimensional framework systems, such as zeolites; two-dimensional layered materials, such as clays, metal oxides, metal phosphates, chalcogenides; and even one-dimensional and zero-dimensional materials, such as $(Mo_3Se_{3-})_n$, chains and clusters.[2] The primary reason for dispersing the particles in a polymer is to improve the polymer properties (e.g., mechanical, electrical, optical). Adding small amounts of nano-clays to polymers have been shown to have significant effects on mechanical, thermal and barrier properties. One of the earliest examples of this was reported by Toyota researchers in which they polymerized ε-caprolactam in the gallery regions of silicate clay to form a hybrid. At only 4.2 wt.% clay loading the modulus doubled, strength increased by more than 50% and the T_g increased by 80°C.

The clays are usually layered materials with thicknesses on the order of 10 angstroms (~1 nm) and width dimensions extending to about 1,000 nanometers. These high aspect ratio materials are attractive for use in nanocomposites because they have high in-plane strength and stiffness. Silicate clays are the primary clays used because their intercalation chemistry can be easily modified to make them compatible with organic polymers and they are abundant in nature and thus are inexpensive.[3]

Clays are typically organophobic. The layers of these silicates are composed of a central sheet of octahedral alumina or magnesia sandwiched in between two tetrahedral sheets. When trivalent magnesium ions (montmorillonite) or divalent magnesium ions are partially replaced by lithium ions (hectorite), the individual layers are anionically charged. Counterions such as Na^+ or Ca^+ are located in the intergallery layers. To facilitate separation and dispersal of the layers, they must be rendered organophilic which is usually done by ion exchange with onium ion such as alkylammonium ions. The clays are then commonly referred to as organically modified silicates (OMS).

The general class of organic/inorganic nanocomposites may also be of relevance to issues of bio-ceramics and biomineralization, in which in-situ growth and polymerization of biopolymer and inorganic matrix is occurring. Finally, lamellar nanocomposites represent an extreme case of a composite in which interface interactions between the two phases are maximized.

Lamellar nanocomposites can be divided into two distinct classes,[2] intercalated and exfoliated.

- In *intercalated nanocomposites* the polymer chains alternate with the inorganic layers in a fixed compositional ratio and have a well defined number of polymer layers in the intralamellar space (Table 1). The intercalated nanocomposites are also more compound-like because of the fixed polymer / layer ratio and possess interesting electronic and charge transport properties.
- With *exfoliated nanocomposites* the number of polymer chains between the layers is almost continuously variable and the layers stand > 100 Å apart. These systems are known for their superior mechanical properties compared to non-exfoliated systems.

There are three common synthetic approaches for these types of inorganic-organic nanocomposites.[3] They are: (a) solution mediation (in which the polymer and clay are mixed in solution), (b) in-situ polymerization, and (c) melt processing.

TABLE 1 Selected minerals showing basal spacing in nanometers

Mineral	Basal Spacing (nm)
Hydrated montmorillonite	1.4–1.54
Dehydrated montmorillonite	0.92–0.95
Hydrated vermiculite	1.16–1.48
Dehydrated vermiculite	0.90
Hydrated kaolin	1.00
Muscovite mica	0.93
Phlogopite mica	0.99

Multilayer Nanocomposites. Multilayer structures are typically formed by gas phase deposition or from the self-assembly of monolayers.[1] Layer-by-layer electrostatic assembly of electrically charged polymers is a versatile technique for the formation of multilayered thin films with a wide range of electrical, magnetic, and optical properties. Multilayers and related composites are more versatile in several respects than other types of film, such as self-assembled monolayers and Langmuir-Blodgett films.

Layer-by-layer assembly has two major advantages. Surface properties of existing objects and devices can be controlled by controlling surface functionality. One can also fabricate thin-film devices in a very controlled way by template-assisted assembly. By controlling the spatial arrangement of functionality in multimaterial-layered nanocomposites, one can engineer the properties of devices. In addition, layer-by-layer assembly is a low-cost, environmentally friendly, wet-bench technique.[4]

Nanoscale multilayer coatings, which consist of alternating layers of materials, further improve the performance of single-layer nanostructured coatings. The types of materials, their bonding characteristics, and crystal structures differentiate multilayer coatings. When properly tailored, nano-multilayer coatings produce super hardness and super modulus effects. Optimum values of mechanical properties can be achieved by varying the number of layers.[5]

Exfoliation

The key to performance of nano-clay based polymers is effective exfoliation. Effective exfoliation and dispersion of nano-clay fillers and additives is essential for achieving maximum performance in any nano-clay based material or application. Early efforts to compound nano-clays into polymer matrices that often failed to achieve proper exfoliation of the clay resulted in limited property improvements and were often very difficult to process. However, recent developments that dramatically improve nano-clay dispersions throughout resin matrices result in dramatic improvements in materials performance and processing. As a result, the latest generations of nano-clay composites have performance characteristics that are more similar to engineered thermoplastics than traditional mineral filled plastics.

Thus, distinct performance advantages of nano-clay composites are becoming evident. These include:

- improved strength and durability (high strength without loss of impact)
- light weight and efficiency
- excellent dimensional stability

- increased heat distortion temperature
- outstanding surface finish for improved aesthetics
- enhanced barrier performance
- improved processability

NANO-MINERALS AND FILLERS—A DILEMMA FOR MINING COMPANIES?

The challenge for most mining companies is that their expertise is mining and production, not research and development or "productization." Thus, market opportunities are often perceived to relate to quantity sold rather than unit value as most historical applications of minerals and fillers have been of a "commodity" as opposed to a "value-added" nature. The emergence of "nano" related materials offers an opportunity for mining companies to realize larger profit margins from such value-added applications.

Most of the work that has been done in the emerging "nanotechnology" area has been done by innovative companies not involved in mining who are focused on technical advantages at very low usage levels. For example, a 1% loading of nano-talc in engineering plastic will impart the same reinforcement as 10% of micronized talc without the loss of impact strength. How the "traditional" mining companies can adapt to the emerging "nanotechnology" industry remains to be seen.

REFERENCES

[1] http://www.azonano.com/details.asp?ArticleID=1147
[2] http://www.azonano.com/details.asp?ArticleID=1283
[3] http://www.tuskegee.edu/Global/story.asp?S=1093138
[4] http://www.chem.fsu.edu/multilayers/C&E%20News%20article%20RE%20 polyelectrolyte%20 multilayers%206%20May%202002.htm
[5] "Nanostructured Coatings for Machining and Wear-Resistant Applications"; Ajay P. Malshe, Wenping Jiang, and Ajit R. Dhamdhere, *JOM*, 2002, *54(9)*, 28.

PART 2

Nanocomposites and Nanofluids

Investigation of Hybrid Organic-Inorganic Systems by the Melt Blending of a Polyhedral Oligomeric Silsesquioxane with Bisphenol A Polycarbonate

David A. Boyles* and Tsvetanka Filipova*

Melt blending of a functionalized polyhedral oligomeric silsesquioxane (POSS) derivative with polymer may afford a facile route to novel hybrid organic-inorganic composite materials. Reactive blending of an octasubstituted POSS derivative, octakis(3-hydroxypropyldimethylsiloxy)octasil-sesquioxane octahedral (OHPS) with bisphenol A polycarbonate using twin screw extrusion was investigated using different temperatures and durations, with comparisons made with and without catalyst. The formulations obtained were examined for morphology, as well as their thermal and mechanical properties. Differences in processing were noted with uncatalyzed versus catalyzed formulations, as well as percentages of OHPS.

INTRODUCTION

Polymeric blends of polycarbonate with a variety of polyesters are well known from the literature [1]. Blending may be used to increase end-use applications of high performance polymers such as polycarbonate. The primary advantage of the technique is to enhance the properties of commercial materials by mixing, obviating the research and development necessary to bring new polymers to market. The technique has been used for polymers which are miscible owing to chain-chain interactions at the molecular level. It has also been used in the case of thermodynamically immiscible polymers to achieve polymers which although phase separated at the molecular level nonetheless are sufficiently stable in the bulk. In many cases compatibilization is obtained upon

* Department of Chemistry, South Dakota School of Mines and Technology, Rapid City, South Dakota

minor chemical modifications at the molecular level which occur during the blending process. Such reactive polymer blending of two or more polymers is, in fact, now a well-established and well-known technique used by the polymer industry to mix materials and introduce property advantages beyond those of the independent components [2,3]. Under conditions used for blending, chain exchange reactions sometimes occur and may generate grafted or cross-linked materials which, whether or not they are inherently miscible, may afford stabilization owing to new block copolymers containing the structural features of the original homopolymers.

Polyhedral oligomeric silsesquioxane (POSS) thermoplastic polymer nanocomposites of polycarbonate produced by melt blending have been known for several years, and mono-substituted POSS derivatives have been used to produce a variety of hybrid polycarbonates, polyesters, and polyurethanes [4,5]. Additionally, there have been investigations on the compatibility of POSS in polycarbonate carried out by researchers at the University of Southern Mississippi [6]. Likewise, the patent literature includes patents on polycarbonate/POSS resin systems [7].

In addition to formulations of POSS with polymers, grafting strategies have been used to modify polymer chains with pendant POSS or to make hybrid inorganic-organic composite materials. Several groups have utilized a variety of POSS derivatives in the development of either composites or grafting from approaches to produce hybrid materials, hybrid in the sense of materials which contain both inorganic POSS cages in organic polymer matrices. For example, Lichtenhan has formed acrylate and styrene hybrids of POSS using POSS as an addition monomer during polymerization [8,9]. Xu et al. have investigated the characterization of poly(vinylpyrrolidone-co-isobutylstyryl) POSS hybrids [10] and glass transition temperatures (T_g) of poly(hydroxystyrene-co-vinylpyrrolidone-co-isobutylstyryl) POSS hybrids [11]. Mather et al. have investigated strain recovery in drawn POSS hybrid thermoplastics [12,13]. Additionally, Schiraldi et al. have investigated thermoplastics with POSS derivatives to form nanocomposites [14]. The properties of POSS systems often show improved properties. For example, when grafted onto POSS, poly(methyl methacrylate) and polyurethanes show increased mechanical strength [15,16,17].

The POSS derivative under investigation in this work is depicted in both valence bond and space-filling representations in Figure 1. It has not been previously formulated with polycarbonate.

With a silicon-oxygen cage diameter of approximately 1.5 nanometers, this material—octakis(3-hydroxypropydimethylsiloxy)octasilsesquioxane (OHPS)—has eight alkoxy arms, each at the corner of an inorganic cubic silsesquioxane cage. It was first synthesized by Laine [19] who has recently prepared a class of epoxy nanocomposites with defined organic/inorganic phases by reacting the material with diaminodiphenylmethane to produce a highly homogeneous nanocomposite having defined inorganic and organic phases [20–23]. Owing to the fact that the introduction of randomly branched cross links into polycarbonate has been shown to substantially improve the melt strength of the resin [24,25], and considering that nanocomposite materials often have increased T_g compared with their pure matrix, we were interested in investigating OHPS melt blending with bisphenol A polycarbonate (PC) to determine if PC/OHPS composites could be formed, and to ascertain processing features involved in melt blending of OHPS with PC under twin-screw extrusion processing conditions.

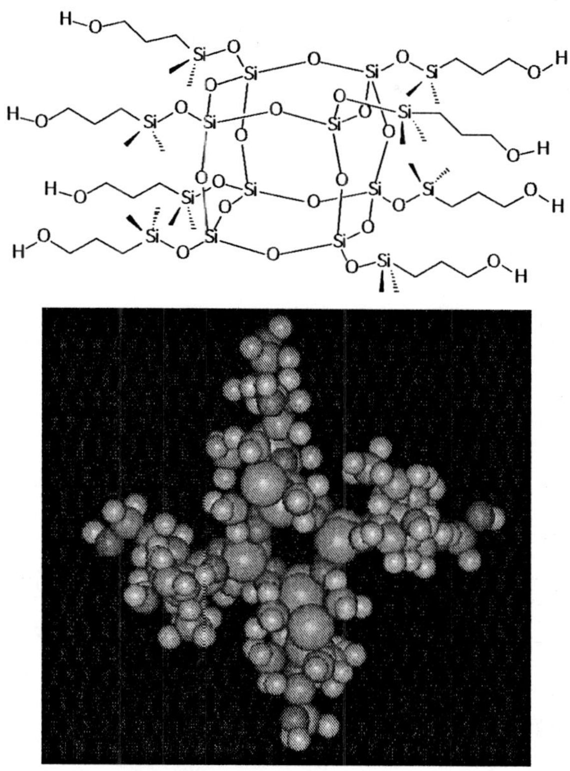

FIGURE 1 Structure of OHPS as valence bond (above) and space filling model (below) [18]

EXPERIMENTAL SECTION

Materials

Lexan® bisphenol A polycarbonate (PC) was supplied by General Electric. The material had a T_g of 149.46°C as measured by differential scanning calorimetry, weight average molecular weight of 29,810 g/mol, number average molecular weight of 18,750 g/mol, with polydispersity of 1.59 as determined by laser light scattering using a Shimadzu HPLC-GPC system consisting of a LC–10AD pump, SIL–10AF autosampler and Wyatt Mini DAWN light scattering detector.

Tetrabutylammonium tetraphenyl borate, 99% (TBATPB, CAS number 15522-59-5) was purchased from Aldrich Chemical Company, Inc. and used directly without purification.

Commerical octakis(3-hydroxypropydimethylsiloxy)octasilsesquioxane (OHPS) was purchased from Aldrich Chemical Company, Milwaukee, WI, as POSS PSS-Octa[(3-hydroxypropyl)-dimethylsiloxy] (CAS Number: 288290-32-4).

Composite Preparation

Composites were prepared using a HAAKE MiniLab (Thermo Electron Corporation), co-rotating, intermeshing twin screw extruder (screw diameter = 5/14 mm, conical). The barrel temperature was 260°C, and the screw speed 80 rpm with a cycle time from 5 minutes to 120 minutes. OHPS was loaded with PC at different weight ratios of 5 weight percent and 30 weight percent. A comparative set of PC/OHPS composites with TBATPB were also compounded.

Characterization

Scanning Electron Microscopy. The PC/OHPS composites specimens were fractured in liquid nitrogen and then coated with carbon in an evaporative coater. The morphology of the composites was examined by a JEOL 840A scanning electron microscope (SEM) with an accelerating voltage of 20 kV. Microanalysis was performed with an Oxford ISIS energy-dispersive X-ray system.

Images were acquired using the CAMEO system (trademark of Oxford Instruments), which gives a complete picture of both the topography and the chemical composition of the sample. The fine detail of the electron image is shown together with corresponding colors showing the composition. Compositional data is automatically incorporated within the image; the color displayed represents sample composition. CAMEO images show the presence of silicon as blue. Carbon, hydrogen and oxygen are red, and low silicon density is green. Thus, high concentrations of OHPS occur in the composites as blue, PC as red, and moderate concentrations of OHPS as green in the CAMEO images.

X-ray Diffraction. Wide angle X-ray diffraction (WAXD) measurements were conducted to estimate the crystallinity of the composite materials from the diffractograms so obtained. WAXD analysis was performed on Rigaku refractometer in the reflection mode using a Cu-Kα X-rays of a wavelength 1.5406 Å. Samples used for analysis were films formed by extrusion. Samples ware scanned from $2\theta = 1°$ to $2\theta = 30°$ with respect to the X-ray beam at a speed of 2.0 deg/min. X-ray analysis was performed on thin films except for the OHPS and TBATPB themselves, which were in powder form.

Infrared Spectroscopy. Infrared analysis was performed using attenuated total reflectance (ATR) with a BIO-RAD FTS-40 spectrophotometer equipped with an UMA-500 microscope used in reflection mode to obtain spectra.

Differential Scanning Calorimetry. Differential scanning calorimetry (DSC) was performed with a DSC Q100 in aluminum pans at heating and cooling rates of 10°C/min from 25 to 250°C under N_2 atmosphere to determine T_g values. The amount of samples used in all measurements was 3–4 mg.

Mechanical Properties. Tensile data was provided using an MTS Q Test 10 Elite instrument equipped with mechanical grips for tensile analysis and a 10N load cell. Cylindrical samples were strained at a rate of 1mm/min until the sample yielded or until stress on the sample exceeded the capacity of the load cell.

RESULTS AND DISCUSSION

The PC/OHPS nanocomposite materials, generated via melt blending in the twin screw extruder, were investigated using different analytic techniques including XRD, SEM, FTIR. In particular, the effect of polymer properties in systems prepared via melts blending of PC with OHPS with and without TBATPB as potential reaction catalyst was studied. TBATPB was chosen for this work since it has been used in polycarbonate work previously [26].

Scanning Electron Microscopy and X-ray Diffraction Analysis

Morphologies of the composite materials were investigated to determine whether OHPS dispersed throughout the polymer matrix. Two factors favor the blending of the inorganic material. These include the melting point of OHPS which was 89–107°C, and the reported solubility of OHPS in common solvents, including acetone, chloroform, dichloromethane, hexane, methanol, and tetrahydrofuran, with water insolubility. Since methylene chloride and tetrahydrofuran are also superior solvents for PC, both OHPS and PC are hydrophobic materials, favoring their miscibility. Additionally, the operating temperature of the twin screw extruder during melt mixing was 260°C, exceeding the melting point of either the OHPS or TBATPB, the latter of which has a melting point of 235–237°C.

SEM photos indicated good dispersion of OHPS at 5 weight percent after 60 minutes of mixing, showing even distribution of material indicating excellent compatibility of OHPS with PC. At 30 weight percent concentrations of silicon are noted at 10 minutes. These appear as blue in the CAMEO image. This indicates that 10 minutes may not be sufficient to fully disperse the OHPS, although even at 10 minutes high concentrations of OHPS have dispersed. Local ordering by intermolecular inorganic-organic or inorganic-organic interactions such as hydrogen bonding may also be responsible for concentrations of silicon-rich regions as indicated by green or blue areas that appear in the SEM.

WAXD indicates fine structure present characteristic of OHPS. Melt blending for as little as 5 minutes results in complete disappearance of the crystalline OHPS. WAXD patterns of all composite materials displayed only a very broad hump centered at $2\theta = 17^0$ unlike the WAXD of pure OHPS, indicating that OHPS crystals melt and neither recrystallize nor sufficiently phase segregate upon cooling. The broader peaks in the composite materials suggest that the silicon-oxygen cage structures remain intact under all conditions studied. This suggests that periodicity remains within the composite systems. From this it is inferred that the octafunctional octahedral silsesquioxane silicon-oxygen cage structure unique to the OHPS is maintained throughout the duration of the blending times. Although blue islands representing high regions of silicon density are seen in the SEM Cameo images of approximately 10 μm in diameter, WAXD does not register these as having a crystalline structure. Figures 2–7 show the SEM and WAXD results.

Infrared Spectroscopy

Figure 8 shows the infrared spectra of OHPS, PC, and composites. Assignments for OHPS given by Laine [19] show bands indicative of OHPS in the composites. A broad band indicating hydrogen bonding of the OHPS hydroxyl groups is seen at 3265 cm^{-1}. The disappearance of this band in the composite materials at 5 weight percent incorporation in composites is attributed to masking by high concentration of polymer. No significant new bands are noted in the composite materials that are not already present in either PC or in OHPS.

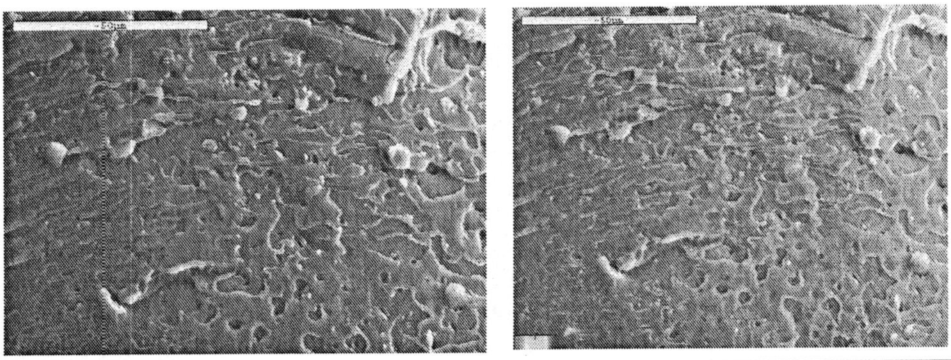

FIGURE 2 SEM photographs of the 5wt% PC/OHPS uncatalyzed composite extruded for 60 min., (a) secondary electron image, (b) CAMEO image

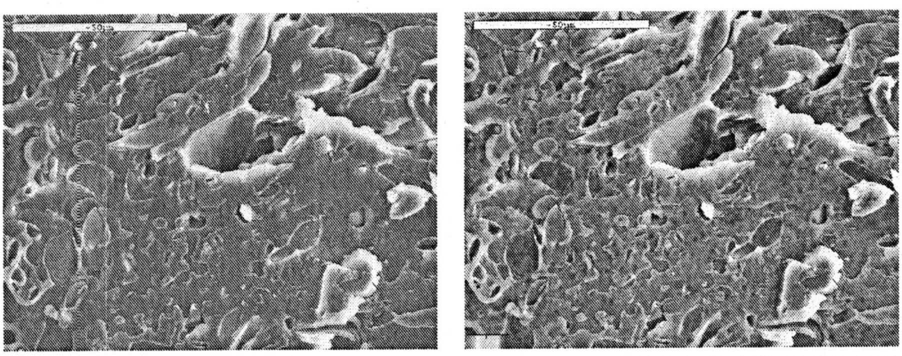

FIGURE 3 SEM photographs of the 30wt% uncatalyzed PC/OHPS composite extruded for 10 min., (a) secondary electron image, (b) CAMEO image

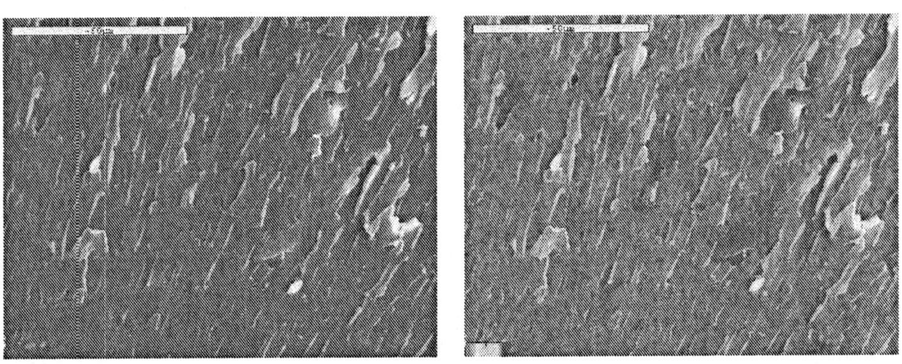

FIGURE 4 SEM photographs of the 5wt% PC/OHPS catalyzed composite extruded for 60 min., (a) secondary electron image, (b) CAMEO image

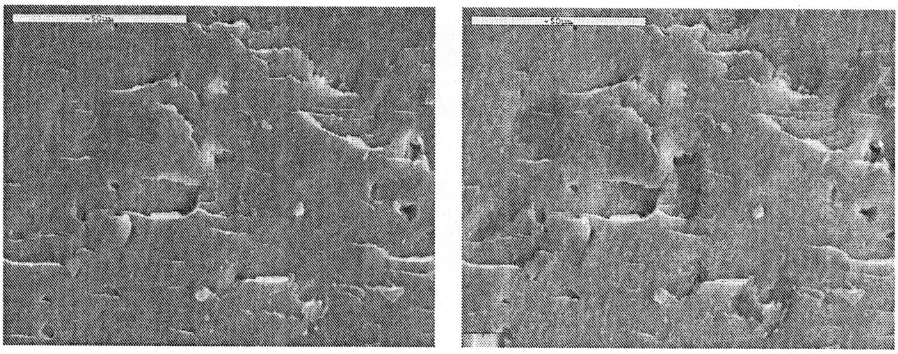

FIGURE 5 SEM photographs of the 30wt% catalyzed PC/OHPS composite extruded for 10 min., (a) secondary electron image, (b) CAMEO image

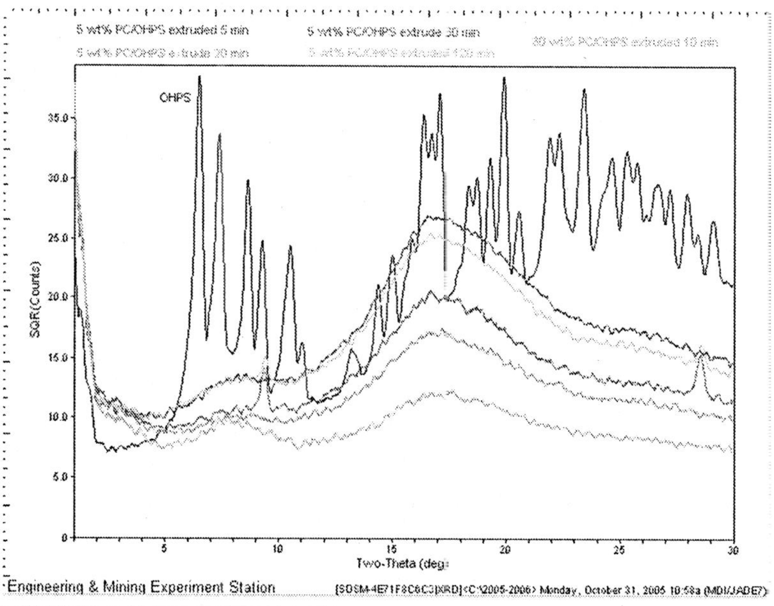

FIGURE 6 WAXD patterns of OHPS and PC/OHPS uncatalyzed composite materials

Glass Transition Temperatures

Glass transition temperatures indicate the onset of longer-range segmental molecular mobility among polymer chains within bulk material. The T_g of unprocessed PC was 149.46°C. Figure 9 shows the unusual behavior between composites formed from catalyzed and the uncatalyzed melt blending. The upper curves show equilibration in the early phases of the blending process. Pure PC shows a slight decrease in the T_g. On the other hand, the uncatalyzed mixture of polycarbonate with 5 weight percent OHPS shows a distinct although slight increase from the initial value at 5 minutes to a value

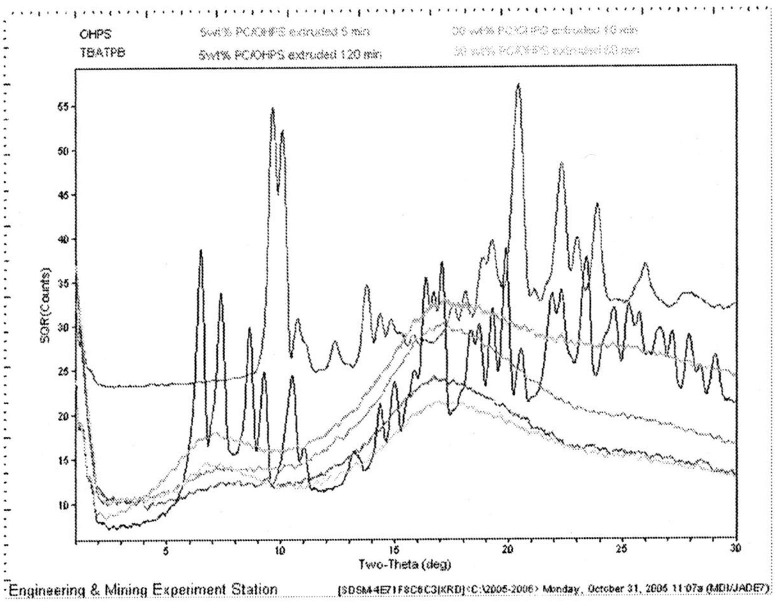

FIGURE 7 WAXD patterns of OHPS, TBATPB, catalyzed PC/OHPS composite materials

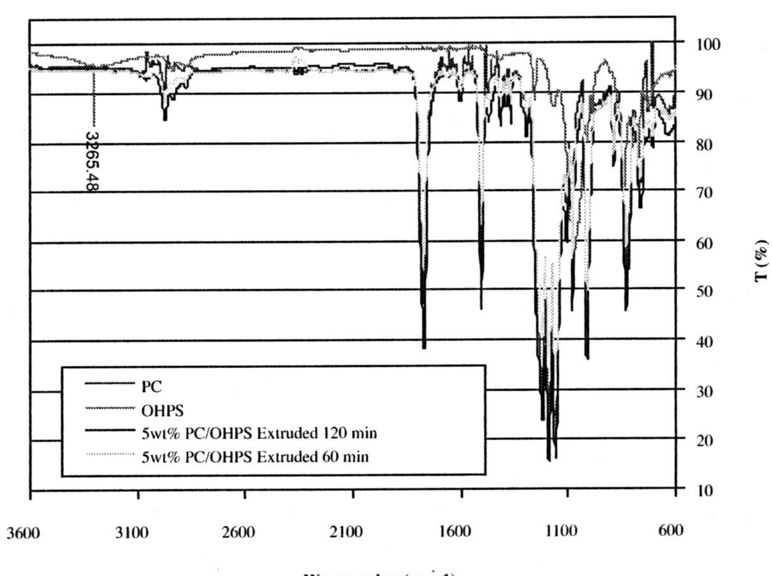

FIGURE 8 IR spectra of OHPS, PC, 5 wt% uncatalyzed PC/OHPS composites extruded 60 min., and 5 wt% uncatalyzed PC/OHPS composites extruded 120 min.

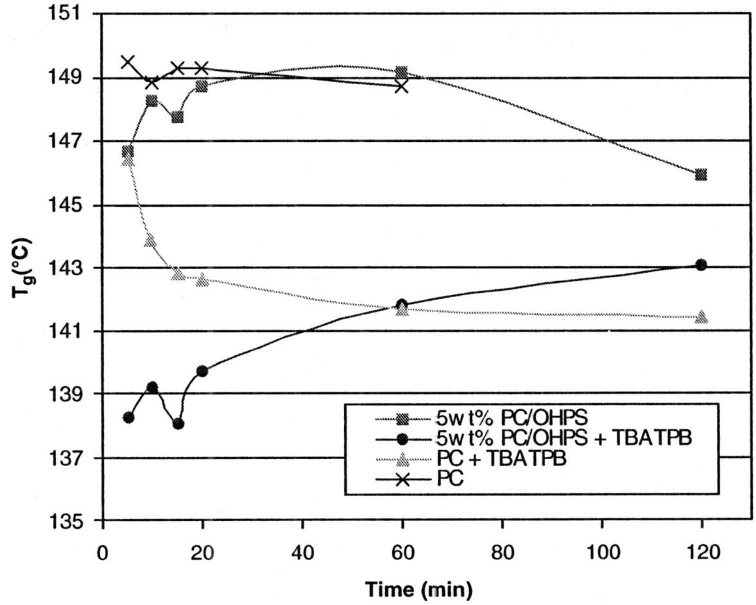

FIGURE 9 DSC data of composite sample plotted versus processing time. (■): 5 wt% uncatalyzed PC/OHPS composite; (×): pure PC; (▲): catalyzed PC; (●): 5 wt% catalyzed PC/OHPS composite

of 149.13°C after 60 minutes. While nanodispersed particles are generally regarded as having a plasticizer effect thereby decreasing the T_g [27] other researchers have noted that POSS incorporation into copolymers affects T_g in a concentration dependent manner, with higher POSS content increasing T_g, an effect attributed to chain aggregation that effectively hinders chain molecular motion. Pittman et al. have equated this to being "akin to attaching an anchor to a jump rope and then trying to use the jump rope" [28].

The T_g maximum of this run was followed by a marked decrease in T_g to 145.93°C after two hours. The T_g of the composites from uncatalyzed processing conditions differ from those of the catalyzed reactions shown in the lower portion of Figure 9. PC with TBATPB and 5 weight percent OHPS shows an increasing T_g from initial fluctuations of earlier samples with a steady increase from 20 minutes at 139.68°C, to 60 minutes at 141.79°C, to a final value at two hours of 143.06°C.

A preliminary period of mixing appears necessary to homogenize the materials thoroughly as evidenced by early fluctuations in T_g values of the composites. In uncatalyzed runs mixing continues for the first hour with a slow enhancement of the T_g followed by a decline in T_g, while in catalyzed runs an initially low T_g at early times is followed by a continuing increase as processing time lengthens. Thus, the uncatalyzed OHPS runs and the corresponding catalyzed runs appear to converge on a similar T_g value, but from different directions. This is attributed to the ionic, hydrophilic nature of TBATPB used in catalyzed runs. It is expected that this material requires a longer time for both melting and dispersion throughout the processing, hydrophobic PC/OHPS material. Likewise, if

TBATPB is involved in catalytic reactions with either PC, OHPS, or both, any such reactions would occur slowly as TBATPB continues to melt and disperse. Since T_g and particularly polymer modulus are susceptible to the influence of impurities and contaminants, the lower moduli seen at earlier processing times tested may be due to TBATPB acting as an ionic impurity and essentially contaminant before it becomes involved in any catalytic role it may exert at later times as processing continues, leading to an increase in T_g and modulus if TBATPB is chemically affecting the system over time.

The maximum T_g followed by its decrease suggests a different mechanism responsible in the uncatalyzed run. While POSS compounds have been reported to hinder the motion of resins formed when they are crosslinked, thus raising their T_g [29–31], no significant T_g increase commensurate with crosslinking in the composite materials is noted, unless it be of a transient nature partially responsible for the initial increases seen in Figure 9. Likewise, chain degradation is known to be among the phenomena responsible for decreases in T_g. Processing PC without addition of either OHPS or TBATPB, however, showed minimal decrease in T_g. It is apparent in PC with TBATPB that the catalyst has a rapid effect on the blending of PC and the T_g of the materials formed. This is expected since TBATPB is used as a redistribution catalyst in commercial polycarbonate processing and causes changes in chain lengths and molecular weight [26]. That blending PC/OHPS with catalyst for more than one hour causes the opposite effect, namely, an enhancement of T_g of the resulting samples versus blending PC with catalyst alone is a remarkable effect, one that results in a reversal of the catalytic effect on the PC. Clearly, therefore, OHPS is having an effect on PC without added catalyst, and OHPS is also having an effect on PC with added catalyst.

Mechanical Testing: Modulus and Elongation at Break

Romo-Uribe et al. have suggested that POSS moieties may introduce interchain interactions, thus retarding polymer chain segmental motion, which leads to higher moduli in POSS-containing systems [32]. In the composite systems of the current study, only decreases in modulus were noted.

The moduli of the five different composite systems are shown in Figure 10. One of the striking features is that in the two different systems of uncatalyzed 5 weight percent OHPS and uncatalyzed 30 weight percent OHPS high moduli are attained which then decrease sharply with increasing processing time. PC with catalyst shows initial fluctuation in modulus with little decrease over two hours, after which the modulus is 91.28 MPa. On the other hand, 5 weight percent OHPS, PC and catalyst shows little initial fluctuation in moduli but rather what appears to be a slow and steady increase of modulus reaching a maximum of 109.27 MPa after two hours.

Several items are particularly noteworthy. First, the moduli of all composites tend to converge on a value of approximately 100 MPa irrespective of catalyst presence or absence, or of percent OHPS. This is similar to the convergence of the T_g values seen in Figure 8. Secondly, the modulus of the 5 weight percent OHPS, uncatalyzed PC composite decreases from a maximum value of 226.84 MPa reached quickly at 20 minutes of processing to 109.27 MPa after two hours. This parallels the rise in T_g followed by the decrease of the same material as depicted in Figure 10. Contrasted to this material, the 5 weight percent OHPS, catalyzed PC composite of Figure 9 shows a steady increase of modulus over a narrow range which parallels the increasing T_g of this system as depicted

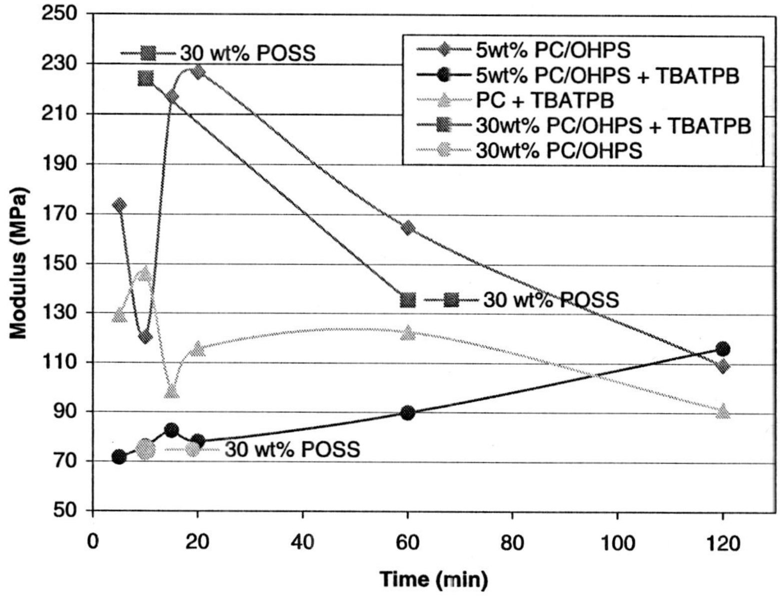

FIGURE 10 Modulus plotted versus processing time. (◆): 5 wt% uncatalyzed PC/OHPS composite; (■): 30 wt% catalyzed PC/OHPS composite; (▲): catalyzed PC; (●): 5 wt% catalyzed PC/OHPS composite; (●): 30 wt% uncatalyzed PC/OHPS composite

in Figure 8. These trends strongly suggest a correlation between changes in moduli and changes in T_g: when the former increase, so do the latter, and vice versa.

In marked contrast to the 5 weight percent OHPS systems, however, the 30 weight percent PC/OHPS systems reverse this correlation: the system without catalyst also has a low initial modulus as does the 5 weight percent system with catalyst. Likewise, the 30 weight percent OHPS, catalyzed PC system has a high initial modulus which decreases with time, similar to the 5 weight percent, uncatalyzed PC system. Thus, considerably higher 30 weight percent concentrations of OHPS affect the modulus in a different manner, and inversely with respect to presence or absence of catalyst, than do the lower, 5 weight percent OHPS concentrations. This may suggest that high concentrations of OHPS may inhibit the effect of the small amounts of TBATPB used, mimicking systems having lower concentrations of OHPS without catalyst.

Finally, it should be noted that with regard to elongation at break data indicates that the systems that exceeded the 90 Newton maximum allowed by the testing equipment and therefore did not break were the 5 weight percent OHPS with PC and pure PC. The 5 weight percent OHPS with catalyst and PC had a lower strain at break than did PC with catalyst control.

CONCLUSIONS

The addition of OHPS to Lexan® has been studied with respect to the effect of different processing times and catalyst addition on morphology, T_g, and moduli of the composite systems. Excellent dispersion of the OHPS was obtained very quickly during processing and all crystalline structure was lost. The structure of the OHPS cubic silsesquioxane was maintained throughout the two hour processing period under all conditions studied while the crystalline character of the bulk material was lost. In the case of the lower weight percent OHPS used in the catalyzed composite processing systems, increases in T_g paralleled increases in modulus over the entire processing time at the selected intervals tested. Likewise, in the case of the lower weight percent OHPS used in the uncatalyzed composite processing systems, initial increases in T_g followed by decreases paralleled similar increases and decline of moduli at the intervals tested. In virtually all cases both T_g and modulus tended to converge on similar values after two hours of processing time, and longer blending time may have resulted in complete convergence at some time. Such behavior may indicate the operation of initial physical mixing in the uncatalyzed systems which with time becomes a chemical reaction, whereas catalyzed systems initiate reactive blending chemical reactions more quickly from the outset. Since the T_g as well as modulus of both catalyzed and uncatalyzed runs are seen to move toward convergence, this suggests the effect of OHPS on PC may be obtained without added TBATPB. Likewise, this suggests the interaction of OHPS on PC at the molecular level is chemical and may be that of a chemical reaction which could be that of polymer chain scission by OHPS, with concomitant grafting of PC chains onto OHPS. OHPS also has an effect on the catalyzed system, an effect which ultimately results in both higher modulus and higher T_g of the composites than result from blending of PC and OHPS alone. The origin of this effect requires further investigation, and work is in progress on molecular characterization of the composites as part of our larger investigation into these unique systems.

ACKNOWLEDGMENT

We gratefully acknowledge the Army Research Laboratory Cooperative Agreement Grant Number DAAD19-02-2-0011 for the support of this work. The views and conclusions contained in this document are that of the authors and should not be interpreted as representing the official policies, either expressed or implied, of the Army Research Laboratory or the U.S. Government. The U.S. Government is authorized to reproduce and distribute reprints for Government purposes notwithstanding any copyright notation hereon. Thanks also go to Dr. John T. Bendler, Physics Department, United States Naval Academy, for the color molecular rendering of OHPS. Additionally, we would like to thank General Electric for gift samples of Lexan®.

REFERENCES

[1] Baker, W.E., Scott, C., Hu, G.H. Reactive Polymer Blending (Progress in Polymer Processing), Hanser Gardner Publications, 2001.

[2] Porter, R.S., Wang, L.H. *Polymer,* 1992, *22*, 2019.

[3] Devaux, J., Godard, P., Mercier, J.P., Touilaux, R., Dereppe, J.M. *J. Polym. Sci. Polym. Phys. Ed.,* 1982, *20*,1881.

[4] Schwab, J.J., Hughes STX, Lichtenhan, M.J., Carr, K., Chaffee, P., Mather, P.T., Romo-Uribe, A. "Polyhedral Oligomeric Silsesquioxanes (POSS): silicon based monomers and their use in the preparation of hybrid polyurethanes," Symposium O, MRS, Spring 1998, San Francisco, CA, 13–17 April 1998.

[5] Zheng, S., Feher, F.J., and Ren-Zhi, Jin. "Polymer nanocomposites via reactive melt blending of thermoplastics with monofunctional POSS monomers," Materials Research Society Symposium Proceedings Series Volume 733E, April 2–4, 2002.
[6] Wu, Chunyong; Wicks, Douglas. *PMSE Preprints,* 2005, *92*, 673–674.
[7] Hideo, S., Masanori, I. JP 2002327062, 2002.
[8] Lichtenhan, J.D., Otonnari, Y.O., Carr, M.J. *Macromolecules*, 1995, *28*, 8435.
[9] Lichtenhan, J.D., Haddad, T.S. *Macromolecules*, 1996, 29, 7302.
[10] Xu, H., Kuo, S-W., Huang, C-F., Chang, F-C. *Journal of Applied Polymer Science*, 2004, 91(4), 2208–2215.
[11] Xu, H., Kuo, S-W., Lee, J-S., Chang, F-C. *Polymer*, 2002, 43(19), 5117–5124.
[12] Mather, P.T., Jeon, H.G., Haddad, T.S. Proceedings of the International Congress on Rheology, 13th, Cambridge, United Kingdom, Aug. 20–25, 2000, ed. Binding, D.M., 2000, 4, 53–55.
[13] Mather, P.T., Jeon, H.G., Haddad, T.S. Polymer Preprints, 2000, 41(1), 528–529.
[14] Yoon, K.W., Polk, M.B., Park, J.H., Min, B.G., Schiraldi, D.A. *Polym. Inter.,* 2005, 54(1), 47–53.
[15] Lichtenhan, J.D., Vu, N.Q., Carter, J.A., Gilman, J.W., Feher, F.J. *Macromolecules,* 1993, *26*, 2141–2142.
[16] Haddad, T.S., Lichtenhan, J.D. *Macromolecules*, 1996, *29*, 7302–7304.
[17] Fu, B.X., Zhang, W.H., Hsiao, B.S., Rafailovich, M., Sokolov, J., Sauer, B.B., Phillips, S., Balnski, R. *High Perform. Polym.* 2000, *12*, 565–571.
[18] The molecular rendering was created by Dr. John T. Bendler using Materials Studio 3.2 with the COMPASS forcefield.
[19] Zhang, C., Laine, R.M. *J. Am. Chem. Soc.*, 2000, *122* (29), 6979–6988.
[20] Laine, R.M., Choi, J., Lee, I. *Adv. Mater*., 2001, *13*, 11, 800–803.
[21] Choi, J., Harcup, J., Yee, A.F., Zhu, Q., Laine, R.M. *J. Am. Chem. Soc.,* 2001, *123*, 11420–11430.
[22] Tamaki, R., Choi, J., Laine, R.M. *Chem. Mater.*, 2003, *15*, 793–797.
[23] Laine, R.M., Tamaki, R., Choi, J. US 6,927,301, 2005.
[24] Boden, E.P., Krabbenhoft, H.O. EP 0 378 858 A2, 1989.
[25] Mobley, D.P., Amaratunga, M.M., Lobos, J.H. US 6,498,271, 2002.
[26] King, Jr., J.A., McCloskey, P.J., Dardaris, D.M. USP 5,259,226, 1995.
[27] Kopesky, E.T., Haddad, T.S., Cohen, R.E., McKinley, G.H. *Macromolecules,* 2004, *37*(24), 8992–9004.
[28] Li, G., Wang, L., Ni, H., Pittman, C.U. *Journal of Inorganic and Organometallic Polymers,* 2001, *11*(3) 123–154.
[29] Jiwon, C., Jason, H., Albert, F.Y., Quan, Z., Richard, M.L. *J. Am. Chem. Soc.*, 2001, *123*, 11420.
[30] Lee, A., Lichtenhan, J.D. *J. Appl. Polym. Sci.,* 1999, *73*, 1993.
[31] Choi, J., Kim, S.G., Laine, R.M. *Macromolecules,* 2004, *37*, 99.
[32] Romo-Uribe, A., Mather, P.T., Haddad, T.S., Lichtenhan, J.D. *J. Polym. Sci. B Polym. Phys.*, 1999, *36*, 1857.

Structural Modification of Clays for Enhancement of Properties of Hybrid Organic-Inorganic Nanocomposites

M. Hassan,[*] H. El-Shall,[†] C. Beatty,[†] and K. Seleem[*]

Layered silicates dispersed as a reinforcing phase in an engineering polymer matrix are one of the most important forms of "hybrid organic-inorganic nanocomposites." In order to make the inorganic clay compatible to the organic polymer, chemicals are inserted or intercalated into the spaces or galleries between the clay layers. Considerable research efforts have been reported in this area. However, research efforts to adequately match the surface polarity of the clay and the polymer are still needed. This must be accomplished in order to allow the polymer to intercalate into the galleries in route to forming the final nanocomposite. Thus, the overall goal of this study is to optimize the conditions that can be used to produce significant improvements in the properties of polymer/clay nanocomposites. This chapter includes a discussion of results of such studies as (a) engineering organically modified clay to optimize their compatibility with a given polymer, and (b) incorporating and dispersing the modified nanolayer silicate in the polymer. In addition, the mechanisms responsible for these improvements are presented.

INTRODUCTION

Naturally occurring silicates, like montmorillonite (MMT), have received much attention as reinforcement materials for polymers because of their potentially high aspect ratio and unique intercalation/exfoliation characteristics (Wei-Bing et al. 2002; Lebaron et al. 1999). Montmorillonite is of particular interest because it has a layered structure, typically about 1 nm in thickness and a high aspect ratio ranging from 100 to 1,500 that,

* Central Metallurgical Research and Development Institute, El-Tibbeen, Egypt

† Department of Materials Science and Engineering, Engineering Research Center for Particle Science and Technology, University of Florida, Gainesville, Florida

with proper exfoliation, can lead to platelets with high stiffness and strength dispersed in the polymer matrix (Lebaron and Pinnavaia 2001; Kawasumi et al. 1997; Tyan et al. 1999; Chen and Yang 2001). Many articles in this field have demonstrated the relationship between the improvement of properties of clay nanocomposites and the complete exfoliation of silicate layers in the polymer matrix. According to many of these papers the degree of exfoliation depends on the structure of the clay, curing temperature, and curing agent, as well as techniques for processing clay-polymers nanocomposites (Yasmin et al. 2003; Isik et al. 2003; Joo Chin et al. 2001; Xu et al. 2002). The commonly used techniques for processing clay-epoxy nanocomposites are direct mixing and solution mixing. Meanwhile, Vaia et al. (1993) suggested that the degree of exfoliation can be improved through the aid of conventional shear devices such as extruders, mixers, ultrasonicators, etc. It has also been reported that the degree of exfoliation depends on the structure of the clay, its functionality, packing density and length of organic modifiers, thermal stability and degradation of the organically modified montmorillonite (Wei Xie et al. 2001; C. Wan et al. 2004).

In this paper, we studied the delamination of montmorillonite using different clays and onium ions. The epoxy-montmorillonite nanocomposites were characterized by X-ray diffraction (XRD), transmission electron microscope (TEM), differential scanning calorimeter (DSC), thermogravimetric (TGA) and dynamic mechanical thermal analysis (DMS).

EXPERIMENTAL

Materials

Na+-montmorillonite with an exchange capacity (CEC) of 84 mmol/100 g was provided by Bentonite Co., Northern Coast of Egypt. Cloisite Na+, cation exchange capacity, CEC = 92 mmol/100 g supplied by Southern Clay Products Inc., USA. Surfactants, dodecylamine hydrochloride (DCAC), citylpyridinium ammonium chloride (CPAC), and hexadecyl ammonium bromide (C16) (HDAB) were provided by Petroleum Research Institute, Egypt.

The epoxy resin was the diglycidyl ether of bisphenol A (DGEBA, epoxide equivalent weight = 178 g, EPON 828) and diethytoluenediamine (EPI-CURE W) was used as curing agent, provided by Shell Chemicals.

Techniques

Twenty grams by weight of MMT dispersed into 500 ml of distilled water. The mixtures of different concentrations of surfactants (10, 5, 2.5 g by weight) and 100-, 25-, and 12.5-ml distilled water were added drop-wise into the MMT suspension. The reaction was maintained for 1 h. The exchanged clays were washed with deionized water until no chloride and bromide were detected with 0.1 N agno3 solutions and dried in the air.

Nanocomposite Preparation

The epoxy resin was mixed with 2%, 5%, 7%, and 10% of homoionic organo-clay (OMMT) in a 100 ml beaker. A magnetic stirrer was used to mix the samples for 3 h. At 70°C, premixing of the clay with the epoxy monomer also was done in the presence of low-boiling point solvents, e.g., acetone, ethanol and isopropyl alcohol. Organo-clays–epoxy–solvent mixture was stirred at 70°C for 3 h. The solvent was evaporated over a 1 h. period at 25°C, 30 min. at 50°C and 1 h. at 75°C under vacuum. Then, the curing agent [diethyltoluenediamine] was added into the epoxy/clay hybrid and mixed thoroughly by the magnetic stirrer. The mixtures were then degassed to remove bubbles using a vacuum oven before they were cast into a mold. Bottom, frame and top cover of mold was made from an aluminum foil covered corning micro slides [75 × 50 mm plain, 0.96 to 1.06 mm thickness]. The samples were cured in a vacuum oven at 70°C for 17 hrs. This last procedure was repeated again at 140°C for 3 hrs.

Characterization Techniques

The change in the interlayer spacing of clay minerals [Na+-montmorillonite and Cloisite Na+] was measured using a Philips PW 1730 X-ray generator with Ni-filtered and Co-K radiation run at 40 kv and 25 mA. The scanning range is from 2° to 12°. The structure of the clay was investigated at different stages during the nanocomposite synthesis. The clay particles were mounted on a sample holder with a large cavity and a smooth surface was obtained by pressing the particles with a glass plate. Analysis of organoclay, swollen in the epoxy resin and in curing agent, was performed by spreading the mixture on a sample holder. The nanocomposite plates produced during the molding process had a fairly smooth surface.

Differential scanning calorimetry (DSC) and thermogravimetric (TG) measurements were carried out using a Satram thermal analyzer to analyze the thermal stability, and onset temperature of degradation of epoxy/OMMT nanocomposite. Samples were heated to 600°C at a scanning rate of 10°C /min under flowing argon.

A Seiko DMS220 interfaced with a Seiko Rheostation model SDM/5600H was used to test dynamic mechanical specimens. Testing was conducted from −120° to 150°C at a heating rate of 5°C/min in dry nitrogen atmosphere maintained at an approximate flow rate of 100 ml min^{-1}. Rectangular samples (20 mm × 10 mm × 1 mm) were cut from the compression molded sheet and tested in bending mode at a frequency of 1Hz.

To verify the nanoscale structure, a Zeiss-EM-10 transmission electron microscope (TEM) operating at an accelerating voltage of 60 kv was also used. Thin TEM specimens with a thickness of 100 nm were cut from nanocomposite blocks using an ultramicrotome, Leica Ultra Cut.

RESULTS AND DISCUSSION

The XRD patterns of the Cloisite Na+ and Na+-montmorillonite (MMT) have a characteristic diffraction peak d001 plane at $2\theta = 8.4–7.8$ based on Bragg equation, corresponding to a gallery spacing of 1.2–1.3 nm respectively. The organoclays (OMMT), with modifiers of different chain lengths (n = 12, 16, and 21), show characteristic diffraction peaks d001 at $2\theta = 5.2–5.9°$ (d-spacing is 1.8–1.9 nm) for DCAC/MMT (Figures 1 and 2). In the case of C16/MMT, the clay was expanded to d-spacing 4.4 nm. However, part of clay layers slightly exfoliated to $2\theta = 6.8$ (d-spacing 1.5 nm), which means that some of MMT layers keep the original crystal structure and exist in primary particles (Figures 3 and 4). According to Shi, Lan, and Pinnavaia 1996, unexchanged Na+ centers in alkylammonium-exchanged clays may act as pinning points to prevent gallery expansion by the polymer

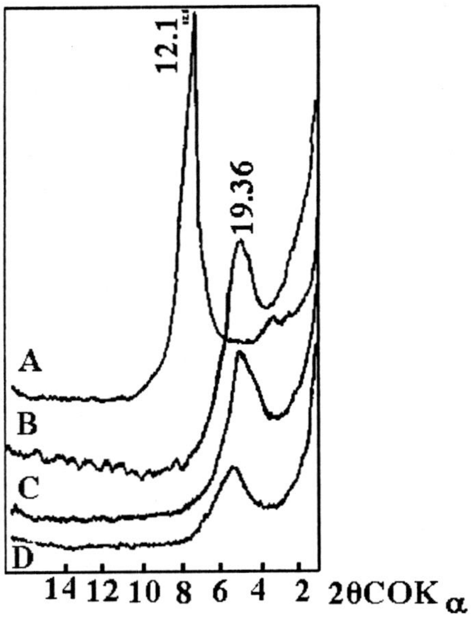

FIGURE 1 XRD patterns of MMT and OMMT; A = Cloisite Na^+; B = DCAC/Cloisite Na^+; C = DCAC/Cloisite Na^+ at 100°C; D = DCAC/Cloisite Na^+ at 200°C

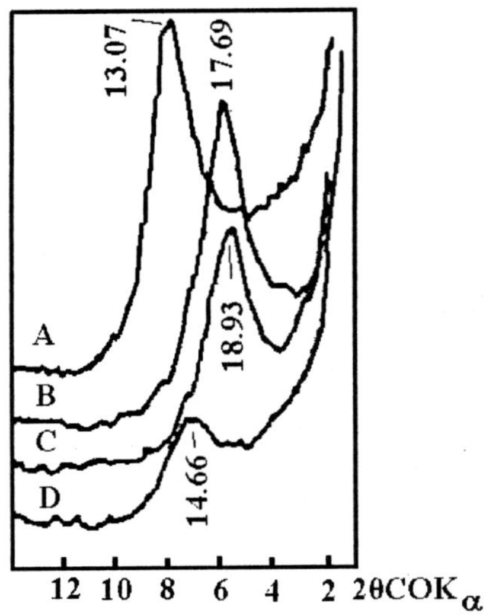

FIGURE 2 XRD patterns of MMT and OMMT; A = Na+-montmorillonite; B = DCAC/Na+-montmorillonite; C = OMMT at 100°C; D = OMMT at 200°C

precursors. Furthermore, the electric constant of the product may be affected by the residual Na+ or other metal ions.

Meanwhile, CPAC/Cloisite Na+ produced characteristic peaks d001 at 2θ =2.1–4.2° (d-spacing 2.3–4.4 nm) compared with d001 at 2θ = 4.95–2.4° (d-spacing 2.0–4.2 nm) for CPAC/Na+-montmorillonite (Figures 5 and 6).

The Effect of Thermal Treatment on OMMT Structure

Alkyl salts are lamellar ionic crystals comprised of alternating layers of paraffinic alkyl chains and of ionic groups and associate water of hydration. Upon degradation around 200°C, the entire structure is rapidly and completely destroyed in our process. In contrast, the layered morphology of the aluminosilicate persists throughout the surfactant decomposition event. X-ray diffraction of the decomposition process up to 200°C yielding the layer repeat distance (d001) and the fwhm of the basal reflection is shown in Figures 1–6. The degree of broadening of the peaks (fwhm) is inversely proportional to the layer registry through the Scherrer equation (Wei Xie et al. 2001). The initial increase in layer repeat distance (d001) and layer registry (≈fwhm −1) from 50–100°C is due to melting of the intercalated surfactant, which creates a fluid-like environment between the layers, enabling local relaxation of residual stress and packing imperfections associated with prior process history. The slight increase in d001 from 100–200°C is attributed to the thermal expansion of the structure normal to the layers.

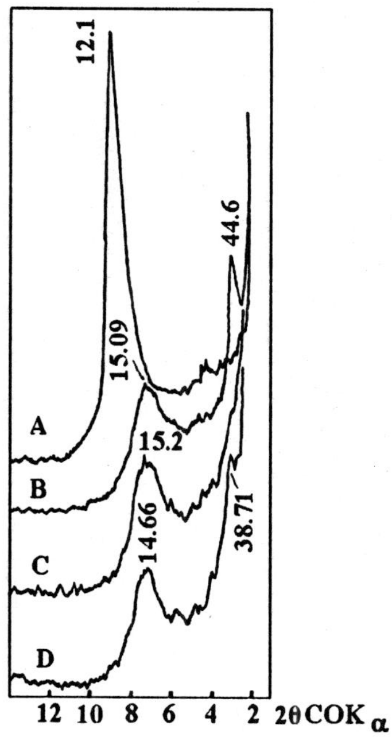

FIGURE 3 **XRD patterns of MMT and OMMT; A = Cloisite Na⁺; B = C16/Cloisite Na⁺; C = C16/Cloisite Na⁺ at 100°C; D = C16/ Cloisite Na⁺ at 200°C**

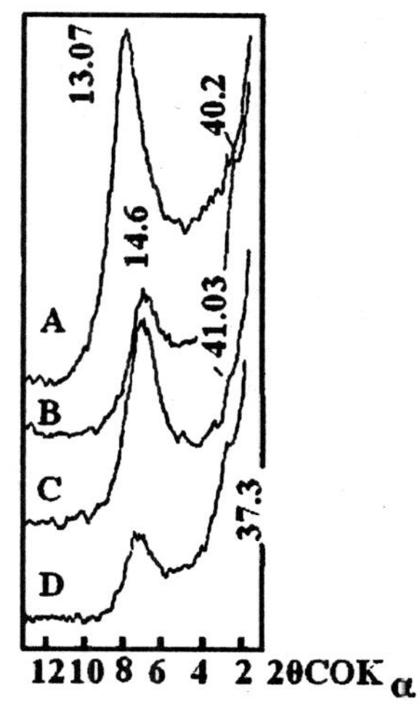

FIGURE 4 **XRD patterns of MMT and OMMT; A = montmorillonite Na⁺; B = C16/montmorillonite Na⁺; C = C16/montmorillonite at 100°C; D = C16/montmorillonite at 200°C**

Effect of Chain Length

As shown in XRD patterns of OMMT, we find that d-spacing of intercalated clay increases with the chain length when the intercalated onium ions are NH_3 $(CH)_{12}$ HCl, $H_{33}C_{16}N^+(C_2H_5)3Br$- and $C_{21}H_{38}Cl$ N^*H_2O. The results agree with published reports (Vaia et al. 1994). In general, as the chain length increases, the intercalated chains adopt a more ordered, liquid-like structure, resulting from a decrease in the gauche/ trans conformer ratio (Vaia et al. 1994). The ordered form will have a larger packing capacity than that of the disordered form. These results may also indicate that the onium ions aggregation in the interlayer has changed from monolayers to bilayers.

Effect of Onium Ions Concentration

The data indicate a correlation between the onium ion concentration (0.11–0.4M) of the bulk solution and the amount of intercalated gallery cations within the clay layers. For example, at 0.16 M of surfactant, Cloisite Na+ has a characteristic diffraction peak d001 at $2\theta = 5.6$ (d-spacing 1.8 nm) as shown in Figure 7. However, the gallery spacing of CPAC/Na+-montmorillonite expands slightly to $2\theta = 7.2$ (d-spacing from 1.2 to 1.4 nm) at the same concentration of surfactants. On the other hand, at (0.15M) concentration of C16, gallery spacing of Cloisite Na+ and Na+-montmorillonite expands to $2\theta = 2.3$–$6.5°$ (d-spacing 1.5–3.7 nm). Meanwhile at (0.14M) concentration of C16, d-spacing of both

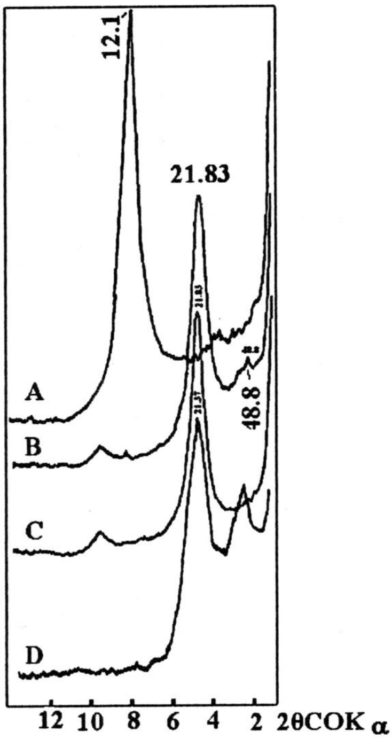

FIGURE 5 XRD patterns of MMT and OMMT; A = Cloisite Na$^+$; B = CPAC/Cloisite Na$^+$; C = CPAC/Cloisite at 100°C; D = CPAC/Cloisite Na+ at 200°C

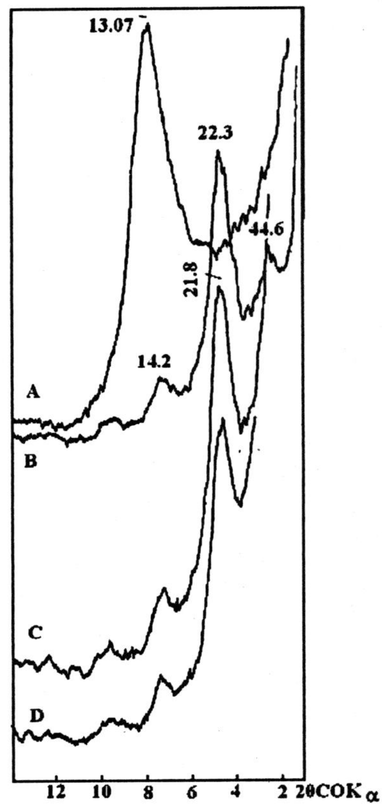

FIGURE 6 XRD patterns of MMT and OMMT; A = montmorillonite Na$^+$; B = Cetylpyridinium ammonium chloride/montmorillonite Na$^+$; C = OMMT at 100°C; D = OMMT at 200°C

bentonites expands slightly to 1.4 nm. The results indicate that the expansion of MMT gallery depends on the type of MMT (CEC of MMT), as well as on the concentration of onium ions. When the onium ion concentration of the bulk solution was lower than 0.0025 M, it was difficult to enlarge the interlayer spacing of the clay as reported by Chen and Yang (2002).

Thermal Properties of Organoclays

The DSC and TG curves of OMMT are present in Figures 8 and 9. The results show that dehydration temperature of OMMT is lower than those of MMT. This means that surfactant plays a role in improving thermal stability of MMT. The dehydration temperature of OMMT depends on the degree of delamination of montmorillonite. Also, the difference of water resistance between modified and unmodified clay can be explained in the same way as the modification of clay surface (Cheng and Yang 2001). The modification by

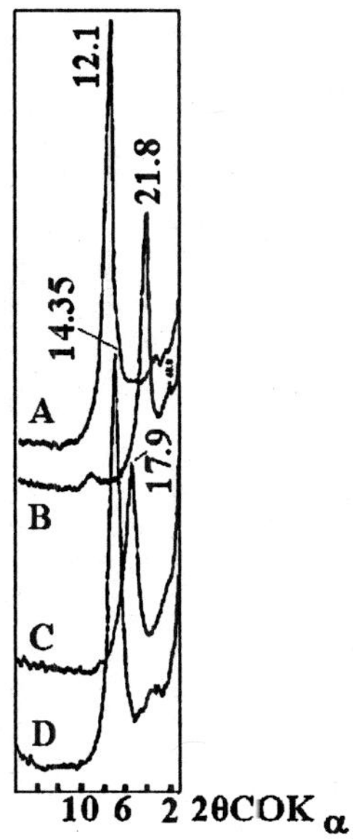

FIGURE 7 **XRD patterns of MMT and OMMT; A = Cloisite Na^+; B = 0.4M CPAC/Cloisite; C = 0.16 M CPAC/Cloisite Na+; D = 0.11 M CPAC/Cloisite Na+**

quaternary ammonium made the clay surface more hydrophobic, because the tail of onium ions was a hydrophobic group. Generally, the weight loss below 200°C is attributed to desorption of water, which includes inter-particles water and interlayer water. The associate weight evident in TGA analysis curves shown in Figure 9 indicate that modification of clay by onium ions was more hydrophobic than was that of unmodified clay. The organic surfactant appears to degrade in several steps, as shown by the changes in the rate of mass loss from the surface. These are (a) the free water region, (b) the region where organic substance evolve, (c) the structural water region and (d) organic carbon reactions. All the quaternary salts of different molecular weights were released in significant quantities during exposure of OMMT to lower temperatures. This study implies that the higher molecular weight material is still evolved, at ~200°C.

Evaluation of Epoxy–Clay Nanocomposite

In general, X-ray diffraction (XRD) has been used to characterize the formation and structure of epoxy-montmorillonite hybrids by monitoring the position, shape, and intensity of the basal reflection from the silicate layers. When insertion of epoxy in the silicate layers occurs, an increase of silicate interlayer volume and corresponding layer spacing is seen, which gives rise to a shift of the (001) diffraction peak to lower angles. No diffraction peaks would be visible in the case of exfoliated structures where silicate layers are completely and uniformly dispersed in a continuous polymer matrix. A broadening of the diffraction profile might indicate partial exfoliation of silicate layers as crystallite size decreases.

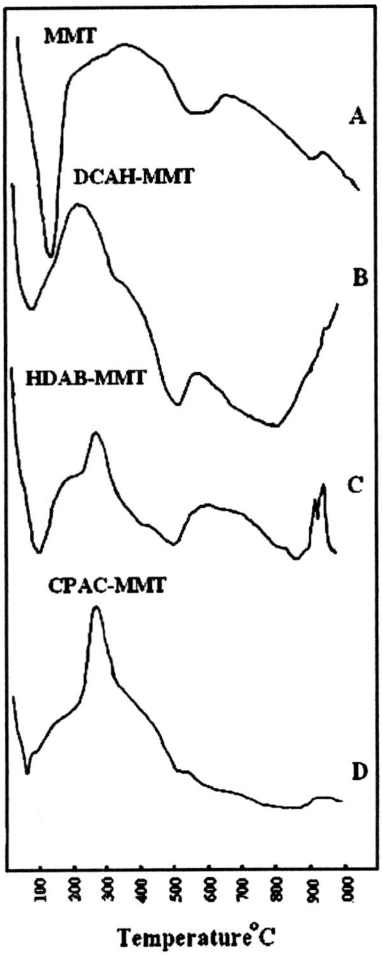

FIGURE 8 DSC thermograms of MMT and OMMT

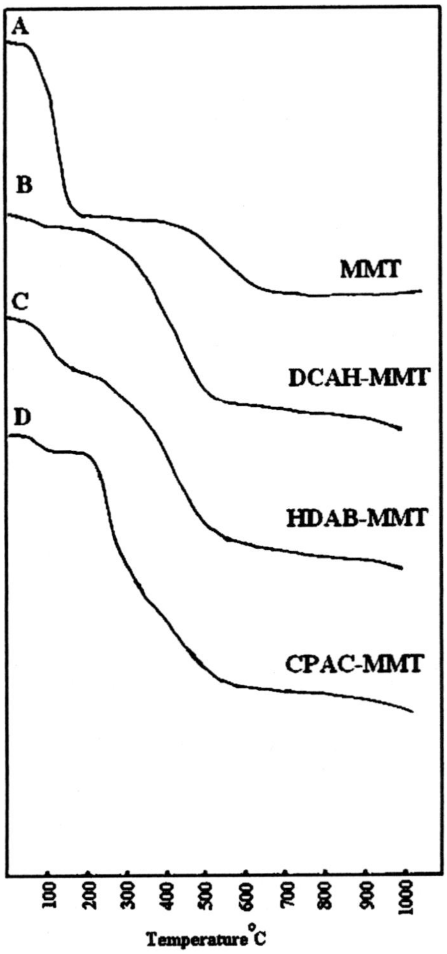

FIGURE 9 TGA thermograms of MMT and OMMT

Figures 10–12 show the X-ray diffraction pattern of (OMMT), and epoxy/OMMT nanocomposite (2, 5, 7, and 10% OMMT loading). The 2θ values along the x-axis can be converted to layer spacing value d through the Bragg relationship (nλ = 2dsinθ). There is an obvious shift of the diffraction peak towards lower angle from that of the corresponding OMMT. However, there is no diffraction peak at 2θ = 4°, even at 2θ = 8° for the epoxy/OMMT nanocomposite. At the low loading (2%) of C16/MMT and DCAC/MMT (Figures 10 and 11), it is difficult to find a prominent peak in a nanocomposite XRD data. This result can be interpreted by the fact that epoxy resin diffused into the galleries and formed intercalated or exfoliated nanocomposite. However, if the nanocomposites are disordered, no peaks are observed in the XRD, due to loss of the structural registry of the layers and/or the large d-spacing (>10 nm), or both. Thus XRD of nanocomposite is limited because a disordered and layered silicate can either be

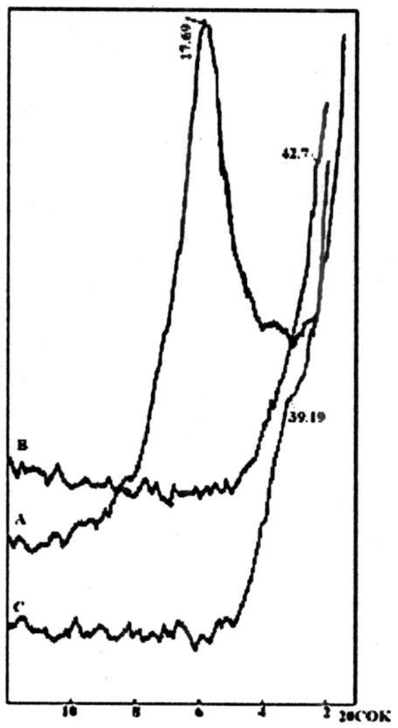

FIGURE 10 XRD patterns of OMMT and epoxy/OMMT, A = DCAC/Na+-montmorillonite; B = epoxy/7% OMMT; C = epoxy/10% OMMT

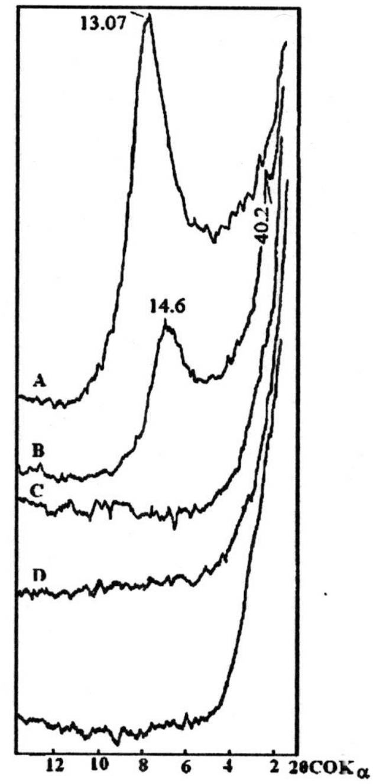

FIGURE 11 XRD patterns of MMT and OMMT and epoxy/OMMT; A = montmorillonite Na^+; B = C16-montmorillonite Na+; C = epoxy/5% OMT; D = epoxy/10% OMMT

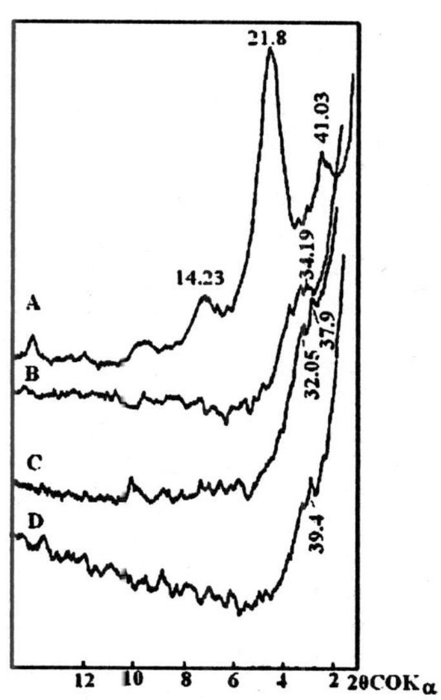

FIGURE 12 XRD patterns of MMT and epoxy/OMMT; A = montmorillonite Na+/Citylpyridnium ammonium chloride; B = epoxy/2% OMMT; C = epoxy/5% OMMT; D = epoxy/7% OMMT

delaminated or intercalated. In such case transmission electron microscopy (TEM) combined with XRD will more accurately characterize these materials.

When the content of OMMT increases further to 10%, however, very broad and very weak d001 reflection is observed at $2\theta = 2.4°$ corresponding to d-spacing 42.7 Å, indicating aggregation of OMMT in the epoxy resin matrix. In spite of this, very broad and spread peaks appeared at $2\theta = 2–2.3°$ corresponding to d-spacing ranging between 3.2–3.9 Å. By increasing OMMT loading, the lattice of 49 Å only appeared. In addition to the increase in gallery height, an increase in breadth of the peak profile was observed. Such broadening of the diffraction peak may indicate that partial exfoliation has occurred as crystal size decreases. The result demonstrates the monomer of epoxy resins or curing agent has diffused into the galleries resulting in the increase of lattice spacing. The data revealed that the sheets of clay were distributed; the d-spacing is independent of OMMT loading. All the epoxy/OMMT nanocomposite developed intercalate/exfoliated structures. We call this nanocomposite wedged nanocomposite. More direct evidence for the formation of a nanocomposite is proved by TEM of an ultramicrotom section. The TEM images in Figure 13 display individual silicate layers apparent as dark lines. Epoxy-OMMT nanocomposite is comprised mainly of face–face groupings of highly exfoliated clay layers separated by 5–15 nm of polymer. A more disordered exfoliated clay phase (i.e., fewer face–face orientations of nanolayers) was found for epoxy-DCAC MMT.

Thermal Properties of Epoxy/OMMT Nanocomposite

TGA were recorded for epoxy/OMMT nanocomposite to monitor the effect of clay on the thermal properties. There is no weight loss up to 250°C, the single step at this point simply corresponding to the polymer decomposition. The results in Figure 14 show that the degradation rate of the nanocomposites become slower compared to that of epoxy, indicating an improvement of thermal stability of epoxy. This could be due to thermal insulation by the inorganic material which limits the further degradation. The epoxy/OMMT (2%wt) nanocomposite onset degradation temperature is at 316°C and is lower than that of the 5% and 7% nanocomposites (331° and 351°C respectively).

The glass transition temperature (T_g) of the epoxy-clay nanocomposite was determined by DSC and DMS. As shown in Figures 15 and 16, there is no glass transition temperature (T_g) up to 99°C, which is characteristic of epoxy. In general, nanocomposite glass transition temperatures (T_g) ranged between 110–140°C, which is higher than that of epoxy with no clay. For all the OMMT, T_g increases in the order 2% > 5% > 7%. The mobility of polymer chains falls due to the interaction between the clay and polymer molecules, resulting in higher T_g. When the OMMT loading is relatively low 2%, the nanocomposite develops an exfoliation dominant structure. The retardant effects of the exfoliated platelets to heat and oxygen in the epoxy matrix is strengthened when the OMMT loading increases since the number of the exfoliated platelets increases with OMMT loading. When the OMMT loading increases further the nanocomposite develops an exfoliated/intercalated structure. The number of exfoliated silicate platelets decreases with the OMMT loading. The tactoids, which are less effective in blocking heat and oxygen than the exfoliated platelet, do, however, increase with OMMT loading. As a consequence, the activation energy decreases when the OMMT loading is higher. As the tactoids still show the retardant effects to heat and oxygen, the activation energy of the sampler with exfoliated/intercalated structures is still higher than that of the neat epoxy resin.

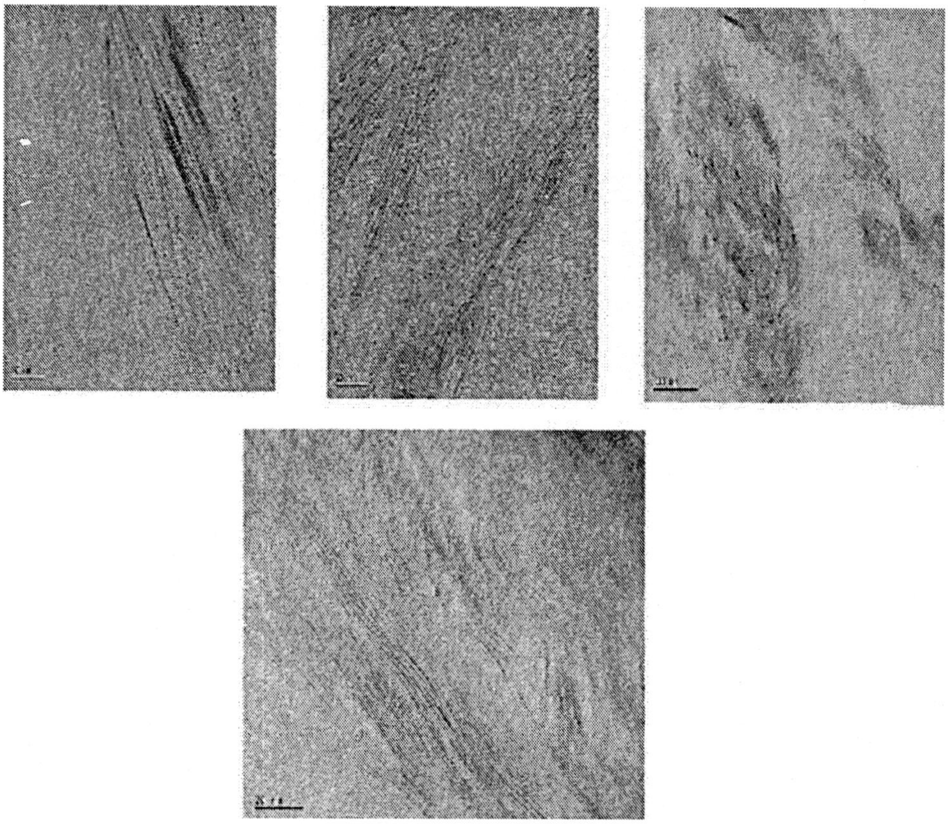

FIGURE 13 TEM photographs of epoxy-clay nanocomposite

CONCLUSIONS

Epoxy-clay nanocomposites were prepared using organically modified sodium montmorillonite by different surfactants. The delamination and exfoliation of the clays was found to depend on the chain length and concentration of the surfactant. Longer chain surfactants were found to enhance the exfoliation of the clays. In general, X-ray diffraction (XRD) has been used to characterize the formation and structure of epoxy-montmorillonite hybrids by monitoring the position, shape, and intensity of the basal reflection from the silicate layers. The data indicate that epoxy resin diffused into the galleries and formed intercalated or exfoliated nanocomposite. Such exfoliation was found to increase as the organic clay loading was increased from 2 to 7% by weight. The presence of such exfoliated clays has been confirmed by the transmission electron microscope images. Most importantly, the composite is found to have higher glass transition temperature and better thermal barrier than the epoxy matrix. Thus, the nanocomposite may be used for the applications that require such thermal barrier properties.

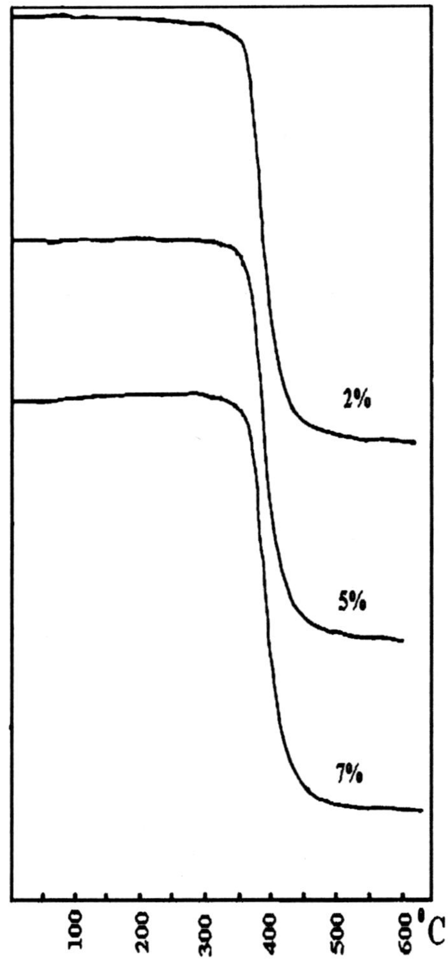

FIGURE 14 TGA thermograms of epoxy/OMMT

ACKNOWLEDGMENT

We are grateful to the Major Analytical Instrumentation Center (MAIC) and the Particle Engineering Research Center (PERC and the Department of Material Science) at the University of Florida for use of their equipment in this investigation. Partial financial support of the PERC, NSF grant # INT0115417 and the industrial partners of PERC, and NSF grant # INT 0211253 is gratefully appreciated.

REFERENCES

I. Isik, U. Yilmazer and G. Bayram. Polymer (2003), 44, 20, pp. 6371–6377.
T. Pinnavaia, T. Lan, Z. Wang, H. Shi and P.D. Kaviratna. ACS Symp Ser 2 (1996), pp. 250–261.
P.C. Lebaron and T.J. Pinnavaia. Chem. Mater. 13 (2001), p. 3760.
C.C. Zeng and L.J. Lee. Macromolecules 34(2001), p. 4098.

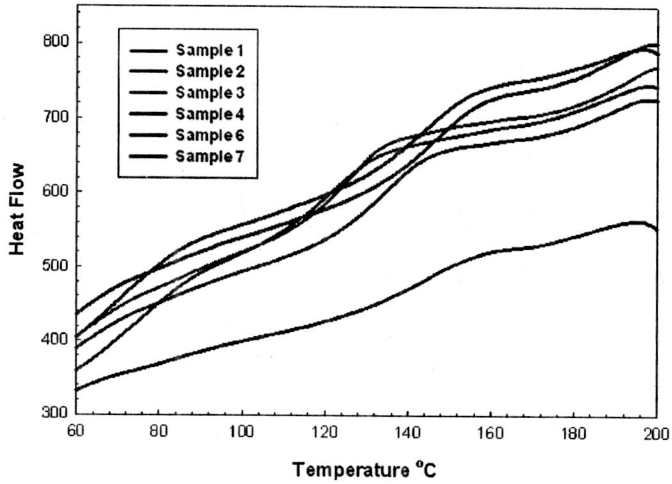

FIGURE 15 DSC thermograms of epoxy/OMMT nanocomposite; S.1 = 10% DCAC-MMT; S.2 = 2% CPAC-MMT; S.3 = 5% CPAC-MMT; S.4 = 2% C16; S.6 = 10% C16; S.7 = 5% C16

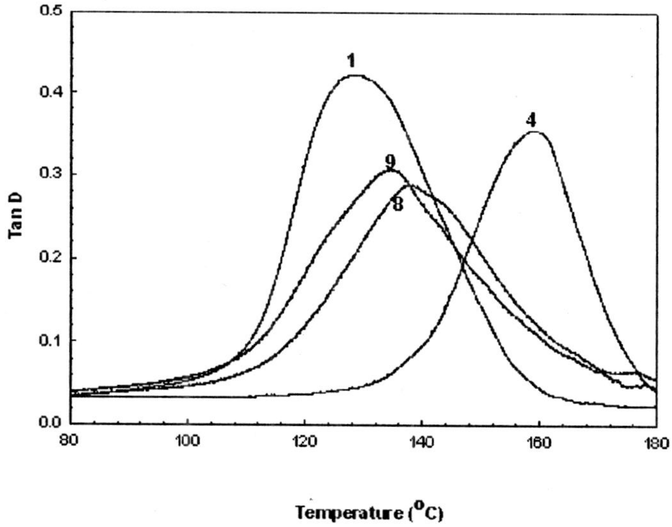

FIGURE 16 DMS of epoxy/OMMT; 1 = 10% DCAC-MMT; 4 = 2% CPAC-MMT; 8 = 10% C16-MMT; 9 = 5% CPAC-MMT

K.H. Chen, S.M. Yang. J. of App. Poly. Scie. 86(2001), p. 414–424.

A. Yasmin, L.J. Abot, and I.M. Daniel. Scripta Materialia 49(2003), p. 81–86.

Wei Xie, Zongming Gao, Wei-Pin Pan, Doug Hunter, Anant Singh, and Riched Vaia. Chem Mater. (2001), p. 2979.

R.A. Vaia, H. Ishii, E.P. Giannelis. Chem Mater 1994;5;1694.

P.C. Lebaron, Z. Wang, T.J. Pinnavais. J App Clay Sci 1999;15;11.

H.L. Tyan, Y.C. Liu, K.H. Wei, Chem Mater.,11,1942 (1999).
M. Kawasumi, N. Hasegawa, M. Kato, A. Usuki and A. Okada. Macromolecules 30 (1997), p. 6333.
K.H. Chen, S.M. Yang. J. of Appl. Poly. Sci. 86, (2001), p. 414–421.
Xu Wei-Bing, S.P. Bao, P.S. He. J. of Appl. Poly. Sci. 84, (2002), p. 842–849.
C. Wan, Y. Zhang, Y. Zhang. Poly. Testing. 23 (2004), p. 299–306.
Joo Chin, T.T. Albrecht, Ho-Cheol Kim, T.P. Russell, J. Wang. Polymer, (2001), p. 5947–5952.

Mica and Its Ability to Be Chemically Exfoliated

L.D. Nielsen,* W.M. Cross,* S.P. Decker,† and J.J. Kellar*

The purpose of this research was to determine the ability of mica to be chemically exfoliated. Toward this end an aqueous lithium treatment was performed, and the treated mica was compared to the as-received mica via a variety of analytical methods including BET surface area measurement, particle size analysis, x-ray diffraction and scanning electron microscopy (SEM). The goal of the research was to attempt to prove conclusively that mica could be chemically exfoliated. Given the sheet nature of mica, and associated complexities, it is difficult to conclusively determine whether exfoliation has been accomplished. In this regard, particle size, x-ray diffraction and SEM analyses were inconclusive in terms of whether exfoliation had occurred. However, BET analysis indicated that the surface area of the mica increased from 19.2 m^2/g to 34.2 m^2/g after surface treatment, indicating possible exfoliation of mica.

INTRODUCTION

In the mid 1990s, Toyota researchers showed that the addition of montmorillonite, with specific surface treatment and polymer compounding conditions, led to a filled polymer with very unique mechanical and physical properties.[1–3] For example, significant improvements in flexural and tensile moduli, and increases in heat distortion temperature under load have been reported with montmorillonite additions to polymers.[1–3] Recent research has shown that the unique properties imparted by the clay addition are due to the nanoscale behavior associated with individual clay platelets. Controlled surface chemistry of the clay leading to dispersion and adhesion with the polymer matrix is

* Department of Materials and Metallurgical Engineering, South Dakota School of Mines and Technology, Rapid City, South Dakota

† Center for Accelerated Applications at the Nanoscale, South Dakota School of Mines and Technology, Rapid City, South Dakota

a necessary prerequisite for effective use of the clay as a nano-scale additive. This research has led to development of a number of new applications of so-called nanocomposites, primarily for the automotive industry.[4] There still is considerable debate about the actual mechanism(s) for property enhancements because of the montmorillonite addition, however, it is generally agreed that the nano-scale clay platelets are responsible for the property changes, as simple rule-of-mixture approaches do not adequately describe the property enhancements found with montmorillonite additions.[3]

The success associated with nano-scale montmorillonite fillers causes one to wonder about other naturally occurring layered minerals and their ability to be exfoliated and utilized for new applications. In this regard, several candidate layered minerals readily come to mind, namely, mica ($K_2Al_4(Si_6Al_2)O_{20}(OH)_4$), molybdenite ($MoS_2$), and hydrotalcites ($[Mg_{8-x}Al_x(OH)_{16}(CO_3)_{x/2}]\cdot nH_2O$).

Molybdenite, along with synthetic $CdPS_3$ and $MnPS_3$, have drawn attention because of their ability to be intercalated with a variety of molecules which might prove useful for magnetic, biological and optical applications.[5-7] Similarly, hydrotalcites are of interest because of the versatility associated with their layered structure.[8-11]

Mica is a term used with 35 layered phyllosilicate minerals. The most common mica is muscovite, the focus of this research effort.

Possible exfoliation of muscovite mica is of interest for several reasons. First, mica can be found and mined in high quality deposits.[12] Second, like the most utilized nano-scale mineral, montmorillonite, it is a layered phyllosilicate.[12] Each sheet, or layer, of mica is estimated to be ~10 Å in thickness.[13] In addition, mica has often been used for a wide variety of uses including the electronic industry, paints/pigments, and fillers for plastics[12-15]; hence, its use is accepted in a number of industries. Given its sheet structure, mica has highly anisotropic properties.[12] Particularly noteworthy in terms of properties is the high modulus and strength of muscovite mica.[12]

Exfoliation of mica can occur in the bulk phase by simply peeling away layers with household tape. Shown in Figure 1 is a scanning electron microscope (SEM) image of macroscopically-peeled mica.

There have been several attempts over many years described in the literature to exfoliate or delaminate mica.[16-26] The most common route explored for exfoliation of mica is by subjecting the mineral to a heated aqueous lithium solution.[26] The reported efforts all seem to agree that surface potassium ions exchange with the lithium ions in solution, however, only a single effort purports to achieve delaminated mica by this treatment.[26] The current study was initiated in an attempt replicate this process and to analyze by a variety of modern analytical methods the efficacy of this approach for delamination/exfoliation of mica.

EXPERIMENTAL PROCEDURE

SEM images were obtained with a JEOL 840A Scanning Electron Microscope and Oxford ISIS Imaging and Microanalysis System. Powder x-ray diffraction measurements were performed using a Rigaku D/Max X-Ray Diffractometer. BET surface areas measurements were performed on a Micromeritics Gemini III 2375 specific surface area analyzer. A Microtrac S3000 was used for particle size analysis. For all analytical methods (surface area, particle size, XRD, SEM) both untreated and treated mica were analyzed. Sample preparation for treated and untreated samples is described below.

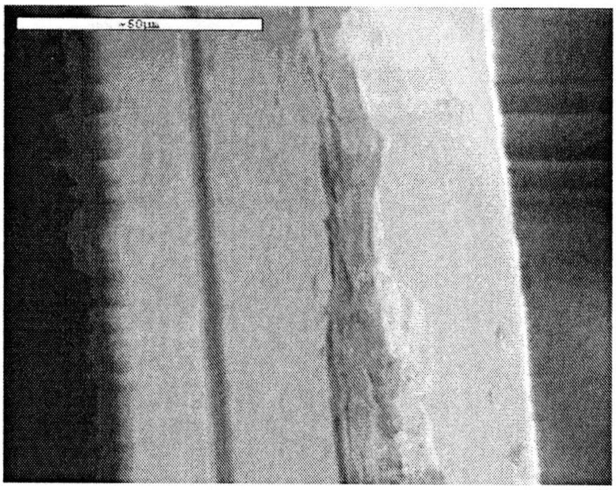

FIGURE 1 SEM image of mica (right-center) remaining on tape after peeling. The adhering mica is roughly 10 microns in thickness.

Samples of mica (muscovite) were hand picked near Hill City, South Dakota. The samples were rinsed with water to remove adhering smaller mineral particulate and then crushed by hand with a mortar and pestle and screened at 200 mesh (74 microns) and the undersize sample used for subsequent analysis. During the comminution process it was noticed that the sample developed significant static charge.

The comminuted samples were analyzed as-received and after a lithium treatment following a slightly modified experimental procedure described by Caseri et al.[26] Specfically, 56.9 grams of lithium nitrate was added to 40 ml of distilled water and 0.21 grams of hand-ground muscovite mica in a round bottom flask. Next, the flask (also containing a magnetic stirrer) was placed in a glass beaker containing silica sand, and a thermometer was positioned in the flask so that the magnetic stirrer did not contact the thermometer. The glass beaker was then placed on a hot plate. A reflux condenser was also placed on top of the flask and the rubber tubing was positioned on the condenser. The sample was slowly heated with the sample being mixed so as to keep the lithium nitrate in solution. The sample was then heated to 110–120°C. The hot plate was positioned so that the sample would stay at a constant temp with a variance of about 5 degrees through the 48 hours that it was heated. After the 48 hour treatment was complete the sample was then removed from heat and allowed to cool. Once the sample cooled to ~40°C the filter process was started. The filter used was a 3 micron cellulose filter, all filtering was performed under vacuum. The filtering process consisted of completely filtering the sample solution twice, washing with ~250 ml of water, and then placing the sample into 50 ml of water and stirring for ~30 minutes. The sample was then completely filtered again and washed with 100 ml of water. After the sample was completely filtered and washed, it was placed in a furnace for approximately 16 hours at temperature between 100–120°C.

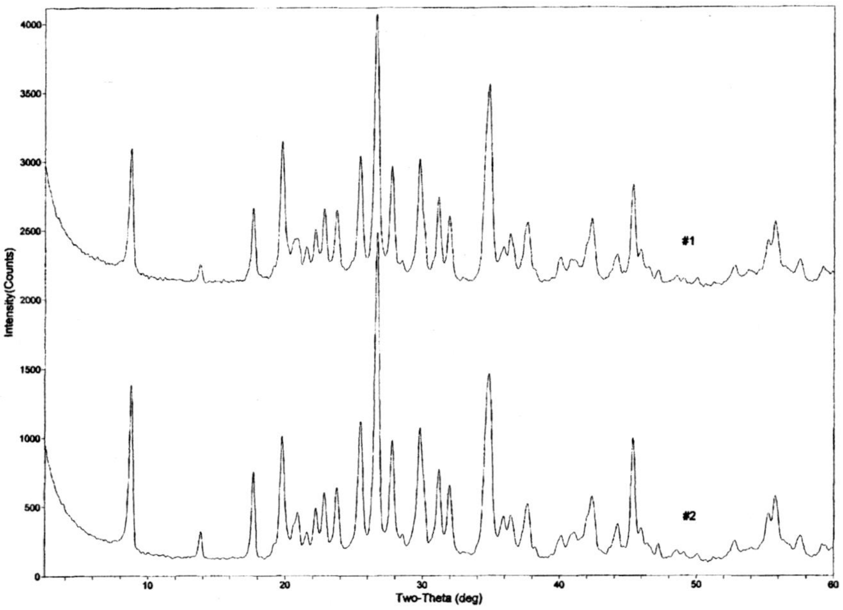

FIGURE 2 Powdered XRD spectrum of the untreated sample (#1-top) and treated (#2-bottom) samples

RESULTS AND DISCUSSION

The purpose of this research is to investigate the effect of a molten lithium salt treatment at elevated temperatures on muscovite mica.

X-ray Diffraction

X-ray diffraction (XRD) was performed on both the as-received (untreated) and treated samples. Shown in Figure 2 is the XRD spectrum for each sample. Both samples were identified to be high-purity muscovite mica, there were no singular differences discernable between the treated and untreated samples. The lack of differences in the XRD spectra between the treated and untreated samples indicate a lack of complete exfoliation of the muscovite mica by the chemical treatment. Probably the best example of exfoliation that has been demonstrated in the literature is that with montmorillonite.[1-4] Here, montmorillonite is surface treated and then subjected to high shear during processing with a polymer (usually thermoplastic) matrix. XRD analysis of such montmorillonite-filled systems often shows an absence of peaks, or weak, broad peaks from the montmorillonite within the polymer matrix. Clearly, the system analyzed here, dry mica not subjected to high shear, is not ideal for promoting exfoliation of the mica sheets. One possibility is that the chemical treatment partially exfoliated some mica sheets but did not fully delaminate all of the mica. In such a case one could not differentiate the untreated and treated samples.

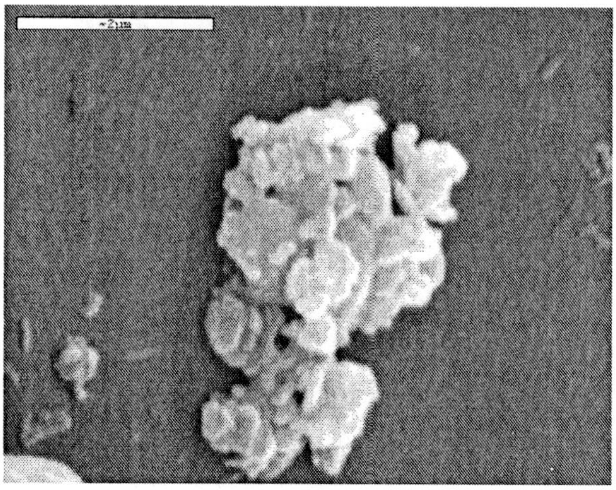

FIGURE 3 SEM image of untreated mica

FIGURE 4 SEM image of treated mica

Scanning Electron Microscopy

Shown in Figures 3 and 4 are representative SEM images of untreated and treated particles. Both images show that there is significant aggregation associated with both samples. In addition, both samples showed particles significantly less than the maximum screen size opening (74 microns). The SEM image of the treated sample confirms the XRD data, namely, that complete exfoliation did not result. Like the XRD data, the SEM images are not able to show partial delamination of the particles.

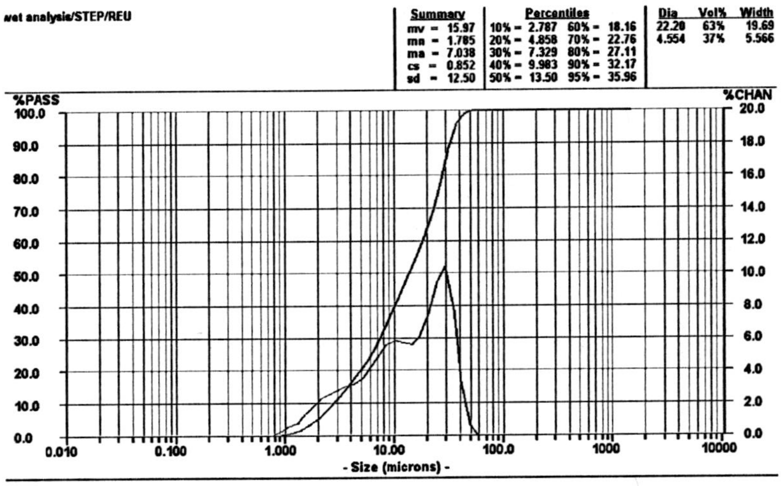

FIGURE 5 Particle size data for untreated mica

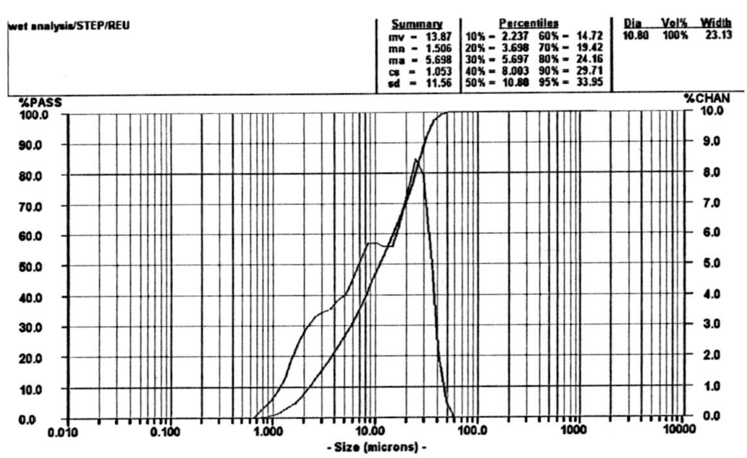

FIGURE 6 Particle size data for treated mica

Particle Size Analysis

Particle size analysis data is shown in Figures 5 and 6 for the untreated and treated samples. As shown in the Figure 5 and 6 data reports the mean diameter (area distribution) was found to be 7.038 and 5.698 microns for the untreated and treated samples, respectively. Three particle size analyses were performed, and results similar to those shown in Figures 5 and 6 were obtained. The average mean diameters for the three runs of the untreated and treated samples was found to be 7.19 and 5.73 microns, respectively. In all cases an increase in finer particles sizes was found for the treated samples. Also, in all cases a trimodal distribution (as shown in Figures 5 and 6) of particles was found for both treated and untreated samples.

Surface Area and Associated Analysis

It was found from BET analysis that the untreated particles had a surface area of 19.2 m^2/gm and the treated samples had a surface area of 34.2 m^2/gm. Two surface area analyses were performed on each sample, and the data reported above is an average of these tests. The surface area data is the most conclusive indication that the chemical treatment resulted in partial exfoliation of the mica. What follows is a discussion of this surface area data in tandem with the earlier described particle size data.

From examining Figures 5 and 6 the treated mica has a smaller mean diameter and more particles at smaller sizes than the untreated mica. The mean area diameter was measured to be 7.2 ± 1.0 micrometers for the untreated mica particles and 5.7 ± 0.3 micrometers for the lithium solution-treated mica.

As discussed above, the BET surface area measurements showed the specific surface area of the untreated mica to be 19.2 m^2/g and 34.2 m^2/g for the lithium-treated mica. This is a 78% increase in specific surface area and about 3 times the increase predicted from the decrease in the mean area diameter. Given the shape of the particles observed by SEM analysis (see Figures 3 and 4), the specific surface area (SSA) is related to the particle thickness (t) by Equation 1.

$$SSA = 2/(\rho^*t) \quad \textbf{(EQ 1)}$$

Rearranging Equation 1 to solve for t and using ρ = 2.82 g/cm^3 for the density of mica, the thickness of the mica is estimated to be 37 nm for the untreated mica and 21 nm for the lithium-treated mica. These numbers seem small given the SEM micrographs in Figures 3 and 4. If the thickness is indeed too small, then the effect of the lithium may have been to partially open the potassium-containing, mica interlayer. The interlayer thickness is less than about 1 nm even in hydrous micas so given the calculated thickness the potassium layer can only be partially open. Our research group is still attempting independent measurements of the particle thickness, as initial efforts to measure particle thickness using scanning electron microscopy have given ambiguous results.

SUMMARY

A variety of analytical methods including BET, particle size x-ray diffraction and scanning electron microscopy (SEM) were used to determine the efficacy of exfoliating mica via a chemical solution treatment. Particle size, x-ray diffraction and SEM analyses were inconclusive in terms of whether exfoliation had occurred. However, BET analysis indicated that the surface area of the mica increased from 19.2 m^2/g to 34.2 m^2/g after surface treatment, indicating possible exfoliation of mica. Future work will involve continued analysis of testing to demonstrate the exfoliation of mica.

ACKNOWLEDGMENTS

The authors would like to thank Ms. Lisa Schlink for performing the BET analysis, Dr. Jan Puszynski for access to the BET apparatus and Mr. Michael Satter for his generous donation of the muscovite mica used in this study. Partial financial support (Ms. Lizabeth Nielsen) for this work was provided by the National Science Foundation through grant #0453216-Materials, Mechanics, and Manufacturing (3M) REU Site.

REFERENCES

1. Usuki A, Kojima Y, Kawasumi M, Okada A, Fukushima Y, Kurauchi T, and Kamigaito O, "Synthesis of Nylon 6-Clay Hybrid," 1992, *Journal of Materials Research* **8** 1179.
2. Liang Y, Omachinski S, Logsdon J, Cho J W, and Lan T, "Nano-Effect in *In Situ* Nylon-6 Nanocomposites," retrieved: October 17, 2005 from: http://www.nanocor.com/tech_papers/antec2001.asp
3. LeBaron P C, Wang Z, Pinnavaia T J, 1999, "Polymer-Layered Silicate Nanocomposites: an Overview," *Applied Clay Science*, **15** 11–29.
4. DeGaspan J, "Feature Focus: Eye on the Future: Nanotechnology-Prospecting Paydirt," retrieved: October 17, 2005 from: http://www.memagazine.org/backissues/april01/features/prospect/prospect.html
5. Westreich P, "The Formation of Novel Layered Compounds by Exfoliation and Restacking of Cadmium Phosphorous Trisulfide with the Biological Molecules Adenosine Monophosphate and Cytinide and Monophosphate Included," 2004, PhD Dissertation, Simon Fraser University.
6. Gordon R A, Yang D, Crozier E D, Jiang D T, Frindt R F, "Structures of Exfoliated Single Layers of WS_2, MoS_2 and $MoSe_2$ in Aqueous Solutions," 2002, *Physical Review B* **65** 125407.
7. Yang D and Frindt R F, "Li-Intercalalation and Exfoliation of WS_2," 1996, *J. Phys Chem Solids,* **57** 1113–1116.
8. Zaneva S, Stanimirova T, "Crystal Chemistry, Classification Position and Nomenclature of Layered Double Hydroxides," Bulgarian Geological Society, Annual Scientific Conference, December 2004, retrieved October 17, 2005 from: www.bgd.gea.uni-sofia.bg/materiali04/Zaneva & Stanimirova.pdf
9. Rey F, Fornes V, Rojo J M, "Thermal Decomposition of Hydrotalcites: An Infrared and Magnetic Resonance Spectroscopic Study," 1992, *J. Chem Soc. Faraday Trans.*, **88**(15) 2233–2238.
10. Valcheva M L, Davidova N P, "Thermal Decomposition of Mg, Al-Hydrotalcite Material," 1993, *Journal of Materials Science*, **28** 2157–2162.
11. Oriakhi C O, Farr I V, and Lerner M M, "Thermal Characterization of Poly(styrene sulfonate)/Layered Double Hydroxide Nanocomposites," 1997, *Clay and Clay Minerals*, **45** 194–202.
12. Xanthos M, "Mica Flakes," in *Functional Fillers for Plastics* (Xanthos M, Editor), 2005, Wiley-VCH, pp. 149–161.
13. Poppe L J, Paskevich V F, Hathaway J C, and Blackwood D S, "A Laboratory Manual for X-Ray Powder Diffraction," 2001, U. S. Geological Survey Open-File Report 01-041, retrieved October 21, 2005: http://www.bucknell.edu/img/assets/6915/USGS%20XRD%20Manual.pdf
14. "Fluorosint: Most Dimensionally Stable PFTE-Based Product," retrieved October 20, 2005 from: http://www.quadrantepp.com/select/fluorosint_europe.html
15. Guseva O, Lusti H R, and Gusev A A, "Matching Thermal Expansion of Mica-Polymer Nanocomposites and Metals," 2004, *Modeling and Simulation in Materials Science and Engineering*, **12** S101–S105.
16. Meier L P, Shelden R A, Caseri W R, and Suter U W, "Polymerization of Styrene with Initiator Bound to High Surface Area Mica: Grafting via an Unexpected Mechanism," 1994, *Macromolecules*, **27** 1637–1642.
17. Barshad I, "Cation Exchange in Micaceous Minerals: II. Replaceability of Ammonium and Potassium from Vermiculite, Biotite, and Montmorillonite," 1954, *Soil Science*, **78** 57–76.
18. Obut A and Girgin I, "Hydrogen Peroxide Exfoliation of Vermiculite and Phlogopite," 2002, *Minerals Engineering*, **15** 683–687.

19. Osman M A, Ploetze M, and Suter U W, "Surface Treatment of Clay Minerals and Thermal Stability, Basal Plane and Surface Coverage," 2003, *J. Mater. Chem*, **13** 2359–2366.
20. Osman M A, Caseri W R, and Suter U W, "H^+/Li^+ and H^+/K^+ Exchange on Delaminated Muscovite Mica," 1998, *Journal of Colloid and Interface Science*, **198** 157–163.
21. Osman M A and Suter U W, "Determination of Cation-Exchange Capacity of Muscovite Mica," 2000, *Journal of Colloid and Interface Science*, **224** 112–115.
22. Graf Von Reichenbach H and Rich C I, "Potassium Release from Muscovite as Influenced by Particle Size," 1969, *Clays and Clay Minerals*, **17** 23–29.
23. Bronson R D, Spain J M, and White J L, "Potassium-Depleted Muscovite. Part I. Preparation using Filtration Process for Treatment with Molten Lithium Nitrate," National Conference on Clays and Clay Minerals, 1959, Pergamon Press, 39–43.
24. Obut A, Girgin I, and Yorukoglu A, "Microwave Exfoliation of Vermiculite and Phlogopite," 2003, *Clays and Clay Minerals*, **51** 452–456.
25. Graf Von Reichenbach, H, "Cation Exchange in the Interlayers of Expansible Layer Silicates," 1968, *Clay Minerals*, **7** 331–341.
26. Caseri W R, Shelden R A, and Suter U W, "Preparation of Muscovite with Ultrahigh Specific Surface Area by Chemical Cleavage," 1992, *Colloid Polym Sci*, **270** 392–398.

. . . .

Novel Organoclay for Rheology Modification and Pigment Dispersion in Hydrophobic Liquids

Ashoke SenGupta* and Ralph Spindler*

Hydrophobically-modified smectite clays, commonly referred to as organoclays, are one of the most widely used thickening agent/rheology-modifier for hydrophobic liquids used in consumer and industrial products. Since the time of their inception, almost fifty years ago, the organoclays have been produced via ion exchange-driven adsorption of quaternary surfactants having long (C_{12}–C_{18}) alkyl chains, onto the face-surface of clay platelets. These organoclays, however, present a number of performance limitations, when used as thickening agents.

Polymer-modified, novel organoclays currently available from AMCOL overcome the drawbacks of traditional organoclays. These new additives exhibit much superior rheological properties, as compared to the traditional organoclays. They also enhance the color properties of color cosmetic products, being able to function as pigment deflocculating agents. The performance properties of these polymer-modified organoclays are described herein.

INTRODUCTION

Numerous color cosmetic products may be simply classified as non-polar dispersions. Despite their varying compositions that can be often rather complex, they are essentially systems where coloring pigments remain suspended in non-polar or hydrophobic solvents. As for most products that are dispersions in their fundamental makeup, the basic

* AMCOL International Corporation, Arlington Heights, Illinois

qualities of color cosmetics, that include color, surface coverage, and ease of application, rely by far on the stability of these dispersions. Once milled to a fine size, the more stable the suspended particles are against flocculation and sedimentation, the better the products are in terms of color and application properties, formulation ease, and shelf life.

Color cosmetic formulations, therefore, must include an ingredient that enables the suspended pigment particles to remain deflocculated by rendering them mutually repulsive. Particles in aqueous dispersions may be stabilized against flocculation by what is known as the electrical double layer (EDL) repulsion if their surfaces bear an electrical charge. In non-polar solvents, however, the EDL repulsion is far too weak for preventing particle aggregation. Flocculation-stability of non-polar dispersions, therefore, would have to be derived utilizing what are described as semi-steric and steric stabilization in colloid literature [1].

Semi-steric stabilization ensues under such conditions where surfactants adsorbed on the particle surface can effectively reduce the van der Waals attraction that tends to draw the particles into flocculation. A method that often proves to be the most effective means of stabilizing suspended particles against coagulation involves the use of polymeric dispersants. These polymers, once adsorbed on the particle surface, with segments dangling out from the particle surface, in the form of loops and tails, can stabilize dispersions by generating the so-called steric repulsion between the particles.

Most pigments used in color cosmetics, including the high-density materials such as various inorganic oxides, tend to settle out easily unless their settling rates are greatly minimized due to an increased viscosity of the dispersion medium. Cosmetic waxes are often used for thickening non-polar dispersions, albeit they pose the threat of marring the pigment color. The other additive used commonly as a thickener for hydrophobic solvents is widely referred to as "organoclay."

Layered silicates, namely the smectite clays such as bentonite and hectorite, are inorganic particulate materials that exist as stacks or aggregates of plate-like, silicate layers, referred to as platelets. These clays can thicken aqueous compositions, the thickening being the outcome of clay platelets forming a particulate network through the entire suspension volume [2]. When suspended in an aqueous medium, the clay platelets stacked in an aggregate are forced apart across their face-surfaces, a phenomenon known as exfoliation of clay platelets. The face-surface of the clay platelets bears an anionic charge for which the adjacent clay platelets in a stack, when wetted with water, repel each other due to the EDL repulsion, resulting in clay exfoliation. Exfoliation releases a large number of platelets in the suspension, which can then form the particulate network.

With weak EDL repulsion in non-polar solvents, the surface of the smectite clays would have to be modified such that the interplatelet repulsion in the form of semi-steric and/or steric repulsion can drive clay exfoliation in these solvents. For traditional "organoclays," the clay surfaces are modified by attaching a long-chain (C_{12}–C_{18}) quaternary surfactant to the clay surface via ion exchange. Such surface-modification renders the clay surface hydrophobic, while enabling the platelets to exfoliate in hydrophobic solvents. Although used widely, these organoclays suffer from many performance limitations.

While many color cosmetic products preferably are viscous when left standing (i.e., under static conditions), it is desirable that their viscosity drops substantially when the product is subjected to shear, and that their initial viscosity is recovered rapidly upon withdrawing the shear. Such shear-thinning properties make the products easier to apply and/or increase the coverage per application stroke of the product.

TABLE 1 Shear-thinning in various solvents due to the polymer-modified organoclays

Organoclay Gel with Clay Content	Diluent	Final Clay Content, %	Brookfield Viscosity	
			rpm	Viscosity, cps
Gel in cyclomethicone	Cyclomethicone, D5	3.9	0.5	50,000
			5	8,000
			10	4,000
Gel in isododecane	Isododecane	3	0.5	20,000
			5	16,000
			10	10,400
Gel in octyl stearate	Capric/Caprylic Triglyceride	3	0.5	488,000
			5	200,000
			10	120,000

The study reported herein involves the evaluation of some novel organoclay materials for their ability to thicken hydrophobic solvents. These organoclays are produced by modifying the surface of bentonite clays, using certain amphiphilic copolymers. Such polymer-modified organoclays, produced in the form of viscous gels in hydrophobic liquids such as cyclomethicone, isododecane, and octyl stearate, were tested for their rheological properties, as well as for their ability to improve the color properties of color cosmetic formulations.

EXPERIMENTAL

Rheological Properties of the Polymer-modified Organoclays

Viscous organoclay gel samples were diluted by mixing them with a given solvent using a dispersion blade agitator. The viscosity of the dilute gel was measured using a Brookfield RVT viscometer. The results of these viscosity measurements are given in Table 1.

Similar viscosity tests were carried out with the octyl stearate-based gel in various solvents including castor oil, mineral oil, and ester solvents. The typical rheological properties demonstrated in these tests include: i) high low-shear-viscosity; ii) high level of shear-thinning; and iii) relatively rapid viscosity-recovery upon withdrawing the shear. Figure 1 shows a comparison of these rheological properties for a polymer-modified organoclay and a traditional organoclay in isopropyl palmitate at two different temperatures. The viscosity measurements were carried out using an automatic, temperature-controlled Brookfield RV viscometer.

Pigment Dispersing Ability of the Copolymeric Surface Modifier

The extent of deflocculation of suspended particles in concentrated dispersions may be assessed from the suspension-viscosity, inasmuch as a lower viscosity indicates a higher level of deflocculation. Accordingly, a copolymer, used herein for the clay surface-modification, was tested for its pigment dispersing ability by measuring the viscosity of concentrated suspensions of iron oxide and titanium dioxide, with and without the copolymer. A Brookfield RVT viscometer was used for measuring the suspension viscosity.

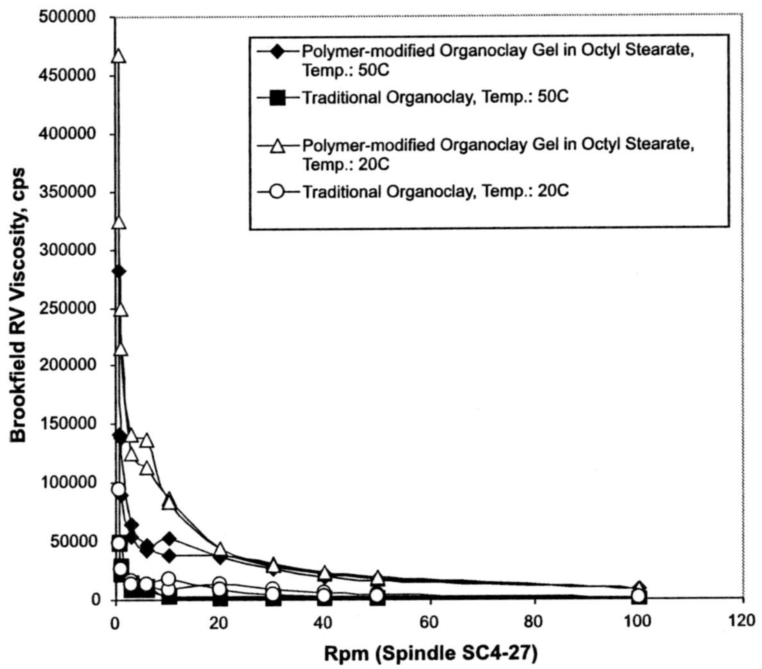

FIGURE 1 Polymer-modified organoclay versus traditional organoclay in isopropyl palmitate at 3% clay solids

A given weight of iron oxide or titanium dioxide was added to a dispersant solution comprising a 60:40 (w/w) mixture of cyclomethicone and C_{12-15} alkyl benzoate, a given amount of the copolymeric dispersant, and an aliquot of a 1:1 mixture (w/w) of propylene carbonate and deionized water. (The organoclay gel samples would invariably contain propylene carbonate and water. These polar materials could potentially affect the dispersing ability of the copolymer. They were added, therefore, to see if they would cause any adverse effects on pigment dispersion.) The resulting slurry was homogenized in a Waring blender at 22,000 rpm for a total mixing time of four minutes. The slurry-viscosity was measured after 15 minutes from the time of completion of mixing. The results of these slurry viscosity tests are summarized in Table 2. In Table 2, the pigment dosage is based on the weight of the slurry (excluding the weight of the copolymeric dispersant), and the dispersant dosage is based on the weight of the pigment.

Color Cosmetic Formulations

Lip color and liquid makeup formulations were prepared with the object of evaluating the color-boosting property of the polymer-modified organoclays. As compared to a traditional organoclay, the polymer-modified organoclay produced deeper shades of color and a more uniform coverage, as shown by the drawdown-films of the aforementioned formulations, presented in Figures 2 and 3. These formulations are given below.

TABLE 2 Pigment dispersing ability of a copolymeric surface modifier

Pigment Material, Dosage %	Copolymer Dosage	Brookfield Viscosity, cps, 10 rpm
Titanium dioxide 38.61	0	15,000
	4	250
	5	100
Iron Oxide 32.61	0	Too viscous
	5	750
	8	260

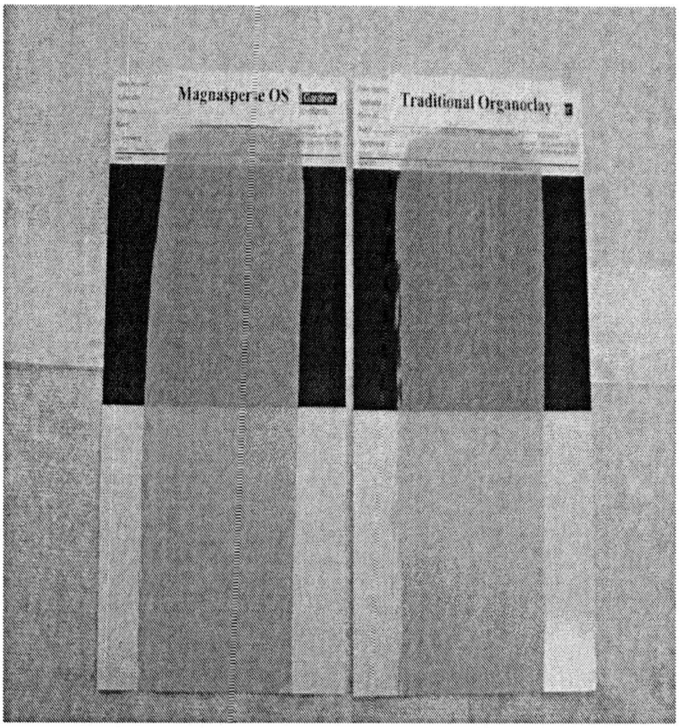

FIGURE 2 Color development in lip color formulation—poly-clay (left) versus traditional organoclay (right)

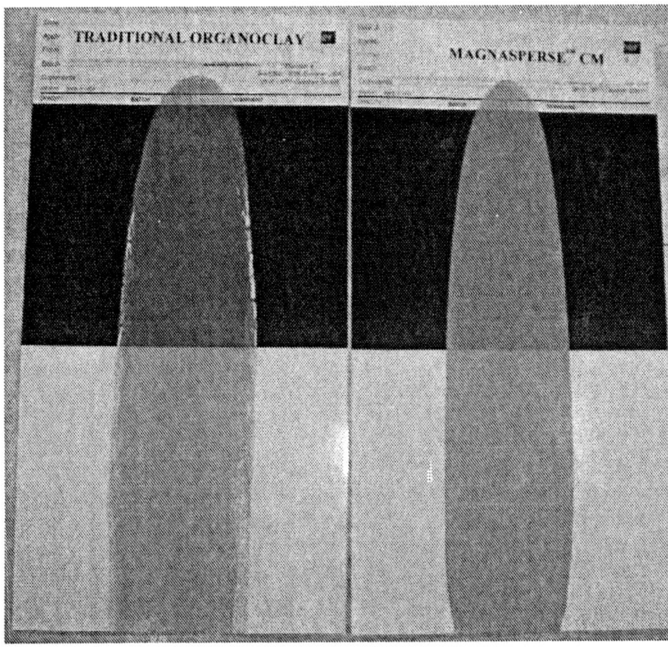

FIGURE 3 Color development in liquid makeup formulation—poly-clay (right) versus traditional organoclay (left)

Lip Color Formulation

Phase	Ingredient	% By Weight	
		Poly-Clay	Organoclay
A	Castor Oil	57.00	57.00
A	Diisopropyl Adipate	11.61	11.61
A	Ethylhexyl Stearate	9.79	18.84
A	Titanium Dioxide	5.00	5.00
A	D&C Red # 7 Ca	0.20	0.20
A	Yellow Iron Oxide	0.10	0.10
A	Red Iron Oxide	0.10	0.10
A	Candelilla Wax	4.00	4.00
A	Propyl Paraben	0.10	0.10
A	Poly-Clay Gel	12.1 (1.2% Clay)	0.00
A	Organoclay	0.00	1.85 (1.2% Clay)
B	Propylene Carbonate		1.2

Liquid Makeup Formulation

Phase	Ingredient	% By Weight	
		Poly-Clay	Organoclay
A	Cyclomethicone	11.5	17.19
A	Diisopropyl Adipate	5.00	5.00
A	Isododecane	5.00	5.00
A	Cetyl-PEG 14 Dimethicone	3.00	3.00
A	Propyl Paraben	0.30	0.30
A	Poly-Clay Gel	8.00 (1% Clay)	0.00
A	Organoclay	0.00	1.54 (1% Clay)
A	Propylene Carbonate	0.77	
B	Titanium Dioxide	5.50	5.50
B	Yellow Iron Oxide	0.70	0.70
B	Red Iron Oxide	0.15	0.15
B	Black Iron Oxide	0.15	0.15
C	Sodium Chloride	0.70	0.70
C	Deionized Water	60.00	60.00

SUMMARY

The study reported herein demonstrates that organoclays may be produced using certain amphiphilic copolymers as the clay surface-modifier. These organoclays show rheological properties characterized by high low-shear-viscosity, high level of shear-thinning, and rapid viscosity-recovery. For color cosmetic products, the benefits of having such rheology include sedimentation-stability of pigments, ease of application of products, and a greater coverage per application stroke of the product.

Equally significant is the prospect of stabilizing pigment particles against flocculation when these polymer-modified organoclays are included in pigment-laden product formulations. The benefits of having such flocculation-stability include the ability to attain a deeper shade of color and/or a greater surface coverage upon product application, for a given pigment dosage. Alternatively, this could mean cost-savings by being able to reduce the pigment dosage.

ACKNOWLEDGMENTS

The authors thank Mr. David McGregor for carrying out most of the developmental work leading to the technology presented herein. We further thank AMCOL Research and Development for permission to publish this work.

REFERENCES

1. Sushumna, I., Gupta, R. K., and Ruckenstein, E., *J. Mater. Res.*, **7-10**, 2891, *1992*.
2. van Olphen, H., *An Introduction to Clay Colloid Chemistry–For Clay Technologists, Geologists, and Soil Scientists*, Krieger Publishing Company, Malabar, Florida, *1991*.

PART 3

Functional Fillers

Thermal and Photo-Induced Wettability for Patterned Titania and Silica

A. Kanta,* R. Sedev,* and J. Ralston*

The surface properties of silica and titania are mainly determined by the presence, density and type of terminal hydroxyl groups (Si–OH "silanol" and Ti–OH "titanol"). Thermal treatment at elevated temperatures causes dehydroxylation on both surfaces, confirmed by streaming potential and ToF-SIMS measurements. The mechanism is condensation of adjacent silanol/titanol groups into siloxane/titanoxane bonds. Ultraviolet light (λ = 254 nm) has little effect on silica but rapidly induces hydrophilicity on titania surfaces.

Patterned inorganic surfaces consisting of silica (SiO_2) and titania can be prepared. The approach is based on a combination of standard photolithography and plasma-enhanced chemical vapour deposition. These surfaces have great potential for high temperature wetting applications, in contact with various media.

INTRODUCTION

The wettability of materials is an important factor in numerous industrial and natural processes. Metal oxides (e.g., SiO_2, TiO_2, etc) are widely popular materials and their wettability is of significant practical interest. The contact angle is the simplest characteristic of the physical interaction between a liquid and a solid surface and is determined by the nature of the three contacting phases (solid, liquid and vapour). The wettability can be considerably modified by varying the composition of the system. However, it is often advantageous to affect the intrinsic wettability by applying an external influence, e.g., light irradiation,[1–3] electric field,[4–6] thermal treatment etc.

The surface of a metal oxide is a high-energy surface (i.e., the surface of hard solid[7]) and therefore should be completely wetted by most liquids.[7,8] This, however, applies for a clean surface only as any adsorption will drastically modify the surface energetics.[8]

* Ian Wark Research Institute, University of South Australia, Mawson Lakes, Adelaide, SA, Australia

Obtaining and keeping a high-energy surface clean under the circumstances of contact angle measurement proved to be a major difficulty in the wetting studies of such surfaces.[9] A contemporary review of the wettability of metals, glass and graphite was published by Schrader.[10] The development of the controversy surrounding the wettability of gold is particularly instructive–it was not until contact angle measurements were performed with carefully purified components and in ultrahigh vacuum that the intrinsic hydrophilicity of gold was firmly established.

A clean silica surface is hydrophilic (water contact angle is 0°) but, when left exposed to the open atmosphere, it gradually turns into mildly hydrophobic (contact angle ~40°[11,12]) due to organic contamination.[13] Such behaviour is typical for high-energy surfaces, e.g., mica, metals, metal alloys,[9] and metal oxides.[13] The advancing water contact angle on metals kept under industrial conditions is typically in the range from 40° to 80°.[14] Titanium dioxide is no exception to this rule.[2]

The surface of silica is probably one of the most extensively studied surfaces.[15–20] Its properties are chiefly determined by the density and type of the hydroxyl groups (Si–OH "silanol") present on the surface. On the average about 5 silanols per nm^2 are found on a fully hydroxylated silica surface. Thermal dehydroxylation occurs at any temperature higher than about 150–200 °C according to the condensation reaction:

$$\underbrace{-\mathrm{Si}-\mathrm{OH} + \mathrm{HO}-\mathrm{Si}-}_{\text{vicinal silanols}} \xrightarrow[-\mathrm{H_2O}]{} \underbrace{-\mathrm{Si}-\mathrm{O}-\mathrm{Si}-}_{\text{siloxane}}$$

Dehydroxylation becomes irreversible above 400 °C and there are less than 1 OH/nm^2 after heat treatment at 800 °C. The surface with many siloxane bridges is much less effective in hydrogen bonding water and the contact angle is about 40°.[11] Rehydroxylation of silica is generally slow.

Hydroxyl groups ("titanol," Ti–OH) are also present on the surface of titania and are responsible for its adsorption and catalytic properties[15,19,21,22]. On the average about 8 titanols per nm^2 are found on the surface of titania (theoretical estimates are even higher: 9–11 nm^{-2} for rutile and 12–14 nm^{-2} for anatase[21]). The chemistry of surface titanols must be analogous to that of surface silanols. In particular, thermal treatment should induce dehydroxylation as follows:

$$\underbrace{-\mathrm{Ti}-\mathrm{OH} + \mathrm{HO}-\mathrm{Ti}-}_{\text{vicinal titanols}} \xrightarrow[-\mathrm{H_2O}]{} \underbrace{-\mathrm{Ti}-\mathrm{O}-\mathrm{Ti}-}_{\text{titanoxane}}$$

After outgassing at 200 °C only three hydroxyls are left per nm^2 and virtually no OH groups are present after heat treatment at about 500 °C.

While high-temperature treatment has a similar effect on silica and titania, the impact of UV irradiation is very different. This is hardly surprising as titania is a semiconductor with a bandgap energy of about 3.2 eV and therefore wavelengths shorter than about 390 nm are capable of exciting electrons. In silica (bandgap energy = 9 eV[23]) a much higher photon energy would be required. Indeed the photo-catalytic properties of titanium dioxide are well documented.[24] UV light has little influence on silica–ultraviolet radiation in the presence of oxygen is used for dry cleaning of silicon wafers from organic contamination.[25] Halfpenny et al.[26] have found that high-intensity UV laser radiation can cause local dehydroxylation on the surface of silica (according to Equation 1) but some damage was also detected.

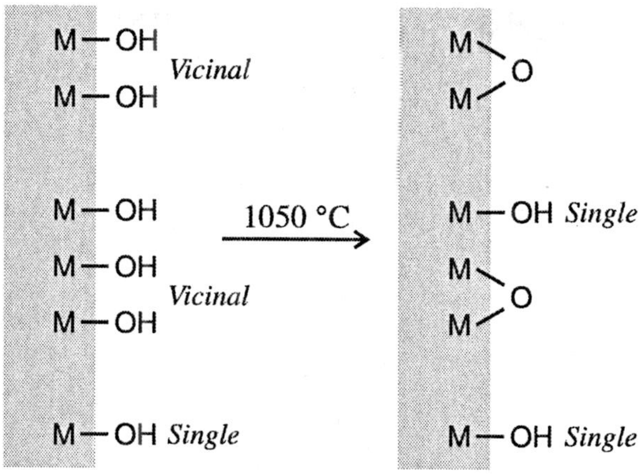

FIGURE 1 Simplified scheme of surface dehydroxylation (M = Si or Ti): silanol/titanol condensation into siloxane/titanoxane bonds

In 1997 Wang et al.[27] announced that UV light generates hydrophilic TiO_2 surfaces and strongly argued that the phenomenon is not directly related to the photocatalytic activity of titania but should rather be attributed to the reversible formation of hydrophilic nano-domains seen with scanning probe microscopy. Contamination was dismissed as a possible explanation of the experimental observations[28] since: (*i*) washing a hydrophobic TiO_2 sample with acetone and then water did not change the water contact angle significantly (51° before and 48° after the cleaning[29]); (*ii*) when a hydrophilic (UV-irradiated) sample was sonicated in pure water the contact angle of water increased from 0° to 11°[29]; (*iii*) in an oxygen atmosphere, as compared with air, the decrease of the water contact angle after UV illumination was slower and the increase after dark storage was faster.[30] Miyauchi et al.[31] investigated the water contact angles and the photo-decomposition of Methylene Blue on a dozen different oxides and concluded that photocatalysis and photo-induced hydrophilicity are uncorrelated.

We have recently shown[2] that titania is easily converted into a hydrophilic state after exposure to UV light and then rapidly reverts to hydrophobic when left in open air. The contact angle change is quite significant (~80°) and reversibly repeatable. The mechanism of the hydrophobic-hydrophilic conversion, however, has not been fully elucidated.

In this investigation we report the wettability changes on silica and titania flat plates after thermal treatment at elevated temperature and irradiation with ultraviolet light. We test the hypothesis that silica and titania surfaces have analogous behaviour when heated (Figure 1). The impact of ultraviolet light is also investigated. The behaviour of the terminal hydroxyl groups is probed with contact angle and streaming potential measurements, ToF-SIMS measurements and AFM imaging.

We then exploit this knowledge to prepare patterned inorganic surfaces of silica and titania with a moderate wettability contrast.

MATERIALS AND METHODS

Silicon wafers, P(100), were kindly provided by Dr. M. Bjelopavlic (MEMC Electronic Materials). The RMS roughness of the wafers was 0.24 nm and the peak-to-valley height was 0.88 nm as obtained with atomic force microscopy (AFM) imaging.

Silicon wafers covered with a thin layer of titanium dioxide were obtained from Philips Research Laboratories (Eindhoven, The Netherlands). Magnetron sputtering was used to deposit 25 nm thick layers of pure and stoichiometric TiO_2 (XPS detection limit was 0.1 at%) in the amorphous state. The root-mean-square (RMS) roughness of the layers was 0.3 nm and the peak-to-valley height was 1.5 nm over an area of 1 μm^2.

Both titania and silica were stored in polypropylene boxes (Fluoroware H22-60) in a laboratory for several months. Samples were cut to size (10×10 mm^2) and subsequently blown with a stream of filtered nitrogen to remove any debris left from the cutting process. These samples are further referred to as "untreated."

Silica and titania samples were subjected to two types of treatment: (*i*) thermal treatment at elevated temperatures (up to 1050 °C) and (*ii*) irradiation with ultraviolet light.

Thermal treatment was carried out for 15 h in air in an oven (Jetflow model K2F, Australia). Samples were placed in a custom-built quartz cell to minimize contamination. After heat treatment samples were steadily cooled down and subsequently loaded into the ToF-SIMS vacuum chamber or the cell of the streaming potential apparatus within 5 min of actual removal from the oven.

Silica and titania samples were irradiated in ambient atmosphere with ultraviolet light of fixed wavelength (λ = 254 nm) using two different sources: a spectral irradiator (MM3, Bunkoh-Keiki Co., Japan, with a 2 kW Xenon arc lamp giving a flux of 13.6 mW/cm^2) or alternatively an UV eraser (Lawtronics ME15). A constant radiant exposure (23.8 J/cm^2) was used throughout the investigation.

Contact angle measurements, streaming potential, time of flight secondary mass spectrometry, and Raman spectroscopy are described elsewhere.[33] X-ray photoelectron spectroscopy, scanning electron microscopy, photolithography and plasma-enhanced chemical vapour deposition of silica were methods and techniques that were also used, as described previously.[34]

RESULTS AND DISCUSSION: INDIVIDUAL SURFACES

The influence of heat treatment on the wettability of titania is presented in Figure 2, which shows the advancing contact angle of water on titania plates before and after baking the sample at various temperatures.

The contact angle on the plates kept in storage (clean polypropylene box in a dark drawer) is (81.0 ± 1.6)° and this remarkable reproducibility is probably one reason for attributing a fundamental significance to the hydrophobicity of titania. Heat-treating the samples reduces the contact angle (measured immediately after the sample has cooled down to room temperature) and if the temperature is higher than 300 °C the samples become totally hydrophilic. Baking the samples for one or 15 hours gave the same result. It appears from these results that the surface was hydrophobic due to adsorbed contamination–after organic components have been pyrolysed, at 300 °C or above, the surface is left clean and therefore hydrophilic.

The titania surface remains hydrophilic even after being treated at 1050 °C (Figure 2) when dehydroxylation must be complete. This is in marked contrast with

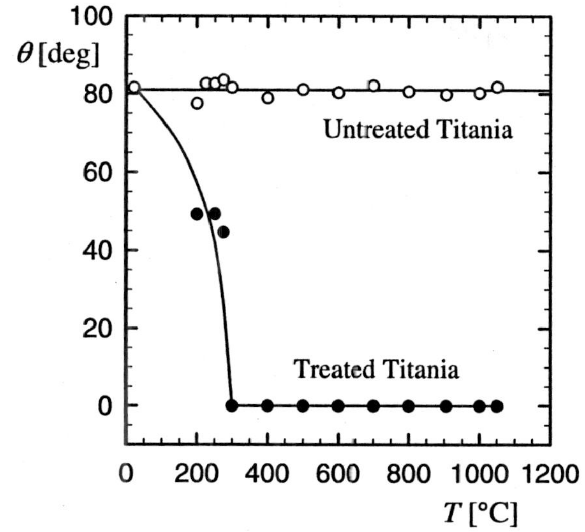

FIGURE 2 Advancing contact angle of water on titania plates: before (○) and after (●) heat treatment at various temperatures. The contact angle is measured immediately after cooling the sample down.

silica, which becomes moderately hydrophobic under similar conditions. The hydrophobicity of dehydroxylated silica is attributed to the significant reduction of the surface density of silanols due to their condensation into siloxane bonds. The ζ-potential is very sensitive to the surface density of OH groups because their dissociation is the underlying mechanism for surface charge creation. The outcome of our streaming potential measurements on silica is given in Figure 3.

The sign and magnitude of the ζ-potential as well as the isoelectric point (IEP ≈ 2) are in line with published data. Note the significant decrease in ζ-potential after the silica samples have been heat-treated at 1050 °C. The widely accepted explanation is that almost full dehydroxylation of the surface has occurred.

Streaming potential measurements were carried out with titania plates in exactly the same manner and the results are shown in Figure 4.

The values of the ζ-potential are smaller and the changes, induced by ionic strength and/or heat treatment, are also smaller when compared with silica. The overall behaviour, however, is the same. The IEP ≈ 4.5 is very close to the value 4.3 reported previously for the same material. This value is also lower than the frequently reported values (≈ 6) for titania particles. It must be noted that the IEP for titania strongly depends on sample preparation and can vary widely.

In full analogy with silica, the IEP of titania is not shifted after heat treatment (Figure 4) implying that the nature of the surface dissociation groups (titanols) has not been radically altered. The decreased magnitude of the surface charge is purely due to the decreased surface density of titanol groups. In other words, at high temperature dehydroxylation occurs on the surfaces of silica and titania in a similar manner (Figure 1).

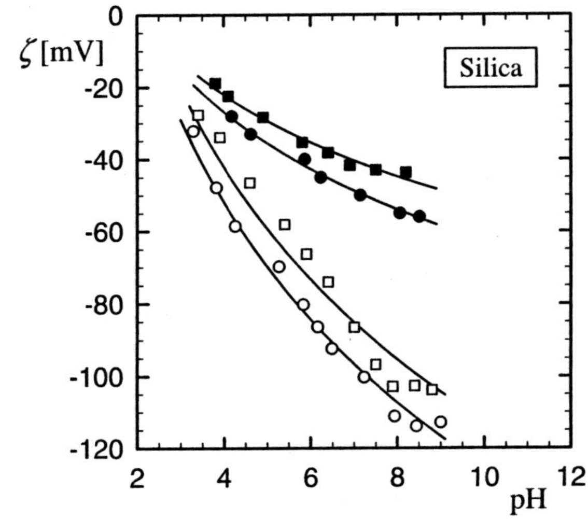

FIGURE 3 Zeta potential on silica as a function of pH (○,●: 10^{-4} M KCl; □,■: 10^{-3} M KCl): before (○,□) and after (●,■) heat treatment at 1,050 °C

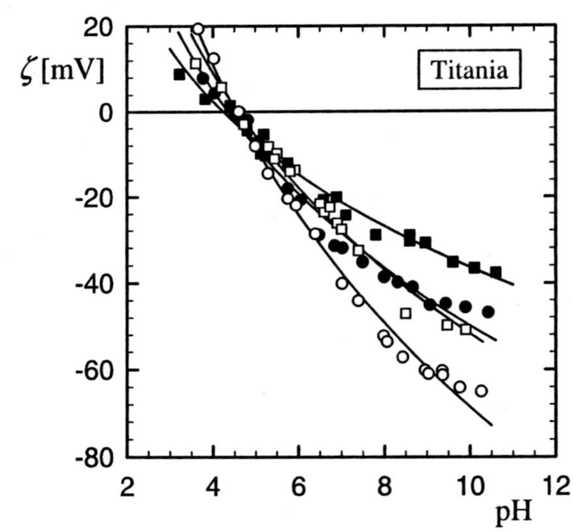

FIGURE 4 Zeta potential on titania as a function of pH (○,●: 10^{-4} M KCl; □,■: 10^{-3} M KCl): before (○,□) and after (●,■) heat treatment at 1050 °C

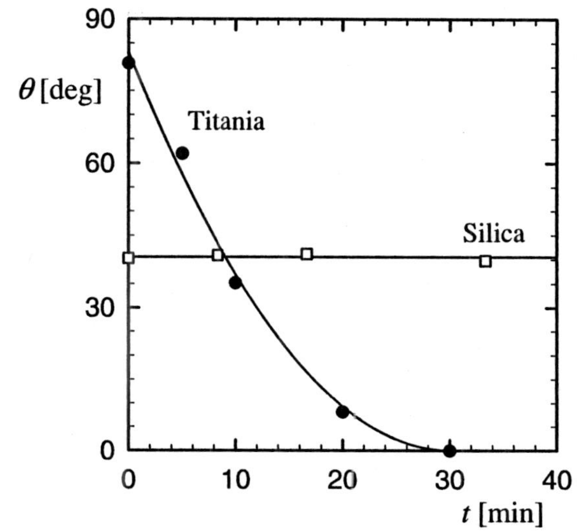

Reprinted from Langmuir, Vol. 21, No. 6, 2400–2407, copyright (2005) with permission from American Chemical Society.

FIGURE 5 Advancing contact angle of water on silica (□) and titania (•) plates after exposure to UV light (λ = 254 nm, 23.8 J/cm²) at room temperature. The contact angle is measured immediately after the illumination.

The influence of UV illumination on the wettability of silica and titania is illustrated in Figure 5.

Since titania is a semiconductor with a moderately wide bandgap energy, excitation with UV light ($\lambda \leq 390$ nm) is possible, and therefore catalytic oxidation of adsorbed organic species will be promoted. The contamination appears to be effectively removed and the intrinsic hydrophilicity of titanium dioxide is rapidly recovered (•, Figure 5). For all practical purposes silica is an insulator and no wettability changes are registered under identical conditions (□, Figure 5).

In order to provide further support for the above hypothesis silica and titania samples were examined with ToF-SIMS. The ratio MOH^+/M^+ is taken as characteristic of the amount of hydroxyl groups on the surface. The $SiOH^+/Si^+$ ratio for the untreated surface is in acceptable agreement with values obtained by others (e.g., 0.04 to 0.25[26]). It decreases when silica is baked at high temperature. This confirms the decrease in the density of hydroxyl groups on the surface, although the ratio cannot be quantitatively related to the area fraction of surface silanols. Even after a prolonged (15 h) heating at 1050 °C some OH groups are still detected on the silica surface. Most probably this ensued from the fact that our samples were treated in air rather than in vacuum. Indeed, when the high-temperature heat treatment was carried out in an evacuated (~10^{-7} torr) quartz furnace, dehydroxylation proceeded further and the $SiOH^+/Si^+$ ratio dropped down to about 0.01. The SIMS results for silica corroborate the conclusion drawn from the electrokinetic measurements.

In the case of titania, however, the trend is more complicated–the surface density of titanols, as represented by the $TiOH^+/Ti^+$ ratio, decreases after thermal treatment but the value for the untreated surface is lower. This is an apparent contradiction as we demonstrate below.

As explained in the introduction clean oxide surfaces are completely wetted by water. In a laboratory air environment, however, re-contamination of high-energy surfaces is unavoidable. Silica and titania plates were pre-cleaned and subsequently exposed to a laboratory (clean room) atmosphere. As organic molecules progressively adsorb the initially hydrophilic surfaces gradually turn into hydrophobic ones.

The higher rate of contamination on titania correlates with the higher surface energy in comparison with silica.[19] Re-contamination is influenced by many factors (temperature and humidity, recent laboratory activities, mode of floor and bench cleaning, etc) and therefore the above example should not be taken as a general rule. The main conclusion is that–under the same conditions–recontamination of titania is much faster (~minutes) than that of silica (~hours).

The wetting behaviour of silica and titania is different (the hydrophobicity of dehydroxylated silica is exceptional among other metal oxides). In an attempt to clarify this difference the work of adhesion of water to silica and titania is estimated below. Following van Oss, Good and Chaudhury[32] the work of adhesion, W_A, is split into two components:

$$W_A = W_A^{LW} + W_A^{AB}$$

The Lifshitz-van der Waals ("LW") term accounts for van der Waals-London (dispersive), Keesom and Debye type of interactions. The acid-base ("AB") term includes the contribution of Lewis type acid-base interactions across the interface, including hydrogen bonding. We consider fully dehydroxylated surfaces, i.e., assume that all surface OH groups have been removed. The Lifshitz-van der Waals component of the work of adhesion, W_A^{LW}, is then calculated as

$$W_A^{LW} = 2(\gamma_S^{LW}\gamma_L^{LW})^{1/2}$$

where γ_S^{LW} is the LW component of the solid surface energy and $\gamma_L^{LW} = 21.8$ mJ/m^2 for water. The γ_S^{LW}, is obtained from the Hamaker constant of the material, A, through

$$\gamma_S^{LW} = \frac{A}{24\pi l_0^2}$$

where $l_0 = 1.57 \pm 0.09$ Å is an average (atomic) equilibrium distance.

The figures for silica and titania are listed elsewhere[33] and shown with black bars in Figure 6. Both values are lower than the work of cohesion for water ($W_C = 2\gamma_L = 2 \times 72.8$ mJ/m^2), shown with a dashed line in Figure 6, and therefore both surfaces should be hydrophobic. The water contact angles can be estimated through the Young equation, written as

$$\cos\theta = -1 + 2\frac{W_A}{W_C}$$

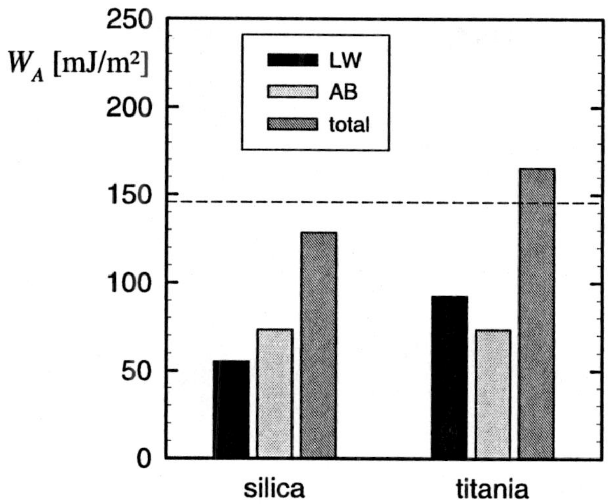

FIGURE 6 Work of adhesion, W_A, of water on silica and titania. The Lifshitz-van der Waals component, W_A^{LW}, is calculated (see text), and the acid-base component, W_A^{AB}, is calculated (see text). The dashed line is the work of cohesion for water ($W_C = 2\gamma_L = 145.6$ mJ/m²).

and the values are unrealistically high (104° and 75° respectively). As aptly described by Laskowski and Kitchener, "all solids would be hydrophobic, if they did not carry polar or ionic groups."

The acid-base component of the work of adhesion, W_A^{AB}, is given by

$$W_A^{AB} = 2(\gamma_S^+\gamma_L^-)^{1/2} + 2(\gamma_S^-\gamma_L^+)^{1/2}$$

where γ_i^+ reflects the acidic (electron-pair acceptor) properties, γ_i^-—the basic (electron-pair donor) properties, and for water $\gamma_L^+ = \gamma_L^- = 25.5$ mJ/m² (by definition). According to van Oss[32] the acidic parameter is rather small for most solids and therefore we have made the simplifying assumption that $\gamma_S^+ = 0$ for both dexydroxylated oxides. The value of γ_S^- for silica was selected such that the contact angle calculated is equal to the experimentally observed value of 40°. The value of W_A^{AB} for titania was taken as equal to the value for silica (see grey bars in Figure 6).

It can be seen clearly that the total work of adhesion (dark grey bars in Figure 6) is lower than the work of cohesion for water in the case of silica and higher in the case of titania. Thus the different wettability of dehydroxylated silica and titania with respect to water is due to their considerably dissimilar Hamaker constants. This nicely illustrates the point that a high-energy surface is wetted by most liquids because of its higher polarisability rather than simply because of its high surface energy.[7]

The chemical difference between the islands (SiO_2) and the matrix (TiO_2) is best seen in the positive SIMS image (Figure 7).

The Si^+ and Ti^+ ion maps are complementary to each other and the thickness of the transition edge between silica and titania is about 0.4 μm.

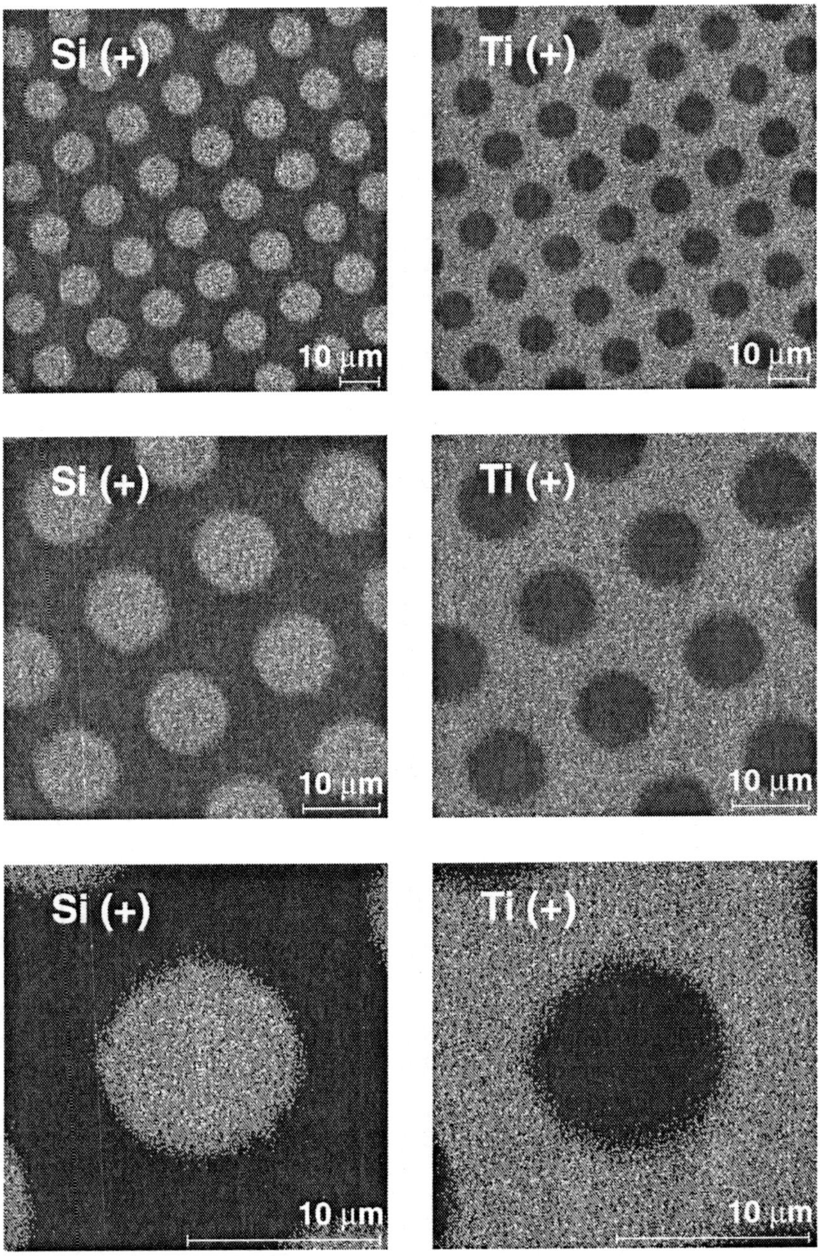

FIGURE 7 Chemical composition of the patterned surface (silica circles on a titania base) as determined by ToF-SIMS: positive ion maps of Si^+ (m/z = 28) and Ti^+ (m/z = 47)

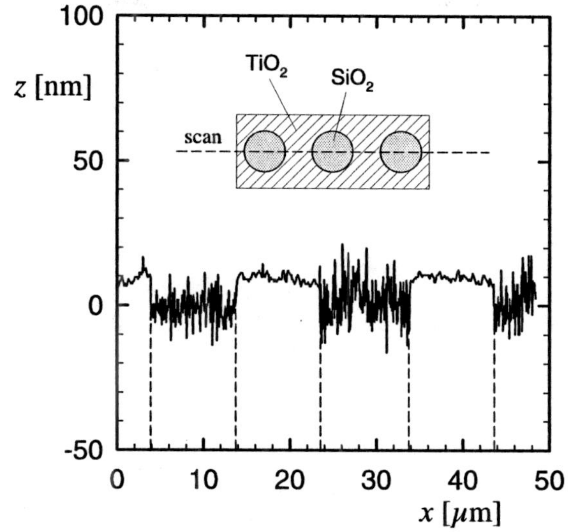

FIGURE 8 Topography of the patterned surface (silica circles on a titania matrix) as determined by AFM (Tapping Mode) scan

The topography of the patterned surface was investigated and an AFM height profile is given in Figure 8.

The two types of patches are clearly distinguished. The rms roughness is different and characteristic of the two materials after high-temperature treatment. The average thickness of the silica layer is about 10 nm.

The patterned silica-on-titania surface is very robust: It can sustain high temperature (at least up to 1,050 °C), UV irradiation or aggressive solutions. Various applications can be envisaged as the properties of SiO_2 and TiO_2 are very different. Thermal dehydroxylation is part of the above preparation protocol and the surface has hydrophilic (TiO_2) and moderately hydrophobic (SiO_2) patches. Another significant difference between the two oxides is that titania, unlike silica, has strong photocatalytic properties.

CONCLUSION

The wettability of titania and silica surfaces is mainly governed by the density of hydroxyl groups (silanols and titanols, respectively). Intensive heat treatment reduces the number of OH groups per unit area. In both cases, the condensation of vicinal silanols/titanols to siloxane/titanoxane bonds is plausible. Silica becomes moderately hydrophobic, while titania remains hydrophilic. UV irradiation has no effect on silica but is remarkably effective in removing organic contamination from titania surfaces. The effect of UV light can be entirely attributed to photocatalytic decomposition, and therefore removal, of organic contamination. The wettability difference between silica and titania can be rationalized with a simple model incorporating van der Waals and Lewis-type acid-base interfacial interactions. The wettability contrast is significant and can be used to produce micro-sized patterns on a flat surface.

ACKNOWLEDGMENT

Financial support for this project from the Australian Research Council through the Special Research Centre Scheme is gratefully acknowledged.

REFERENCES

(1) Abbott, S.; Ralston, J.; Reynolds, G.; Hayes, R. *Langmuir* **1999**, *15*, 8923.
(2) Stevens, N.; Priest, C. I.; Sedev, R.; Ralston, J. *Langmuir* **2003**, *19*, 3272.
(3) Richards, N.; Ralston, J.; Reynolds, G. *Contact Angle, Wettability and Adhesion* **2003**, *3*, 361.
(4) Quinn, A.; Sedev, R.; Ralston, J. *J. Phys. Chem. B* **2003**, *107*, 1163.
(5) Peykov, V.; Quinn, A.; Ralston, J. *Colloid Polym. Sci.* **2000**, *278*, 789.
(6) Schneemilch, M.; Welters, W. J. J.; Hayes, R. A.; Ralston, J. *Langmuir* **2000**, *16*, 2924.
(7) de Gennes, P. G. *Rev. Modern Phys.* **1985**, *57*, 827.
(8) Zisman, W. A. *Adv. Chem. Ser.* **1964**, *43*, 1.
(9) White, M. L. In *Clean Surfaces: Their Preparation and Characterization for Interfacial Studies*; Goldfinger, G., Ed.; Marcel Dekker: New York, 1970; pp 361.
(10) Schrader, M. E. In *Modern Approaches to Wettability: Theory and Applications*; Plenum Press: New York, 1992; pp 53.
(11) Lamb, R. N.; Furlong, D. N. *J. Chem. Soc., Faraday Trans. 1* **1982**, *78*, 61.
(12) Yang, J.; Duan, J.; Fornasiero, D.; Ralston, J. *J. Phys. Chem. B* **2003**, *107*, 6139.
(13) Takeda, S.; Fukawa, M.; Hayashi, Y.; Matsumoto, K. *Thin Solid Films* **1999**, *339*, 220.
(14) Horsthemke, A.; Schroeder, J. J. *Chem. Eng. Fund.* **1983**, *2*, 66.
(15) Hair, M. L. *Infrared Spectroscopy in Surface Chemistry*; Marcel Dekker: New York, 1967.
(16) Iler, R. K. *The Chemistry of Silica: Solubility, Polymerization, Colloid and Surface Properties, and Biochemistry*; Wiley: New York, 1979.
(17) Legrand, A. P. *The Surface Properties of Silicas*; John Wiley: New York, 1998.
(18) Vansant, E. F.; van der Voort, P.; Vrancken, K. C. *Characterization and Chemical Modification of the Silica Surface*; Elsevier: Amsterdam, 1995.
(19) Parfitt, G. D. *Pure Appl. Chem.* **1976**, *48*, 415.
(20) Zhuravlev, L. T. *Colloids Surfaces A* **2000**, *173*, 1.
(21) Parfitt, G. D. In *Progress in Surface and Membrane Science*; Cadenhead, D. A.; Danielli, J. F., Eds., 1976; Vol. 11, pp 181.
(22) Textor, M.; Sitting, C.; Frauchiger, V.; Tosatti, S.; Brunette, D. M. In *Titanium in Medicine*; Brunette, D. M.; Tengvall, P.; Textor, M.; Thomsen, P., Eds.; Springer: Berlin, 2001; pp 171.
(23) Sze, S. M. *Physics of Semiconductor Devices*, 2nd ed.; Wiley: New York, 1981.
(24) Fujishima, A.; Rao, T. N.; Tryk, D. A. *J. Photochem. Photobiol. C* **2000**, *1*, 1–21.
(25) El-Kareh, B. *Fundamentals of Semiconductor Processing Technologies*; Kluwer Academic Publishers: Boston, 1995.
(26) Halfpenny, D. R.; Kane, D. M.; Lamb, R. N.; Gong, B. *Appl. Phys. A* **2000**, *71*, 147.
(27) Wang, R.; Hashimoto, K.; Fujishima, A.; Chikuni, M.; Kojima, E.; Kitamura, A.; Shimohigoshi, M.; Watanabe, T. *Nature (London)* **1997**, *388*, 431.
(28) Sun, R.-D.; Nakajima, A.; Fujishima, A.; Watanabe, T.; Hashimoto, K. *J. Phys. Chem. B* **2001**, *105*, 1984.
(29) Sakai, N.; Wang, R.; Fujishima, A.; Watanabe, T.; Hashimoto, K. *Langmuir* **1998**, *14*, 5918.
(30) Wang, R.; Sakai, N.; Fujishima, A.; Watanabe, T.; Hashimoto, K. *J. Phys. Chem. B* **1999**, *103*, 2188.
(31) Miyauchi, M.; Nakajima, A.; Watanabe, T.; Hashimoto, K. *Chem. Mater.* **2002**, *14*, 2812.
(32) Van Oss, C. J. *Interfacial Forces in Aqueous Media*; Marcel Dekker: New York, 1994.
(33) Kanta, A.; Sedev, R.; Ralston, J. *Langmuir* **2005**, 21, 6, 2400–2407.
(34) Kanta, A.; Sedev, R.; Ralston, J. *Langmuir* **2005**, 21, 13, 5790–5794.

Characterization of Surface Free Energies of Talc Powders Using the Thin Layer Wicking Technique

Emilio Lobato,* Jinhong Zhang,† Ismail Yildirim,† Roe-Hoan Yoon,† and Jorge Yordan‡

Thin layer wicking technique was used to determine the contact angles (θ) of different liquids on powdered talc samples. The measured contact angles were used to calculate the surface free energies (γ_S) of talc and its components, which included dispersion (γ_S^d), acidic (γ_S^+), basic (γ_S^-), and acid-base (γ_S^{AB}) surface free energies, using the van Oss-Chaudhury-Good (OCG) equation. The surface free energy characterization was conducted on both the basal and edge surfaces. The results showed that talc is a low surface energy solid, and that its basal surfaces are basic, while its edge surfaces are acidic.

It has also been found that the hydrophobicity of talc increases with decreasing particle size, indicating that fracture occurs preferentially along the basal surface of talc. As the particle size is reduced beyond a certain limit, however, talc becomes increasingly hydrophilic, indicating that talc begins to fracture along the edge surfaces, thereby exposing more hydrophilic surfaces.

INTRODUCTION

Talc ($Mg_3Si_4O_{10}(OH)_2$) (Orowan, 1933) is a natural mineral of which the elementary sheet is composed of a layer of magnesium-oxygen/hydroxyl octahedral sandwiched between two tetrahedral silica layers (Evans and Guggenheim, 1988; Gehring, et al.,

* Industrial Minerals Division, Magnesita S.A., Praca Louis Ensch, 240 – Contagem, MG – Brazil

† Center for Advanced Separation Technologies, Virginia Tech, Blacksburg, Virginia

‡ Luzenac America, Englewood, Colorado

1988). Its main surface, the face surface, of an elementary sheet shows a high hydrophobicity with no hydroxyl groups or active ions; whereas the edge surface shows some hydrophilicity because of the existence of pH dependent SiOH and MgOH sites (Jenkins and Ralston, 1998; Michot, et al. 1994). This dual surface property of talc plays an important role in many of the industrial applications, including paper coatings, pitch control, ceramics manufacture, and as filler in the plastics, polymer, paint and cosmetics industries.

In the paper industry, for the pitch and sticky control applications, the hydrophilic property of the edges allows the particles to be dispersed in aqueous media, while the hydrophobic property of the basal surfaces attract the sticky hydrophobic substances present in wood pulp. For filler applications, proper control of the adhesion between talc filler and the matrix is essential in controlling the property of the composite material. The strength of adhesion depends on the surface properties of the filler and of the matrix (Fourchet, 1995). In general, strong filler-matrix interactions result in improved processability, impact strength, and surface quality, while interactions that are too weak lead to decreased strength and increased deformability of the composite (Pukansky, 1995).Therefore, determination of surface properties both at the basal and edge surfaces of talc is essential for understanding the origin of molecular interaction and strength of adhesion between two interacting surfaces, i.e., talc and polymers, stickies, etc. It will also be helpful for the proper selection of surfactants used in the surface treatment of talc. Thus, it is crucial to determine the surface free energy and its components at the basal and edge surfaces of talc individually.

In this communication, the thin layer wicking technique is applied to determine the surface free energy of talc powders. The relationship between solid surface free energy and particle size is thus drawn based on the wicking results. Further, the flow microcalorimetry is utilized to determine the basal/edge surface area ratio of talc powders. Thus, solid surface free energy components on basal and edge of talc surfaces are accordingly inferred based on the unique technique developed in present work.

THEORETICAL BACKGROUND

Acid/Base Theory

Determination of surface free energies of solids has been a topic of heated discussion for a long time, and there is not a generic consensus as yet. The acid/base theory developed by van Oss et al. (1986, 1987, 1988, and 1994) is widely used at the present time. The theory is briefly described as follows.

Fowkes (1962, 1963) proposed that the surface free energy of a material i can be subdivided into separate components:

$$\gamma_i = \sum_j \gamma_i^j \qquad \textbf{(EQ 1)}$$

where j refers to types of surface free energy component, e.g., dispersion, H-bonding, and metallic interactions. Van Oss et al. considered the surface tension as a sum of apolar Lifshitz-van der waals, γ^{LW}, and polar acid/base interactions, γ^{AB}. Thus,

$$\gamma_i = \gamma_i^{LW} + \gamma_i^{AB} \qquad \textbf{(EQ 2)}$$

where γ_i^{LW} and γ_i^{AB} refer to the non-polar and polar (acid-base) surface free energy components, respectively. The former can be represented by the Lifshitz-van der Waals (or

LW) interactions that include the dispersion (London), induction (Debye) and orientation (Keesom) components. The latter represents interactions between Lewis acids (electron acceptor) and bases (electron donor) on the surface.

van Oss et al. used the geometric mean combining rule to define γ_i^{AB} as follows:

$$\gamma_i^{AB} \equiv 2\sqrt{\gamma_i^+ \gamma_i^-} \tag{EQ 3}$$

where γ_i^+ is the acidic component of the surface tension and γ_i^- is the basic component. Therefore, Eq. [2] becomes

$$\gamma_i = \gamma_i^{LW} + 2\sqrt{\gamma_i^+ \gamma_i^-} \tag{EQ 4}$$

It can be readily shown also that the free energy change (ΔG_{SL}) associated with the adhesion of a liquid on a solid in the air is given by:

$$\Delta G_{SL} = -2\left(\sqrt{\gamma_S^{LW} \gamma_L^{LW}} + \sqrt{\gamma_S^+ \gamma_L^-} + \sqrt{\gamma_S^- \gamma_L^+}\right) \tag{EQ 5}$$

where γ^{LW} is the Lifshitz-van der Waals component of the surface energy, γ^+ the acidic component, and γ^- is the basic component of the interacting solid (subscript S) and liquid (subscript L) surfaces.

Combining this with the Dupre equation:

$$\Delta G_{SL} = \gamma_{SL} - \gamma_S - \gamma_L \tag{EQ 6}$$

one obtains:

$$\gamma_{SL} = \gamma_S + \gamma_L - 2\left(\sqrt{\gamma_S^{LW} \gamma_L^{LW}} + \sqrt{\gamma_S^+ \gamma_L^-} + \sqrt{\gamma_S^- \gamma_L^+}\right) \tag{EQ 7}$$

Combining Eq. [7] with Young's equation,

$$\gamma_L \cos\theta = \gamma_L - \gamma_{SL} \tag{EQ 8}$$

where θ is contact angle, one obtains the van Oss-Chaudhury-Good (OCG) equation:

$$(1 + \cos\theta)\gamma_L = 2\left(\sqrt{\gamma_S^{LW} \gamma_L^{LW}} + \sqrt{\gamma_S^+ \gamma_L^-} + \sqrt{\gamma_S^- \gamma_L^+}\right) \tag{EQ 9}$$

which is used to characterize a solid surface in terms of its surface free energy components, i.e., γ_S^{LW}, γ_S^+, and γ_S^-. To determine these values, it is necessary to determine contact angles of three different liquids of known properties (i.e., γ_L^+, γ_L^-, γ_L^{LW}) on the surface of the solid of interest. One can then set up three equations with three unknowns, which can be solved simultaneously to obtain the values of γ_S^{LW}, γ_S^+, and γ_S^-. Table 1 gives a list of liquids that can be used for the contact angle measurements, along with the values of γ_L, γ_L^{LW}, γ_L^+, and γ_L^-.

Thin Layer Wicking

In practice, however, it is impossible to directly measure the contact angle of liquid on powdered solids, especially for clay particles. Thin layer wicking technique is a convenient way to measure contact angles of liquids on powdered materials (Yildirim, 2001; Lobato, 2004) by determining the capillary rise rate of the wicking liquids in packed

TABLE 1 Surface tension components (mN/m) and viscosities (mPa·s) of the wicking liquids at 20°C used in thin layer wicking measurements

Liquid	γ_i^{LW} (mN/m)	γ_i^{AB} (mN/m)	γ_i^{+} (mN/m)	γ_i^{-} (mN/m)	η (mPa·s)
Heptane	20.3	0	0	0	0.409
Octane	21.6	0	0	0	0.542
Decane	23.8	0	0	0	0.907
Dodecane	25.35	0	0	0	1.493
1-Bromonaphthalene	44.4	0	0	0	4.89
Methylene iodide	50.8	0	0	0	2.82
Water	21.8	51.0	25.5	25.5	1.08
Formamide	39.0	19	2.28	39.6	3.76

beds of those particulate materials. The wicking process follows the Washburn's equation (Washburn, 1921).

$$\frac{2\eta h^2}{t} = R_{eff} \cdot \gamma_L \cdot \cos\theta \qquad \textbf{(EQ 10)}$$

in which η is the viscosity of the liquid, h is the height to which wicking liquid has risen at time t, R_{eff} is the effective interstitial pore radius of packed powder bed, γ_L is the surface tension of wicking liquid and θ is the contact angle.

The parameter R_{eff} of Eq. [10] is generally determined by wicking with low-energy spreading alkanes, for which the contact angle θ is assumed to be zero. Thus, Eq. [10] is reduced into the following equation:

$$\frac{2\eta h^2}{t} = R_{eff} \cdot \gamma_L \qquad \textbf{(EQ 11)}$$

After wicking with several low energy alkanes, one can draw a plot of $2\eta h^2/t$ vs. γ_L. The slope of the straight fitting line passing through origin gives the effective interstitial pore radius R_{eff}.

The contact angle of high energy liquid, such as water, on powder surface can be obtained by wicking with the liquid on thin powder layer. Using the R_{eff} value obtained from wicking result with low energy alkanes, one can get the water contact angle by solving Eq. [10], in which contact angle is the only unknown parameter.

Free Energy of Basal/Edge Surfaces

Flow microcalorimetry has been shown to be a powerful tool for characterizing solid surfaces and their interactions with liquids. It was also used to estimate the percentage of hydrophilic (polar) and hydrophobic (nonpolar) surface sites on powdered solid surfaces (Groszek, 1962, 1966, 1970). In this method, a talc sample is equilibrated initially with heptane and then contacted with a 2 g/l butanol solution while measuring the calorimetric heat effect. The heat effect is due to the butanol displacing heptane from the hydrophilic edge surface of the talc sample. Since heptane is hydrophobic, it will be adsorbed strongly on the hydrophobic basal surface but not on the hydrophilic edge surface. By dividing the heat effect with the total surface area of talc sample used in the measurement, one can obtain the molar heat of adsorption ($-\Delta H_{ad}$) of butanol in units of mJ/m^2. Further, by dividing this value with the heat of adsorption of butanol on silica ($197\ mJ/m^2$), one obtains the fraction of the talc surface area that is hydrophilic, i.e., fraction of edge surface area, (f_e). Also, one can get the basal surface area fraction, f_b, which equals one minus f_e.

In order to determine the surface free energy components at the basal and edge surfaces of talc, surface free energy characterization studies must be conducted at two different sizes, D_1 and D_2 (Yildirim, 2001; Lobato, 2004). The solid surface free energy (γ_s) and its components (i.e., γ_s^{LW}, γ_s^+, and γ_s^-) of each sample are obtained using the above described thin layer wicking technique. From the known f_e and f_b values obtained from flow microcalorimetry, one can then set up the following equation (Yildirim, 2001; Lobato, 2004):

$$\gamma_s^{D_1} = f_b^{D_1}\gamma_s^b + f_e^{D_1}\gamma_s^e \qquad \text{(EQ 12)}$$

in which $\gamma_S{}^b$ is γ_S at the basal surface and $\gamma_S{}^e$ is γ_S at the edge surface. Eq. [12] suggests that the value of γ_S is a composite of $\gamma_S{}^b$ and $\gamma_S{}^e$ in proportion to $f_b^{D_1}$ and $f_e^{D_1}$. Similarly, one can set up the following equation at size D_2:

$$\gamma_s^{D_2} = f_b^{D_2}\gamma_s^b + f_e^{D_2}\gamma_s^e \qquad \text{(EQ 13)}$$

An explicit assumption in writing Eqs. [12] and [13] is that $\gamma_S{}^b$ and $\gamma_S{}^e$ are intensive properties of talc and do not change with particle size, which is a reasonable assumption. By solving Eqs. [12] and [13] simultaneously, one can determine the values of $\gamma_S{}^b$ and $\gamma_S{}^e$. Likewise, one can set up simultaneous equations for $\gamma_S{}^{LW}$, $\gamma_S{}^{AB}$, $\gamma_S{}^+$, and $\gamma_S{}^-$, and obtain the respective values at the basal and edge surfaces of talc.

EXPERIMENTAL

Materials

Nanopure water was obtained by Nanopure (Barnstead, IA) water system. The conductivity of the water was 18.2 MΩ/cm at 23°C and the surface tension was 72.5mN/m at 22°C. Pre-cleaned glass microslides (75×25 mm, Corning, NY) were used without further cleaning.

Wicking liquids used in the present work, Alkanes, such as Heptane, Octane, Decane and Dodecane (Aldrich Chemical Company) were HPLC grade (>99.5% purity). The surface tension and viscosity data of these wicking liquids are listed as Table 1. HPLC-grade acetone and butanol were purchased from Fluka and both had purities greater than 99.5%.

Powdered talc samples, labeled Luz A and Luz B, were kindly provided by Luzenac America (Englewood, CO). Talc samples, labeled Mag A to Mag G were obtained from Magnesita (Brumado, Brazil). The Mag series samples were all from the same mine; however, they were of different particle sizes as obtained by using different grinding technologies.

Thin Layer Wicking Technique

A known amount of a given talc powder was dispersed in deionized water to make a 4% (w/v) suspension. Aliquots of 4.5 ml suspensions were withdrawn with a pipette and evenly distributed onto a microscopic glass slide, which was kept strictly horizontal. The water was allowed to evaporate overnight at room temperature and further dried in an oven at 150°C overnight to remove residual pore water. It was then cooled down in a vacuum desiccator before being used for wicking experiments.

The wicking experiments were performed by means of a horizontal chamber which was originally designed for thin layer chromatography (DS-II chamber, Chromdes, Poland), following the procedure of Chibowski and Holysz (1992). All the experiments were carried out in a closed chamber. The glass slide was placed horizontally in the chamber with the talc layer facing up. The probe liquid was placed in one of the edges of the chamber and brought to contact with the adsorbent layer by means of another glass slide. A scale was fixed outside the glass cover and carefully aligned with the edge of the slides to set the zero position. The time for the probe liquid to travel through every 5 mm interval was recorded using a stopwatch. The wicking was stopped by moving back the glass slide of the liquid container. With each sample, the tests were repeated at least three times for each probe liquid and the values reported here were the arithmetic averages over three sets of experiment data.

The value of the parameter R_{eff} of Eq. [11] was determined using liquids of low surface free energy (heptane, octane, decane and dodecane), all of which were found to completely spread over the solid surface (i.e., $\theta=0$). Experimental results showed that the $2\eta l^2/t$ vs. γ_L yielded a straight line passing through the origin. This finding confirms that the Washburn equation, Eq. [11] is applicable under the wicking conditions in present work.

Flow Microcalorimetry

Heats of adsorption measurements were conducted using a Model 3V flow microcalorimeter (FMC) from Microscal, United Kingdom. Figure 1 shows a schematic diagram of the calorimeter. In each measurement, a talc sample was dried overnight in an oven at 110°C. A known amount (usually 0.05–0.15 gram) of the dried sample was placed in the calorimeter cell and degassed for at least 30 minutes under vacuum (<5 mbar) at ambient temperature. The vacuum system consisted of a vacuum pump and a liquid nitrogen vapor trap. The solvent was then introduced to the calorimeter cell at a steady flow rate of 3.3 ml/h by means of a syringe pump, and two Valco Instrument changeover valves for switching between solvent and solution. The heat effect was recorded by means of a strip chart recorder and a PC. After the thermal equilibrium was reached, which usually took 8 to 30 minutes, the solvent was switched over to the solvent containing adsorbate and the solution was introduced to the cell at a flow rate of 3.3 ml/h. The heat of interaction of an adsorbate on the solid was also recorded using a strip chart recorder and a PC. The recorded experimental data was analyzed using the Microscal Calorimeter Digital Output-Processing System (CALDOS), which enables the analysis of calibration and experimental data and converts the raw downloaded data into the heat of immersion (h_i) and heat of adsorption (h_{ads}) values.

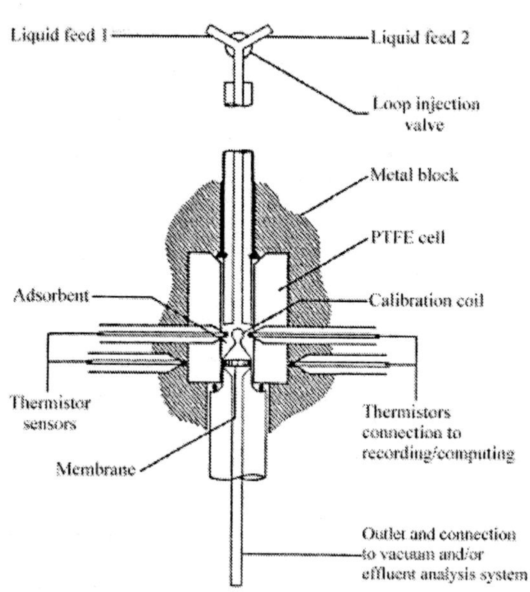

FIGURE 1 Diagram of the flow micro-calorimeter used for the adsorption studies

RESULTS AND DISCUSSION

Effective Pore Radius

Figure 2 shows the results of the thin layer wicking experiments conducted with four different non-polar alkane liquids, namely Heptane, Octane, Decane and Dodecane, on the glass slides coated with Luz A talc powders, which has an average particle size, D_{50} of 7 μm. It is clearly seen from the figure that there is a very good linear relationship between the wicking distance square h^2 and wicking time t.

Figure 3 shows the thin layer wicking experiment results of four different alkanes on Luz B coated glass slides. The average particle size, D_{50} of Luz B was 3.2 μm. As shown in Figure 1, the wicking distance square h^2 and the wicking time t follow a good linear relationship.

The good linearity of h^2 vs. t plots shown in Figures 2 and 3 showed that Eq.[11] can be used for talc. When low surface free energy alkanes are used as wicking liquids, they spread automatically on solid surface and form zero degree contact angles; thus the viscosity of the wicking liquid was the only parameter that determined the wicking velocity.

The wicking experiment results shown in Figure 2 and Figure 3 were fitted linearly and the slope, which equals h^2/t, was inputted to the left of Eq.[11] to obtain the value of $2\eta h^2/t$ for each wicking liquid. In Figure 4, $2\eta h^2/t$ vs. γ_L of each wicking liquid was plotted for the Luz A and Luz B samples. It is clearly seen from the figure that for each talc sample, $2\eta h^2/t$ vs. γ_L follows a good linear relationship and the fitting line passing through the origin, which indicates that the contact angles of these wicking liquids are zero and Eq.[11] is applicable under the present experiment conditions. Figure 4 shows that the slope of $2\eta h^2/t$ vs. γ_L, or the effective pore size (R_{eff}) for Luz A, which is a coarser talc sample, is larger than that of the Luz B talc sample, which is of a smaller particle size.

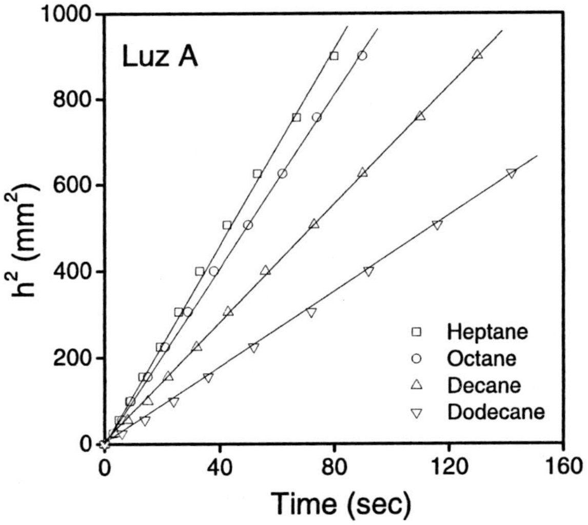

FIGURE 2 h^2 **vs. time of Luz A in thin layer wicking using alkanes**

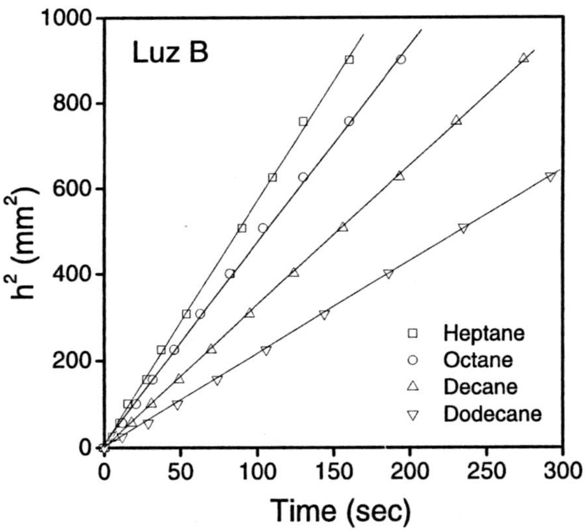

FIGURE 3 h^2 **vs. time of Luz B in thin layer wicking using alkanes**

Effect of Particle Size on Pore Radius

In present work, besides Luz A and Luz B samples, we obtained the effective pore sizes (R_{eff}) of Mag series samples of different particle sizes. The effective pore size (R_{eff}) and corresponding powder size (D_{50}) for each of the talc samples tested are listed as Table 2 and plotted in Figure 5. It is clearly seen from Figure 5 that R_{eff} increases with D_{50}. That all the data points are almost on a same straight line suggests a strong linear relationship between R_{eff} and D_{50}. Consecutively there comes a relationship between R_{eff} and D_{50}, which writes as follows:

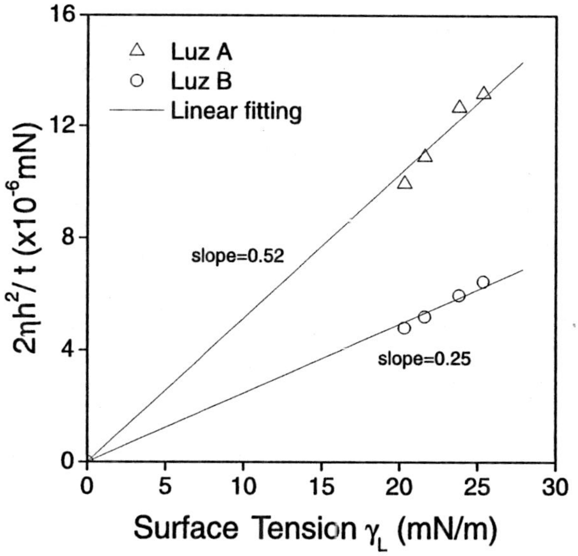

FIGURE 4 $2\eta h^2/t$ **vs. surface tension of the wetting liquids for Luz A and Luz B talc samples. The slope gives effective pore radius.**

TABLE 2 Effective pore size, R_{eff} of talc obtained with the thin layer wicking technique using low surface free energy liquids

Sample	D_{50} (μm)	R_{eff} (μm)
Mag A	7.5	0.456
Mag B	4.49	0.295
Mag C	1.53	0.129
Mag D	1.35	0.076
Mag E	1.15	0.091
Mag F	0.49	0.055
Mag G	0.39	0.046
Luz A	3.2	0.25
Luz B	7.0	0.52

$$R_{eff} = k \cdot D_{50} \quad \text{(EQ 14)}$$

where k is a constant that relates to the packing method of particles in a powder bed.

The packing of a powder bed is generally determined by the aspect ratio (l_l/l_w) of the powders, where l_l and l_w are respectively the dimensions of the particle in length and in width. As shown in Figure 5, all the talc samples have a nearly constant k value suggesting that the aspect ratio (l_l/l_w) of all of the samples should have similar values, which in turn results in a similar basal/edge surface area ratio, f_b/f_e. We will further discuss this in a later section.

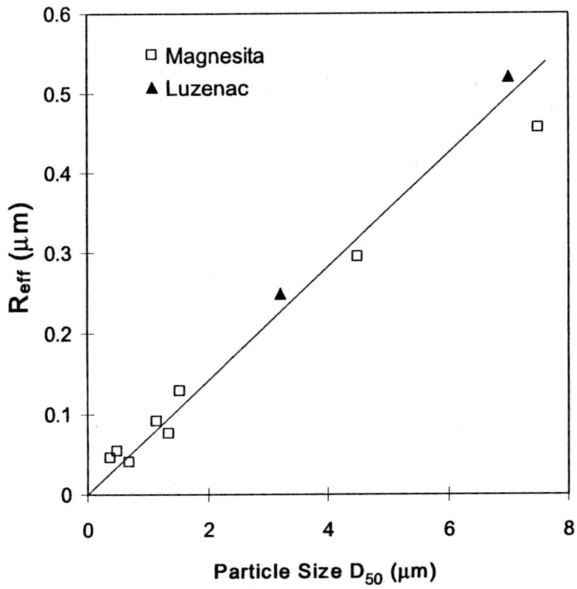

FIGURE 5 ***R_{eff}* vs. particle size *D_{50}* of different talc powder samples using thin layer wicking technique**

Surface Free Energy Components versus Particle Size

Table 3 shows the contact angles of various liquids on powdered talc samples measured with thin layer wicking technique. These contact angle data are further used to obtain the value of the various surface free energy components, i.e., γ_S^{LW}, γ_S^{AB}, γ_S^+, γ_S^- and γ_S, using the combinations of Eq. [9]. The calculation results are listed as Table 4 and also plotted as a function of average particle size (D_{50}) in Figure 6. The results demonstrate that the surface of talc is predominately basic, though there are few acidic sites. For example, the γ_S^- on Mag D is 6.3 mJ/m^2; whereas γ_S^+ is only 0.08 mJ/m^2. Another finding is that the value of γ_S^- increases with decreasing particle size down to the average particle size of 1.5 μm after which it starts to decrease again. The opposite is true for the γ_S^+, i.e., it first decreases down to the particle size of 1.5 μm and then increases again. The values for γ_S^{AB} are nearly the same down to the particle size of 1.5 μm due to the balance in the γ_S^-/γ_S^+ parameters.

However, there is a clear trend that γ_S^{AB} increases with decreasing particle size (E, F and G). Also shown, the decrease in the value of γ_S is primarily attributed to the decrease in the value γ_S^{LW} down to the limiting size of 1.5 μm. After that, γ_S tends to increase as a result of increasing γ_S^{AB}.

The results show that an increase in the percentage of basal planes would result in an increase in γ_S^- and a decrease in γ_S^+. This was indeed what happened down to the particle size of about 1.5 μm. On the contrary, an increase in the edge surfaces would give a higher value of γ_S^+ and lower γ_S^-, which was observed for particle size finer than 1.5 μm. It suggests, therefore, that there is a limit, i.e., 1.5 μm, in terms of delamination that can be achieved using the grinding technology used in the present work. Beyond that breakage begins to occur across the basal plane generating more hydrophilic edge surfaces.

TABLE 3 Contact angles of various liquids on powdered talc samples measured with thin layer wicking technique

Talc Sample	D_{50} (μm)	Contact Angle (θ)			
		W	FO	1-Br	DI
Mag A	7.5	85.87	51.70	36.33	55.05
Mag B	4.49	83.48	60.15	55.05	57.76
Mag C	1.53	85.73	63.85	44.81	54.55
Mag D	1.35	80.92	60.96	36.47	29.32
Mag E	1.15	82.96	52.75	38.89	45.14
Mag F	0.49	78.30	50.95	40.15	43.60
Mag G	0.39	76.07	45.61	43.35	50.58

W = water; FO = Formamide; 1-Br = 1-bromo-napthalene; DI = Diiodomethane

TABLE 4 Surface free energy and its components of powdered talc samples obtained with the thin layer wicking technique

Talc Sample	D_{50} (μm)	Surface Free Energy (mJ/m²)				
		γ_s^{LW}	γ_s^-	γ_s^+	γ_s^{AB}	γ_s
Mag A	7.5	36.28	0.78	1.63	2.25	38.53
Mag B	4.49	31.73	3.93	0.7	3.32	35.05
Mag C	1.53	32.44	3.85	0.27	2.03	34.47
Mag D	1.35	36.13	6.3	0.08	1.4	37.53
Mag E	1.15	35.1	2.07	1.38	3.38	38.48
Mag F	0.49	34.55	4.19	1.41	4.86	39.42
Mag G	0.39	33.11	3.9	2.72	6.51	39.62

Fraction of Hydrophilic Surface on Talc Sample from Flow Microcalorimetry

Table 5 shows the fraction of hydrophilic surface on the Mag series talc samples obtained with a flow microcalorimetry. One can see from the table that for the largest sample (Mag A), the fraction of hydrophilic surface is 25.7%. When the particle size decreases to 1.5 μm, the fraction of hydrophilic surface also decreases to 17.42%. When the particle size further decreases to 0.15 μm, the hydrophilic surface fraction, however, increases to 22.61%. That is, when particle size is larger than 1.5 μm, the fraction of hydrophilic surface decreases with D_{50}, which is due delamination along the basal plane; when particle size is smaller than 1.5 μm, the fraction of hydrophilic surface increases with decreasing D_{50}, which is due to the fracture across the basal plane. This finding correlates to the one obtained with surface free energy analysis.

Surface Free Energy Components at the Basal and Edge Surfaces of Powdered Talc Samples

In order to obtain the surface free energies of the basal and edge surfaces individually, it is necessary to know the area fractions of the basal surface (f_b) and that of the edge surface (f_e) at two different size fractions, as well as the total (or composite) surface free

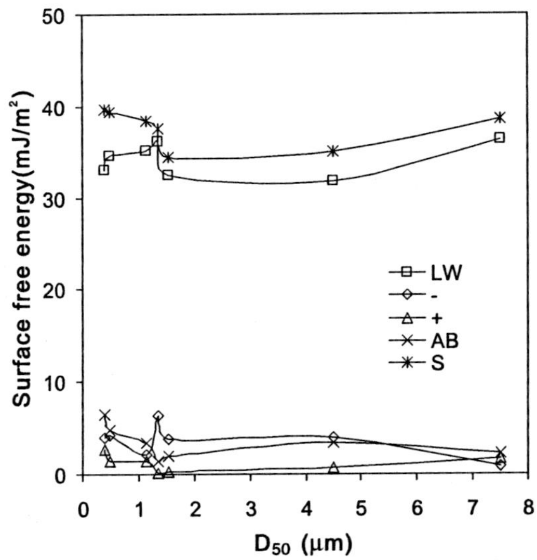

FIGURE 6 Effect of particle size on surface free energy (S) samples, i.e., London-van der Waals component (LW), basic component (–), acid component (+), and acid-base components (AB)

TABLE 5 Percentage of hydrophilic surface of the talc samples obtained from butanol-in-heptane method using flow microcalorimetry

Talc Sample	D_{50} (μm)	$-\Delta H_{ads}$/BuOH from n-Heptane on Talc (mJ/m^2)	% Hydrophilic Surface
Mag A	7.5	50.54	25.66
Mag B	4.49	43.38	21.54
Mag C	1.53	34.38	17.42
Mag D	1.35	44.21	22.45
Mag E	1.15	36.75	18.66
Mag F	0.49	47.97	24.53
Mag G	0.39	44.53	22.61
Silica	—	197	100

energies (γ) of the two talc samples of different size fractions. In the present work, two talc samples, i.e., Mag A (D_{50} = 7.5 μm) and Mag E (D_{50} = 1.15 μm), were chosen for sample calculations. These two samples were chosen because of the relatively large difference in particle size, which should give sufficient differences between the two samples in surface area fractions and surface free energies. The values of f_b and f_e for the two samples are given in Table 6. Using the uncorrected data, the surface free energy components of the basal and edge surfaces have been calculated by solving Eqs. [12] and [13] simultaneously, and are given in Table 7. Despite the negative surface free energies obtained, which must be incorrect, the results show important points, that is, the basal surface is more basic than acidic, and the edge surface is more acidic than basic. This

TABLE 6 Corrected and uncorrected fractions of the basal and edge surfaces of selected talc samples

	Uncorrected		Corrected	
Talc Sample	f_b	f_e	f_b	f_e
Mag A (D_{50} = 7.46 μm)	0.743	0.257	0.303	0.697
Mag E (D_{50} = 1.15 μm)	0.813	0.187	0.813	0.187

TABLE 7 Surface free energy components (mJ/m²) at the basal and edge surfaces of powdered talc samples obtained using flow microcalorimetry

	Uncorrected		Corrected	
Surface Free Energies	Basal Surface (mJ/m²)	Edge Surface (mJ/m²)	Basal Surface (mJ/m²)	Edge Surface (mJ/m²)
γ_s^{LW}	51.59	−26.98	34.67	36.99
γ_s^-	5.84	−14.16	2.54	0.34
γ_s^+	−4.08	23.07	1.29	1.78
γ_s^{AB}	3.27	1.84	3.79	1.58
γ_s	54.86	−25.14	38.46	38.56

finding is reasonable because the basal surface of talc can be considered as Lewis base because its basal surface consists of oxygen atoms only without cations. The basal surfaces of talc are made up of fully-compensated oxygen atoms, linked together by siloxane (Si-0-Si) bonds, while the edge surfaces are composed of MgOH, SiOH and the other substituted cations, e.g., Al^{3+}, Fe^{2}+. The oxygen atoms can donate electron pairs to a base, thereby meeting the definition of Lewis base. On the other hand, the edge surfaces, can readily dissociate to give up protons, thereby acting as Bronstead acid.

The most likely reason that Eqs. [12] and [13] gave the negative surface free energy components was the inaccuracies associated in determining the values of f_b and f_e. In the present work, these values were determined using the heats of adsorption technique, which is used in the talc industry. In this technique, heat of adsorption is measured by immersing a dry talc sample in a 10 g/l butanol-in-heptane solution, and the measured value is divided by the heat of adsorption of silica in the same solution (Groszek, and Partyka, 1993). This method assumes that butanol adsorbs only on the edge surface and that the heat of adsorption on talc is the same as on quartz. A problem with this technique is that the % hydrophilic surface (f_e) does not change significantly with particle size. As shown in Table 5, a talc sample with D_{50} = 0.49 has f_e = 0.226 while a talc sample of D_{50} = 7.5 μm has f_e = 0.256. It appears, therefore, that the technique is insensitive to particle size or aspect ratio. In principle, f_e should increase with increasing particle size. In this regard, a value of f_e = 0.697 was arbitrarily chosen (which gave f_b = 0.303) for the talc sample with D_{50} = 7.5 μm. Eqs. [12] and [13] were solved again using these "corrected" values of f_b and f_e, and the results are given in Table 7. All of the surface tension values are positive and appear reasonable. The corrected values of the surface free energies and their components show that the basal surface is basic and the edge surface is acidic.

CONCLUSION

In present work, thin layer wicking technique was used to determine the surface free energy components on powdered talc samples. The results showed that under the experimental conditions employed in the present work, the penetration of liquids in powder layer follows the Washburn's equation and the effective interstitial pore radius (R_{eff}), which is obtained by wicking with low-energy alkenes, varies linearly with particle size (D_{50}).

The results also show that there is a relationship between particle size and the basal-to-edge surface area ratios. As the particle size decreases, the basal-to-edge surface area ratio increases, indicating that the mineral fractures preferentially along the basal surface. When the particle size becomes smaller than approximately 1.5 μm, however, the basal-to-edge surface area ratio begins to decrease, indicating that the fracture occurs preferentially across the basal surface. The particle size of 1.5 μm, therefore, is the limit of effective delamination under the grinding conditions employed in the present work. These results are consistent with the changes in surface free energies measured using the acid-base-theory.

The basal-to-edge surface area ratios of talc samples have been determined from the heats of adsorption of butanol on talc. Using these values and the surface free energies measured using the acid-base theory, the surface free energies of the basal and edge surfaces of the mineral have been determined. The results show that the basal surface is basic and the edge surface is acidic.

REFERENCES

Chibowski, Emil, and Holysz Lucyna, *Langmuir*, 1992, 8(2), 710–716.

Chibowski, Emil, Gonzalez, Caballero Fernando, *Langmuir*, 1993, 9(1), 330–340.

Evans, B. W., and Guggenheim, S., *Rev. Min.*, 1988, 19, 225.

Fourchet, G., *Polymer Eng. Sci.*, 1995, 35, 957.

Fowkes, F. M., *J. Phys. Chem.*, 1962, 66, 382.

Fowkes, F. M., *J. Phys. Chem.*, 1963, 67, 2538-2541.

Gehring, A. U., *Clay Minerals*, 1988, 33, 661–664.

Groszek, A. J., and Partyka, S., *Langmuir*, 1993, 9, 2721–2725.

Groszek, A. J., *Carbon*, 1987, Vol. 25, 6, 717–722.

Groszek, A. J., *Chemistry and Industry*, Oct. 1966, 1754-1756.

Groszek, A. J., *Nature*, 1962, 196, 531.

Groszek, A. J., *Proc. Roy. Soc.*, A, 1970, 314, 473–498.

Holysz, Lucyna, Chibowski, Emil, *Langmuir*, 1992, 8(2), 717–721.

Jenkins, P., and Ralston, J., *Colloids and Surfaces A*, 1998, 139, 27–40.

Lobato, E., Master Thesis, Virginia Tech, November 2004.

Michot, L. J., Villiéras, F., Franqois, M., Yvon, J., Dred, R. Le, and Cases, J. M., *Langmuir* 1994, 10, 3765–3773.

Orowan, E., *Z. Phys.*, 1933, 82, 235.

Pukansky, B., in *Polypropylene: Structure, Blends and Composites*, ed: J. Karger Kocsis, Vol. 3, pp. 1, Chapman and Hall, London, 1995.

van Oss, C. J., "*Interfacial forces in aqueous media*," Marcel Dekker, 1994.

van Oss, C. J., Chaudhury, M. K., and Good, R. J., *Adv. in Colloid Interface Sci.*, 1987, 28, 35.

van Oss, C. J., Chaudhury, M. K., and Good, R. J., *Colloid and Interface Science*, 1988, 128, 313.

van Oss, C. J., Good, R. J., and Chaudhury, M. K., *Colloid and Interface Science*, 1986, 111, 378.

van Oss, C. J., Good, R. J., and Chaudhury, M. K., *Langmuir*, 1988, 4, 884.

Washburn, E. W., *Physical Review*, 1921, 1, 273.

Yildirim, I., Ph.D. Thesis, Virginia Tech, April 2001.

The PolyMag™ Process—A Novel Approach to Plastics Recycling

Michael J. Mankosa* **and Gerald H. Luttrell†**

With the increasing cost of natural resources, recovery and recycling of raw materials is becoming a necessity in the plastics industry. Unfortunately, injection molding processes can often have reject rates up to ten percent. Since an effective means of separating multi-component assemblies is currently not available, the rejected products are typically ground and used in low value secondary markets or simply disposed to a landfill. In light of this problem, a new method has been developed that makes it possible to selectively separate plastics by including a functional filler during the molding process. The filler enhances the magnetic susceptibility of one or more of the plastic components so that the manufacturing rejects can be granulated and separated using a specially designed magnetic separator. This article provides a brief overview of the PolyMag™ process and presents results from an investigation of the effect of the filler on plastic properties. Experimental results from separation testing of granulated, injection-molded components are also presented.

INTRODUCTION

Plastics have found wide acceptance in industry and are used to make many articles that have a variety of applications. Hundreds of types of plastics are routinely used for commercial and consumer products such as containers for household cleaners and detergents, industrial chemicals, automobile components, children's toys and various other consumer goods (Muccio, 1994a; 1994b). With the exception of a small percentage of household consumer products, a great majority of plastic polymers are simply discarded due to a lack of acceptable recycling techniques. Unfortunately, disposal of these plastics contributes to a continuous accumulation of consumer refuse and represents a vast

* Eriez Magnetics, Erie, Pennsylvania

† Mining and Minerals Engineering, Virginia Tech, Blacksburg, Virginia

waste of natural resources. The growing stacks of plastic articles at waste disposal facilities are monuments being built to our failure to effectively recycle these plastics.

Considerable research has been conducted over the past several decades with regard to material recycling. Limited success has been achieved in recycling of post consumer waste in which plastics are hand sorted from other refuse by consumers and disposed of using curbside collection facilities. Separation of the various types of plastics (i.e., PVC, PET, etc.) is accomplished by hand sorting at material recycling facilities. In some instances, whole bottle sorting machines have been employed in place of hand sorting. In addition, industrial materials such as rubber tires, chopped wire and cable, and various metals from automobile shredding are currently reclaimed with commercial success using different physical processing techniques (Ura, 1994; Frisch and Razim, 1995; Sendijarevic et al., 1997). For example, rubber tires are currently shredded to liberate nylon cord and metal belting. The nylon and metal are subsequently removed using a combination of screening, air classification and magnetic separation (Shah, 1998). The final granulated rubber product has sufficient purity to be recycled into various components such as paving materials and play area ground coverings. As another example, wire and cable are extensively recycled with the intent to recover the copper and/or aluminum wire and the plastic insulation. A typical communications cable consists of approximately 80–85% copper and various types of plastics. The cable is typically sheared into smaller pieces suitable for shredding/granulation. The granulated product (particle size of approximately 6 mm and smaller) is air-classified to remove dust/fluff and separated into metallic (copper/aluminum) and non-metallic (plastic) components using gravity and electrostatic separation methods. This approach produces a high-quality metal product that can be sold directly as a commodity. Unfortunately, the plastic by-product from this approach is often not of an acceptable quality for re-use since it consists of a mixture of components such as PVC, PET, rubber and other plastics. Each plastic component must be refined to an acceptable quality to be considered for recycling by mixing with virgin components. Further refining of the different plastic components, however, has proven a difficult task.

Various attempts have also been made to recycle mixed plastic components using gravity and electrostatic separation processes. Gravity-based processes rely upon the inherent density differences between various plastic components. For instance, process specifications may require separation of two components having a density difference of less than 0.05 specific gravity units. Separation of components with nearly the same density requires an extremely precise process that, while possible, is not practical in a continuous commercial process. Furthermore, control of medium density at such a low value requires the use of dissolved salts. This presents two additional problems, i.e., high corrosion of process equipment and the need to rinse and dry the final products, which add cost. To avoid these problems, electrostatic separation processes have been employed to separate various plastics. Triboelectrostatic separation depends upon the electrical properties of the plastic surface. Specifically, the work function of one plastic must differ substantially from the second plastic component so as to acquire an opposite electrical charge. Unfortunately, this approach suffers from an extremely low throughput capacity and an inability to produce a high purity product. Additionally, since the separation mechanism is surface dependent, coatings, contaminants and changes in atmospheric conditions (i.e., temperature and humidity) can affect the quality of the separation. Gravity- and electrostatic-based processes have had limited success since these techniques cannot achieve the desired product purity at an acceptable product recovery rate.

The PolyMag™ Process

It is apparent that the current methods for recycling plastic are inadequate to accomplish efficient separation of high purity plastics of the type needed for most recycling applications. In light of this shortcoming, the PolyMag™ process has been developed as a cost-effective method to efficiently separate plastic components. In this approach, the properties of the plastics are re-engineered using a function filler. Specifically, fillers are selected to manipulate the magnetic susceptibility of one or more of the plastic polymers used to formulate final products. The magnetic susceptibility of a plastic polymer can be altered by the inclusion of a magnetic filler during the manufacturing process. The amount of magnetic material added is sufficient to influence the magnetic susceptibility of the plastic without affecting other physical properties such as tensile strength, elongation or hardness. In a typical case, the amount of filler required ranges from 0.05% to 5% by weight of the plastic formulation. Once modified, plastics with different magnetic susceptibilities can then be selectively recovered using efficient, low-cost, high-capacity magnetic separators that are well suited for the separation of granular-size materials.

Figure 1 provides an example of a process flowchart for the separation of plastic scrap using the PolyMag™ process. In the first step, the mixed plastics are passed through a size reduction process which may include crushing, shredding, grinding and granulating of the plastic scrap. The liberated particles may then be classified using one or more of the steps of dedusting, screening and washing to removing surface coatings. The prepared particles are then passed through a bulk scalping magnet to remove any tramp metal which may have been incorporated into the stream of plastic particles. The particles are then fed to a series of one or more magnetic separators arranged in order of increasing magnetic strength. The first magnet is specifically configured to remove only plastic particles with the greatest magnetic susceptibility (P1). This separation step may, for example, be performed using a low-intensity magnetic drum separator. The drum magnet will keep the particles with the desired susceptibility in contact with the drum surface while discharging the low-intensity particles to the non-magnetic fraction. The non-magnetic fraction from this first pass separation is then transferred to a second magnetic separator with a stronger magnetic field. This separator will remove a second group of plastic particles with lower magnetic susceptibility (P2). Theoretically, this procedure can be repeated, as necessary, for as many plastic materials as are desired to be separated. Practically speaking, however, systems with more than three components are rarely encountered. If required, the magnetically sorted particles may undergo secondary separation processes (e.g., color sorting, electrostatic sorting, density separation, etc.) to provide recycled plastics of any desired purity.

POLYMAG™ PROPERTIES

Many different additives are used with plastic resins to impart a change to various physical properties such as density, conductivity and strength. For example, calcium carbonate, talc, kaolin and other mineral fillers are often used (Wypych, 1999). Table 1 shows a list of additives that may be used in the plastics industry.

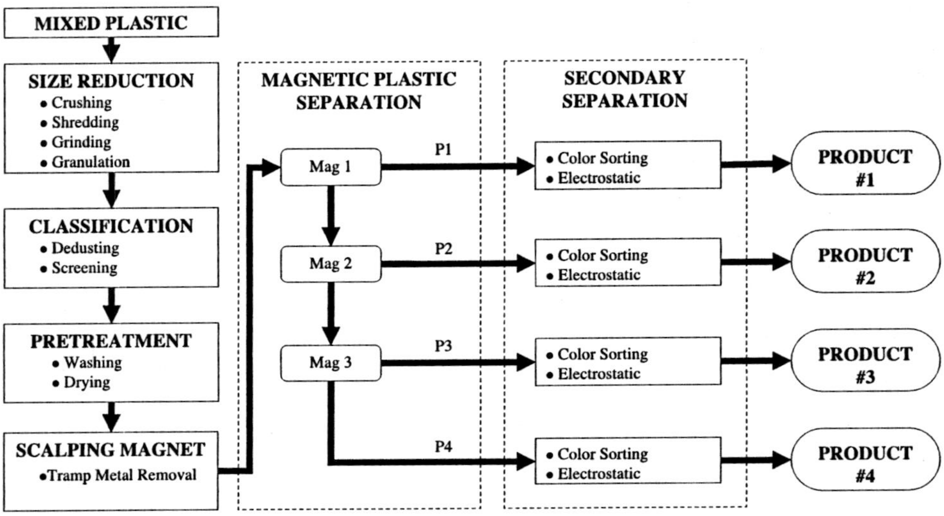

FIGURE 1 Flowsheet for the PolyMag™ process for separating mixed automotive plastic scrap

TABLE 1 Typical mineral fillers used in the plastics industry

Name	Chemical Composition
Aluminum	Al
Barium Sulfate	BaS04
Bentonite	$(Na,Ca)(Al,Mg)_6(Si_4O_{10})_3(OH)_6.nH_2O$
Calcium Carbonate	$CaCO_3$
Carbon Black	C
Dolomite	$CaMg(CO_3)_2$
Graphite	C
Silica	SiO_2
Talc	$Mg_3Si_4O_{10}(OH)_2$
Titanium Dioxide	TiO_2

As one might expect, however, any type of contamination (including fillers) in a plastic component will have an effect on the physical properties of the material. This is significant in that the property change may have a detrimental impact on the final part that is manufactured from this material. For instance, a decrease in the yield strength of the plastic may result in premature failure of a component. Therefore, before proceeding to industrial implementation of the PolyMag™ process, a detailed material characterization was conducted.

Evaluation of Physical Properties

The additive for the PolyMag™ process consists of a high-concentration pellet that contains as much as 50 percent (by weight) of a magnetic component and a polyolefin binder that is compatible with various resins such as thermoplastic elastomers (TPE) and

TABLE 2 Physical property comparison for low durometer TPE

Parameter	Specification	Neat	w/PolyMag™
UTS (psi)	500 min.	748	757
EB (%)	290 min.	414	418
MI100 (psi)	230–350	279	289
Hardness (Zwick)	57–63	57–63	61.5

TABLE 3 Physical property comparison for high durometer TPV

Parameter	Specification	Neat	w/PolyMag™
UTS (psi)	2,000 min	2,346	2,563
EB (%)	500–660	591	628
MI100 (psi)	890–1130	1,000	1,003
Hardness (Zwick)	90–96	94.1	93.8

polypropylenes (PP). The high-concentration pellet is combined with the base resin prior to entering the injection molding machine. Axial mixing conditions within the extruder thoroughly homogenize the additive with the host resin prior to injection into the mold cavity.

In this evaluation, a TPE formulation was used to create various test specimens. The specimens were evaluated for ultimate tensile strength (UTS), elongation to break (EB), modulus at 100% elongation (M100), and Zwick hardness. Details of these standard test methods are available elsewhere (Shah, 1998). Tests were conducted with both the virgin base material (neat) and with the magnetic additive. Two different tests were conducted in parallel using a low- and high-durometer material.

Test results for the key physical properties specified above are shown in Tables 2 and 3. In each case, a direct comparison can be made to the neat material and the standard product specifications. The latter are typically provided by the resin manufacturer as expected minimum values. Results indicate that the PolyMag™ additive has little effect on ultimate tensile strength, elongation to break, modulus at 100 percent elongation, and hardness.

Evaluation of Over-Molding Properties

The second aspect of the physical evaluation examined the effect of the PolyMag™ process as it applies to the injection over-molding and blow-molding processes. In these processes, a material such as TPE is either injection molded over an existing component made from a different material such as polypropylene, or co-injection blow-molded. An example of the former process is the soft "rubbery" handle that overlays a typical household appliance or hand tool. The bond between the two materials is an important characteristic that should not diminish when the magnetic additive is included for material separation. Two process methods comparing bond strength were evaluated: blow molding and insert injection molding.

Using the insert injection molding process, components were injection molded individually onto T-bars. The bars were then pulled using a Monsanto T-10 tensometer and the resulting peel strength recorded. Using the blow molding process, injection molded

TABLE 4 Comparison of pull strength for injection- and blow-molded components

Test Method	Durometer	Neat	w/PolyMag™
Insert Injection Molding (lbs)	Low	15	17
	High	51	60
Blow Mold Molding (lbs)	Low	14	17
	High	42	53

plaques were inserted into a blow mold and encapsulated by a polypropylene parison. Individual one-inch wide pull bars were cut from these components and again pulled using the tensometer. The results, shown in Table 4, indicate that the magnetic additive does not have a detrimental effect on the adhesion characteristics of injection molded components.

In general, the detailed material tests conducted in this evaluation indicate that the PolyMag™ additive has little or no effect on the physical properties of the thermoplastic elastomers studied.

Evaluation of Magnetic Susceptibility

The effectiveness of the PolyMag™ process relies on the use of a functional filler that enhances the magnetic susceptibility of one or more of the components in a plastic component. Particle fillers may be classified into one of four groups based upon the manner in which such particles respond to a magnetic field: ferromagnetic, paramagnetic, diamagnetic, and non-magnetic. Ferromagnetic materials are those that have a very high susceptibility to a magnetic field and are strongly influenced thereby. Paramagnetic materials are those that have a low magnetic susceptibility and are weakly influenced by a magnetic field. Diamagnetic materials have a negligible magnetic susceptibility and, for all practical purposes, may be considered non-magnetic. Non-magnetic materials have no susceptibility to magnetic fields.

Ferromagnetic materials and to a lesser extent, paramagnetic materials, become magnetized when placed in a magnetic field. The amount of magnetization induced on the particle depends on the mass, the magnetic susceptibility of the particle, and the intensity of the magnetic field that is applied. The induced magnetization, M, of a particle can be expressed as follows:

$$M = mXH \quad \textbf{(EQ 1)}$$

where m is the mass of the particle, X is the specific magnetic susceptibility of the particle (a constant for any particular material), and H is the magnetic field intensity. By definition, a separation process relies upon the ability to make one type of particle move relative to another by applying an external force. In this instance, a magnetic force is applied that is sufficient to move particles of high magnetic susceptibility relative to all other material. Likewise, particles exhibiting a low magnetic susceptibility (i.e., paramagnetic materials) can be moved relative to non-magnetic and diamagnetic particles.

The magnetic force acting on the particle depends upon the induced magnetization, M, as well as the field gradient. The magnetic field gradient refers to the rate of change or the convergence of the magnetic field strength. This concept is illustrated in Figures 2(a) and 2(b). Figure 2(a) shows a uniform pattern of flux lines without gradation. A magnetic particle entering this field will be attracted to the lines of flux and remain

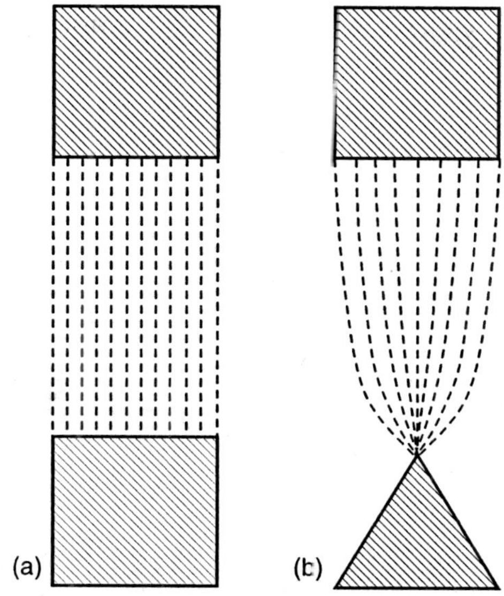

FIGURE 2 Magnetic fields illustrating (a) uniform pattern of flux lines without gradation and (b) converging pattern of flux lines displaying a high gradient

stationary without migrating to either pole piece. Figure 2(b) shows a converging pattern of flux lines displaying a high gradient. As these lines pass through a smaller area, there is a significant increase in the magnetic field intensity. A magnetic particle entering this field configuration will not only be attracted to the lines of flux but will also migrate to the region of highest flux density, which occurs at the tip of the bottom pole piece. In simplified terms, the magnetic field intensity holds the particle, while the magnetic field gradient moves the particle.

From Equation (1), the magnetic attractive force acting on a particle is the product of the particle magnetization and the magnetic field gradient and can be expressed as:

$$Fm = mXH\frac{dH}{dx} = M\frac{dH}{dx} \quad \textbf{(EQ 2)}$$

where dH/dx is the magnetic field gradient. It can be seen from Equation (2) that for a fixed magnetic field strength and gradient, relative particle motion can only be achieved through a difference in magnetic susceptibility. To achieve separation of a mixture of polymers requires that the susceptibility of one type of plastic be substantially different from another. Magnetic susceptibility (X) is a fundamental bulk property of a material and is quantified using a vibrating magnetometer. The response curve, shown in Figure 3, provides the magnetization M (emu/cc) as a function of the applied magnetic field that is measured in kilo-Oerstads (kOe). The magnetic susceptibility is determined from the initial slope of the response curve and is reported as emu/cckOe. Using this procedure, the response curves for plastics containing different amounts and/or types of additives can be determined. Furthermore, as shown in Figure 3, it is clear that the magnitude of the susceptibility can be adjusted by varying the amount of additive. Given an appropriate magnetic field gradient, a separation can be achieved based on the difference in magnetic susceptibilities.

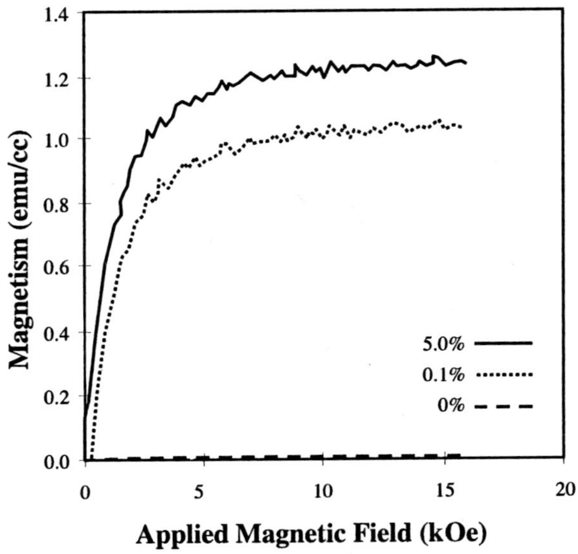

FIGURE 3 Response curve showing magnetization M (emu/cc) as a function of the applied magnetic field (kOe) for samples containing different amounts of paramagnetic filler

SEPARATION RESULTS

Laboratory Testing

Several series of small continuous-scale tests were conducted to evaluate the impact of adding parametric fillers to plastic particles. The tests were carried out using a predefined mixture of synthetic particles containing 0%, 0.1%, 0.5%, 1% and 5% ferric matter by weight. The synthetic particles were color coded using dyes so that the effectiveness of the separation could be easily quantified by simple hand sorting of the colored particles. The test runs were performed by passing the entire lot of mixed plastic particles through a drum-type magnetic separator. After each pass, the separated fractions of each type of particle were individually collected, weighed, and analyzed for purity. This procedure was repeated several times at different magnetic field strengths to generate recovery-quality curves for each type of plastic particle.

Figure 4 compares the results obtained using different amounts of the functional filler. As expected, the entire lot of plastic particles containing 5% ferric filler reported to the magnetic concentrate at the lowest field. Likewise, the bulk of particles containing 1%, 0.5% and 0.1% filler reported to the magnetic concentrate as the field strength was increased. These small differences make it possible to efficiently segregate each of the four test materials into high-purity groups through the use of properly designed, multistage, magnetic separators. Also, it is important to note that none of the nonmagnetic particles (0% ferric filler) reported to the magnetic concentrate, which indicates that very high purity levels can be maintained between particles with and without ferric filler using just a single magnetic separator.

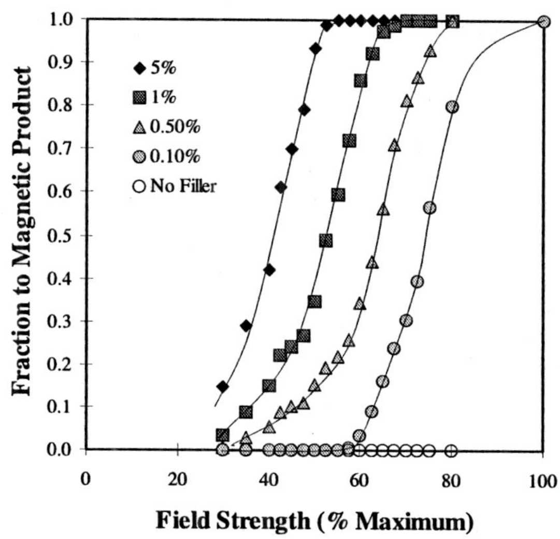

FIGURE 4 Separation curves obtained for plastic containing 0–5% functional (ferric) filler

Industrial Applications

Although the PolyMag™ process can be used in various applications, the initial effort is directed towards recovery of rejects that occur during the manufacturing process. For example, a typical injection over-molding process may have a reject rate as high as 5–10%. Additionally, scrap material is also generated during the initial set-up period for a new molding process. Since plastic resins are derived from petroleum products and/or natural gas, recovery of manufacturing rejects is becoming critical due to the cost of raw materials. Three different applications were evaluated during this investigation: injection over-molding (mechanical bond), injection over-molding (chemical bond), and an injection blow-molded assembly.

Injection Overmold (Mechanical Bond). Many plastic assemblies consist of simple mechanically locked components. A typical example is the two-component wheel assembly shown in Figure 5 which consists of a thermoplastic elastomer tire injection molded over a nylon hub. In this case, the TPE containing the magnetic additive is linked to the nylon through a mechanical bond. As a result, both components are completely liberated during the granulation process. The wheel assemblies were ground to a size of 90% passing 6-mm and subjected to separation testing using a 4-inch diameter laboratory rare-earth roll magnetic separator. In this example, the TPE was successfully separated from the nylon. Final product purity was greater than 99 percent for both products. It should be noted that in most cases a small amount of cross-contamination is allowable. The one exception is contamination of light colored materials with darker particles.

Injection Overmold (Chemical Bond). Depending on the raw materials, injection over-molding can result in a chemical bond between components. For instance, a polypropylene substrate over-molded with TPE results in a chemical bond at the interface of the two components. An example of this type of application is shown in Figure 6. In this case, the softer TPE is used on the border of the main component as a weather seal for an automotive application. During granulation, this material parallels a binary

FIGURE 5 Magnetic separation of mechanically bonded TPE/PP wheel assembly

mineral system. Locked particles are created as a result of the boundary condition between the two materials. Additional size reduction will minimize this condition. From a practical perspective, however, the minimum particle size is limited to a few millimeters due to downstream handling requirements.

The component shown in Figure 6 was granulated to approximately 90% passing four millimeters and processed using the rare earth roll separator described above. In this case, the TPE component was only 5.5 percent of the weight of the original part. Since the magnetic additive was included with the TPE, separation testing resulted in a pure polypropylene fraction. Although the locked particles had a lower magnetic signature, they ultimately reported to the TPE product. The final purity of the TPE product was approximately 85% due to misplacement of the PP from the middlings particles. In this application, recycling of the separated components resulted in a PP contaminant level in the TPE feed of 0.3–0.5%. This is considered acceptable since TPE and PP are chemically compatible and this small amount of carryover will not affect the physical properties of the final part. Since TPE tends to be substantially more expensive, the full value of the majority of the material is recovered.

Injection Blow-Molded (Chemical Bond). The third part evaluated in this investigation is a two-component, blow-molded hose assembly for an automotive application (Figure 7). In this case, the material composition of the wall transitions from pure TPE to PP several times over the length of the hose to allow for flexibility in the accordion joints of the product. After granulation, locked particles are created from the overlapping materials in the transition zone. In this case, however, the percentage of locked particles is substantially higher due to multiple material transitions along the length of the hose. The current recycling practice is to manually cut the hose using a band saw. By adding the magnetic filler to the TPE, the plastic components can be magnetically separated after granulation. Separation of the granulated material produced a TPE component with a purity of 85% at a recovery of 95%. Likewise, the PP was greater than 99% pure

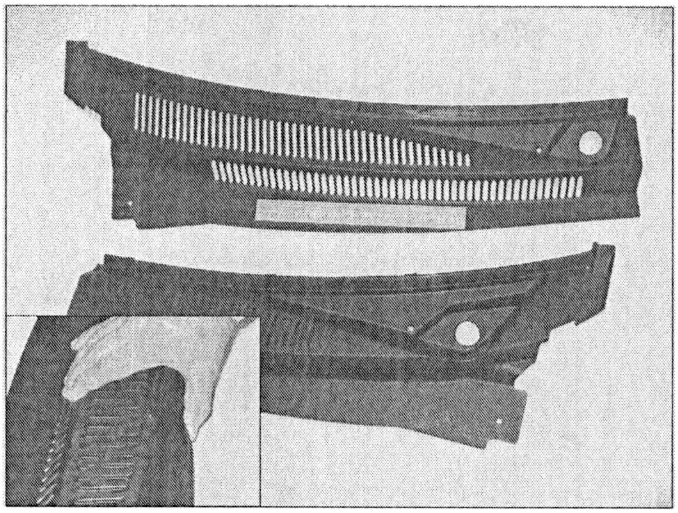

FIGURE 6 Chemically bonded injection over-molded TPE/PP component

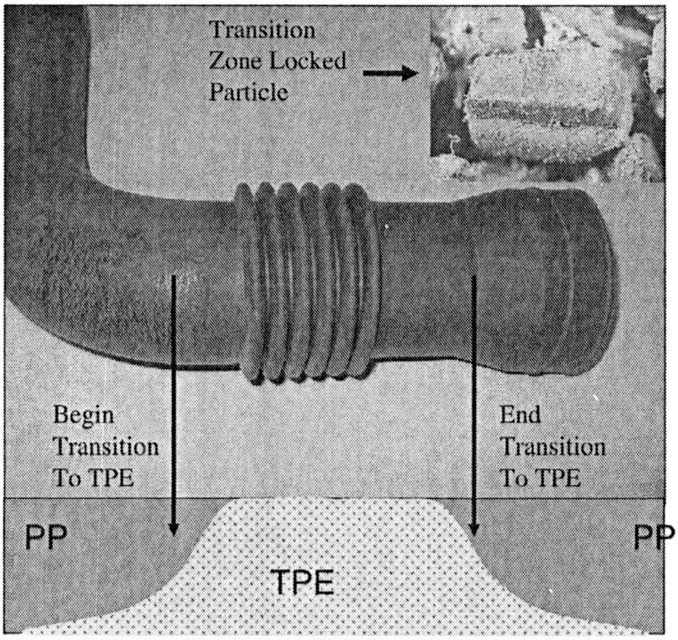

FIGURE 7 Chemically bonded, blow molded hose assembly

TABLE 5 Summary of separation test results

	PP		TPE		Nylon	
Component	Rec. (%)	Purity (%)	Rec. (%)	Purity (%)	Rec. (%)	Purity (%)
Mech. Bonded Wheel Assem.	n/a	n/a	>99	100	100	>99
Chem. Bonded Auto. Comp.	98.9	>99	100	85	n/a	n/a
Blow Molded Hose Assem.*	90	>99	95	85	n/a	n/a

* Two-stage separation

with a recovery of 90%. A summary of all of the results from the three examples above is provided in Table 5.

CONCLUSIONS

The PolyMag™ process makes use of functional fillers to enhance the magnetic susceptibility of one or more plastic articles to enable their separation from a manufacturing reject stream using magnetic separators. Results from tests conducted on parts manufactured using this process indicate that the addition rate of filler is sufficient to influence the magnetic susceptibility of the plastic, but small enough to avoid adversely impacting the desirable properties associated with the functionality of the plastic article. Results from physical property testing of plastics incorporating the filler indicate that the changes to ultimate tensile strength, elongation to break, and hardness were within normal tolerances for the material studied. In order to demonstrate the impacts of the filler on plastics separations, several different industrial trials were conducted in both injection- and blow-molding applications. The separation tests indicate that high-purity products (>99%) can be obtained at very high recovery values using a single pass magnetic separator. An economic evaluation of the process indicates a favorable payback for the process in plastics recycling applications.

REFERENCES

Frisch A., Razim C., 1995. "Materials Technology in Automotive Recycling," *Advances in Materials*, Vol. 7, No. 6, pp. 513–518.

Muccio, E.A., 1994a. "*Plastics Processing Technology*," Materials Information Society, ASM International, Materials Park, Ohio, 320 pp.

Muccio, E.A., 1994b. "*Plastic Part Technology*," Materials Information Society, ASM International, Materials Park, Ohio, 304 pp.

Sendijarevic V., Pokorski B., Klempner D., Frisch K.C., 1997. "Recent Developments in Shredder Downstream Separation Processes and Recycling Options for Automotive Shredder Residue," Society of Automotive Engineers, *Special Publications* 1997, SP-1253 (New Plastics Applications for the Automotive Industry), pp. 163–171.

Shah, Vishu, *Handbook of Plastics Testing Technology*, John Wiley & Sons, Inc., 1998.

Ura, M., 1994. "Recycling Technologies in Automobiles," *Transactions of the Material Resources Society of Japan*, Vol. 18A (Ecomaterials), pp. 759–61.

Wypych, George, *Handbook of Fillers*, 2nd Ed., ChemTec Publishing, 1999.

Functional Fillers for Dimensional Stability

W.M. Cross,[*] B.D. Henderson,[*†] W.C. Weyer,[‡] C. Kroetch,[‡] L. Kjerengtroen,[‡] J. Welsh,[**] and J.J. Kellar[*]

The purpose of this research was to investigate the effect of zirconium tungstate (ZrW_2O_8) particles in a polymer matrix on the coefficient of thermal expansion (CTE) of the resulting composite. In addition, surface modification of the ZrW_2O_8 particles with γ-aminopropyltrimethoxysilane (silane) and its effect on the composite CTE was examined. With the addition of 50.2 ± 3.0% particles by volume, the CTE of the neat CTD-403 resin was decreased by approximately 84%. The silane treatment had little or no effect on the overall composite CTE.

INTRODUCTION

Polymer matrix composites are very useful because of their lightweight and relatively high strength. Although polymer matrix composites exhibit several highly desirable characteristics, these materials often expand and contract considerably when exposed to temperature changes. As most polymers have a relatively large (>50 ppm/°C) coefficient of thermal expansion (CTE), a material with a negative CTE (these materials shrink as the temperature rises) can be added to the polymer matrix to aid in reducing the CTE. Theoretically, this approach could lead to a dimensionally stable material with a CTE

---

* Department of Materials and Metallurgical Engineering, South Dakota School of Mines and Technology, Rapid City, South Dakota

† Now: Senior Associate Engineer, Materials Technology Group, Caterpillar Specialty Products, Mossville, Illinois

‡ Department of Mechanical Engineering, South Dakota School of Mines and Technology, Rapid City, South Dakota

** Air Force Research Laboratory Space Vehicles Directorate, Kirtland AFB, New Mexico

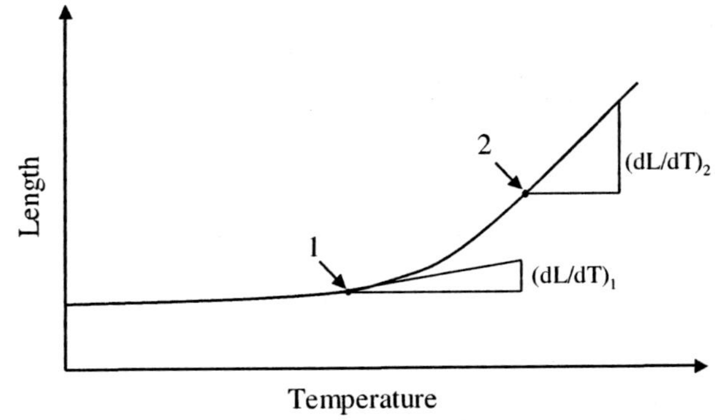

FIGURE 1 Plot of length versus temperature showing derivatives. Adapted from ISO 11359-2:1999(E).[2] The derivative is shown at points 1 and 2.

approaching 0 ppm/°C. Zirconium tungstate (ZrW_2O_8) is very promising in this respect as it has a large negative CTE over a wide temperature range.[1]

The purpose of this research was to investigate the effect of the incorporation of ZrW_2O_8 particles into a polymer matrix. The effect of the proportion of ZrW_2O_8 particles on the CTE of the composite material was investigated. In addition, the effect of a particle surface treatment on the composite CTE was examined.

Coefficient of Thermal Expansion

The coefficient of thermal expansion is a material property that defines the extent to which a material expands upon heating. From an atomic perspective, thermal expansion occurs because of an increase in the average distance between atoms. Heating to successively higher temperatures raises the vibrational energy of the material, which, in turn, raises the average interatomic distance between atoms. This phenomenon causes most materials to expand upon heating.

The CTE can be expressed as a linear quantity, describing expansion in one direction, or a volumetric quantity, which takes into account the expansion of the material in all directions. The linear CTE is most commonly used, because it is more easily measured. Equation 1 describes the general equation for linear expansion,

$$\frac{\Delta L}{L_0} = \alpha \Delta T \qquad \textbf{(EQ 1)}$$

where L_0 is the initial length of the sample, ΔL is the change in length of the sample, ΔT is the change in temperature, and α is the CTE.[2] For this work, the instantaneous CTE, in which ΔL and ΔT are replaced by their infinitesimal counterparts dL and dT, will be used.

To calculate the instantaneous CTE, a plot of temperature versus length, as shown in Figure 1, is used. This derivative, which is the slope of the line, must be calculated and the CTE is given by Equation 2.[2] In Figure 1, the derivative is demonstrated at two points.

$$\alpha = \frac{dL}{dT} \cdot \frac{1}{L_0} \qquad \textbf{(EQ 2)}$$

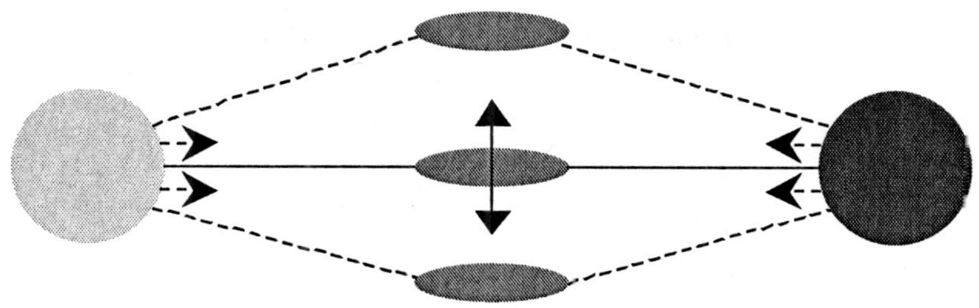

FIGURE 2 Zr-O-W bond responsible for the negative CTE of ZrW_2O_8. Left circle = Zr, right circle = W, center ellipses = O.

Zirconium Tungstate

Zirconium tungstate (ZrW_2O_8) is a very unique material. It possesses a negative CTE that is isotropic and has a relatively large magnitude over a wide range of temperatures (−273°C to 777°C).[3] This makes ZrW_2O_8 very useful in applications where a reduction in the CTE of a material is desired.

The unusual properties of ZrW_2O_8 are due to its unique crystal structure. The structure consists of ZrO_6 octahedra at the origin and face center positions with WO_4 tetrahedra in between. The octahedra and tetrahedra are linked by corner sharing.[4]

The mechanism responsible for contraction with increasing temperature is demonstrated in Figure 2. At lower temperatures, the oxygen atom situated between the zirconium and tungsten atoms has a bond illustrated by the solid lines. As the temperature is increased, the oxygen atom, having a high degree of mobility compared to the ZrO_6 octahedra and WO_4 tetrahedra, vibrates between the other two atoms, causing the effective bond length to be shortened.[5] Recall that the zirconium atom is part of an octahedron and is connected to five other oxygen atoms while the tungsten atom is part of a tetrahedron and is connected to three other oxygen atoms. These polyhedra do expand when heated, as expected, but this expansion is minimal compared to the contraction caused by the polyhedral vibrations and therefore the structure, as a whole, contracts upon heating.

This negative thermal expansion behavior was shown to occur from 0.3 to 1,050 K by Sleight and his co-workers in 1996.[1] Because of its cubic crystal structure, the contraction of ZrW_2O_8 is isotropic, which makes it even more useful, because the thermal expansion will be equal in all three directions.[1]

Other materials have been shown to have negative thermal expansion. For example, SiO_2 has a negative CTE when it is in the faujasite structure, but only from 20 to 500 K.[4] The isostructural $A_2M_3O_{12}$ family, where M is W^{6+} or Mo^{6+} and A^{3+} can range from Al^{3+} to rare earth cations, has shown highly negative CTEs. This includes the record value, to date, for negative thermal expansion of −11 ppm/°C shown in $Sc_2W_3O_{12}$, but this behavior is not isotropic.[4] The cubic phases of the type $ZrP_{2-x}V_xO_7$ have shown negative thermal expansion, but only at temperatures above 100°C.[1] Negative CTE has been shown in AgI, when in the metastable cubic form, but is confined to a region no wider than 200°C.[1] Gold nanoparticles have been shown to have a negative CTE over a relatively wide temperature range, and are receiving more attention.[6] ZrW_2O_8, however, has been the most promising negative CTE material because it has a large negative CTE over a wide temperature range and, because of its cubic crystal structure, the effect is isotropic.[3] Hafnium tungstate, HfW_2O_8, is isostructural to ZrW_2O_8 and therefore has similar

TABLE 1 Phases and phase transitions in zirconium tungstate[12–16]

ZrW_2O_8 Phase	Crystal Structure	Space Group	CTE (ppm/°C)
α	cubic	$P2_13$	−8.7
β	cubic	$Pa\overline{3}$	−4.9
γ	orthorhombic	$P2_12_12_1$	−1

Transition	Temperature (°C)	Pressure (psi)	Pressure (MPa)
α/β to γ	25	29,000	199.95
γ to α	117	14.7	0.101
α to β	155–175	14.7	0.101

thermal expansion characteristics. However, the negative thermal expansion coefficients for HfW_2O_8 are slightly smaller in magnitude than for ZrW_2O_8, thus leaving the focus for negative CTE squarely on ZrW_2O_8.[7]

Zirconium Tungstate Phase Transformations

Zirconium tungstate shows several phases in temperature regions of interest to this research. The phases and phase transformations of zirconium tungstate are shown in Table 1. The α to β transition is reversible upon cooling, whereas the γ phase is metastable upon the release of the pressure and exists until heating above 117°C at ambient pressure.[3,8–11] Phase behavior is especially important as the CTE for the various phases can vary by nearly an order of magnitude. This could have a considerable effect on the CTE of a composite material. The α-γ transformation begins at approximately 200 MPa and may be of importance in these composite systems.

The α-γ transition is also accompanied by a 5% reduction in volume.[12] In addition, Dunand's group showed that thermal residual stresses in a Cu-ZrW_2O_8 composite were sufficiently large to trigger a phase transformations in the ZrW_2O_8.[3,8,13] They showed that hydrostatic pressures in the range of 200–300 MPa may form the undesirable γ-ZrW_2O_8 phase. The γ-ZrW_2O_8 phase has a coefficient of thermal expansion of ~−1 ppm/°C, whereas the desired α-ZrW_2O_8 and β-ZrW_2O_8 phases have a thermal expansion coefficient of ~−9 and −5 ppm/°C, respectively.

Uses of Low/Zero CTE Materials

Table 2 lists the CTEs of some commonly used materials.[17] In general, ceramics and metals have relatively low CTEs, whereas polymers have high CTEs. The CTE of a material is a very important structural parameter. For instance, CTE mismatch between the two constituents in a composite can create thermal stresses, as the expansion of one material is restricted by its bond to another material. It is possible, though, to tailor the CTE of the composite by controlling the proportion and distribution of reinforcement in the matrix. This concept is being used in the electronics industry, where the CTE of high temperature computer chips needs to match the CTE of their substrates.

Zirconium tungstate is being evaluated for use in fiber optics also. One example is a fiber Bragg grating, which, when placed in an optical fiber, reflects a very narrow wavelength of light centered around the Bragg wavelength.[18] This has led to their use as filters, laser source tuners, and stabilizers in fiber optics applications.[19] These reflection

TABLE 2 Coefficient of thermal expansion for some common materials

Materials	CTE (ppm/°C)
Metals	
Aluminum	23.6
Iron	11.8
316 Stainless Steel	16.0
Ceramics	
Alumina (Al_2O_3)	7.6
Soda-lime glass	9.0
Borosilicate (Pyrex) glass	3.3
Polymers	
Polypropylene	145–180
Teflon®	126–216
Nylon 6,6	144

Abridged from reference 17

gratings are temperature-sensitive, though, causing wavelength changes with temperature. To overcome this, they can be bonded to a zirconium tungstate package structure to account for any temperature changes, therefore remaining consistent over a temperature range.[20]

The negative CTE zirconium tungstate possesses can be used to counteract the positive CTE of the matrix material and give a composite with a low or zero CTE. There would be many uses for such a "dimensionally stable" composite with zero CTE. One possible use is a dimensionally stable mirror backing structure or support system for a telescope. Zero CTE in the structure would eliminate the need for adjustments due to changes in curvature of the mirror during the considerable temperature fluctuations that occur in space. As a spacecraft orbits the earth, the temperature varies from roughly −100°C to 100°C, depending on whether the craft is in direct sunlight or the shadow of the earth. The weight saved by using composites over other materials, with low CTEs (like metals), is very important in space-based applications. As of the year 2000, the average cost for putting a pound of payload into geosynchronous orbit was $11,729.[21]

EXPERIMENTAL PROCEDURE

Measurement of the CTE was accomplished through thermomechanical analysis (TMA) according to ISO 11359-2:1999(E).[2] In TMA, a non-negligible load is applied to the sample while the sample is heated/cooled at a constant rate as the dimension change is recorded. For this work a TA Instruments Q400EX Thermomechanical Analyzer was used. A load of 0.2 N (6.8 kPa) was applied via a macroexpansion probe and the temperature ramp rate was 5°C/min.

Composite samples were manufactured by mixing 10 grams of zirconium tungstate particles (Wah Chang, Albany, OR) with CTD-403, an aromatic cyanate resin from Composites Technology Development, Inc. (Lafayette, CO) to achieve approximately 70 weight% particles. After mixing, the samples were subject to a vacuum environment for at least 5 minutes to remove air from the mixture. Next, the composite mixture was poured into a cylindrical mold 5/32" in diameter and 1.282" deep. The zirconium

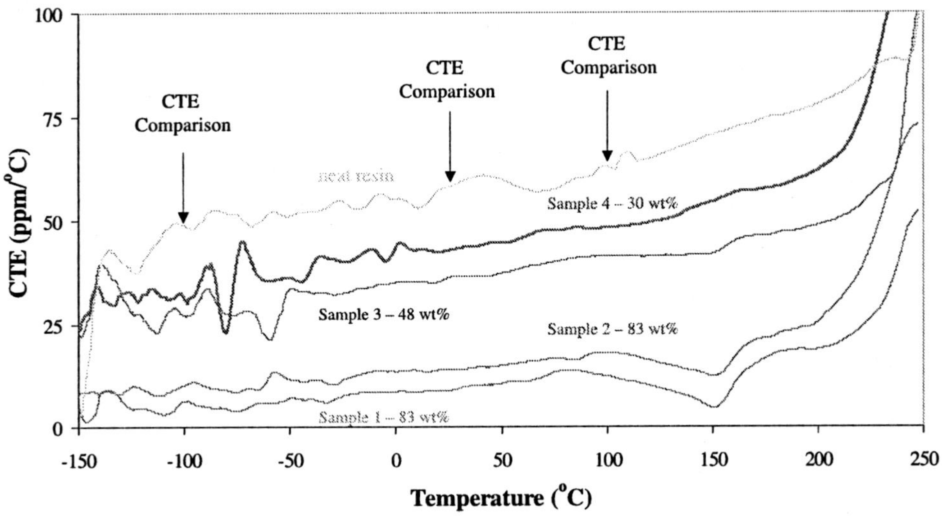

FIGURE 3 CTE of the four sections from a representative specimen compared with the CTE of neat CTD-403. The arrows indicate the temperatures at which further comparisons were made. The volume percentages of particles in each section are listed in the legend.

tungstate particles were allowed to settle as the composite underwent the manufacturers recommended cure schedule. Following cure, the composites were removed from the mold, cut into four samples and polished to obtain two parallel sides for TMA testing. Surface treatment of the zirconium tungstate was performed by treating the particles in a 1% aqueous γ-aminopropyltrimethoxy silane solution at pH 4.5.

Specific details of the experimental procedure can be found in reference 22. Particle size data and more comprehensive examination of the volume fraction and particle size in the four cut composite samples are found in references 22 and 23.

RESULTS AND DISCUSSION

The purpose of this research is to investigate the effect on composite material CTE of adding zirconium tungstate (ZrW_2O_8) to a polymer. Therefore, samples were prepared with different weight percentages of ZrW_2O_8 to investigate the effect on the composite's CTE.

Introduction to Data Collected and General Trends

The CTE data obtained from TMA measurements of a representative sample is given in Figure 3. For comparison, the CTE data for the neat CTD-403 resin is also included in Figure 3 for comparison.

From Figure 3, the addition of ZrW_2O_8 lowered the CTE in all cases. Also, the addition of more ZrW_2O_8 caused a greater reduction in CTE. For a more thorough analysis, the instantaneous CTE of every sample was examined at three different temperatures: −100°C, 25°C, and 100°C. Another important point is the CTE slopes upward with increasing temperature for all of the samples. This is due to an increase in thermal expansion of the CTD-403 with increasing temperature. The CTE of the ZrW_2O_8 particles

TABLE 3 Data from TMA experiments on composite samples made with untreated particles and silane treated particles. The variation shown is one standard deviation.

A. Untreated Particles, Average of 10 Samples					
Section	Volume %	Weight %	CTE −100°C, (ppm/°C)	CTE 25°C, (ppm/°C)	CTE 100°C, (ppm/°C)
1	50.2 ± 3.0	80.4 ± 1.8	6.9 ± 1.4	9.2 ± 0.5	11.4 ± 0.8
2	47.3 ± 3.9	78.5 ± 2.7	8.5 ± 1.7	12.2 ± 2.0	14.8 ± 2.2
3	16.2 ± 1.4	44.1 ± 2.5	31.9 ± 4.0	36.7 ± 1.3	41.6 ± 1.6
4	8.6 ± 1.5	27.7 ± 4.0	35.5 ± 3.7	44.1 ± 0.9	49.8 ± 1.2
B. 1% Silane Treated Particles, Average of 10 Samples					
Section	Volume %	Weight %	CTE −100°C, (ppm/°C)	CTE 25°C, (ppm/°C)	CTE 100°C, (ppm/°C)
1	41.7 ± 2.2	74.5 ± 1.7	11.3 ± 3.1	16.2 ± 1.3	18.0 ± 2.7
2	34.1 ± 5.0	67.6 ± 5.1	12.6 ± 3.8	19.8 ± 5.2	24.5 ± 5.2
3	23.8 ± 4.5	55.7 ± 6.3	21.4 ± 5.6	30.2 ± 4.1	34.2 ± 4.7
4	11.2 ± 3.2	33.4 ± 7.6	27.9 ± 6.3	42.1 ± 4.9	47.3 ± 4.1

TABLE 4 Comparison of instantaneous CTE of untreated ZrW_2O_8 filled section 1 to CTD-403 CTE

Sample	Volume %	Weight %	CTE −100°C, (ppm/°C)	CTE 25°C, (ppm/°C)	CTE 100°C, (ppm/°C)
CTD-403	0	0	49.12	58.07	63.04
Section 1 Untreated	50.2	80.4	6.9	9.2	11.4
Percentage reduction in CTE			86%	84%	82%

was found by Sleight et al.[1] to be relatively constant at temperatures less than the α to β phase transition at 155°C.

The sections with a high loading of ZrW_2O_8 particles (sections 1 and 2) have a dip in the CTE data at this phase transition temperature (155°C). The CTE curves from the sections with very little ZrW_2O_8 (sections 3 and 4) don't show this dip clearly, demonstrating that the dip in the CTE data is caused by the ZrW_2O_8 particles, not the CTD-403 resin. The fact that this dip occurs at 155°C indicates that it is likely due to the transition of the particles from the α to β phase. Additionally, Yamamura et al. used X-ray diffraction to examine the lattice parameter of ZrW_2O_8 as a function of temperature and their data showed a similar drop in CTE around this temperature, which they attributed to the α to β phase transition.[7]

The thermal expansion data for samples fabricated with untreated and silane treated ZrW_2O_8 particles is summarized in Tables 3 and 4. Table 3 contains the data from TMA experiments run on ten different samples that were made with untreated ZrW_2O_8 particles and particles treated in a 1% by weight solution of γ-aminopropyltrimethoxysilane. In Table 4, the instantaneous CTE data for neat CTD-403 resin is compared to the data for section 1 of the samples prepared with untreated ZrW_2O_8. Section 1 of the untreated particle samples was chosen, because it contained the largest amount of ZrW_2O_8 particles, and thus provided the maximum CTE reduction of all the composite samples fabricated. The CTE of the neat CTD-403 resin was lowered significantly, about 84%. The percentage reduction in CTE was nearly the same at all three temperatures.

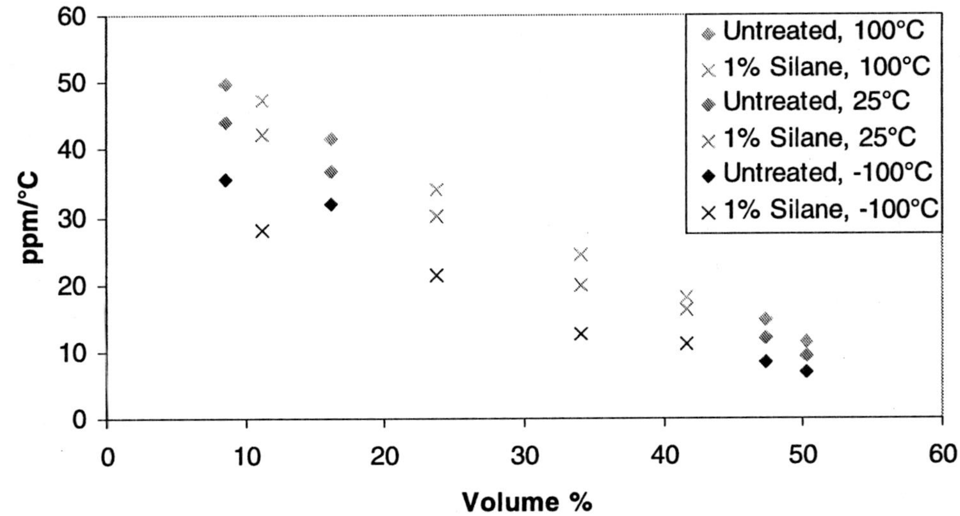

FIGURE 4 Overview of all of the TMA data for samples with treated and untreated particles at 3 different temperatures

In both the untreated and treated particle samples, the coefficient of variation of the instantaneous CTE measurements is larger for section 2 than for all of the other sections. This section is second from the bottom, and contains a mixture of both small and large particles indicating a transition in the sample. The way in which the small and large particles stack and interact is likely the factor that is increasing the variation in these samples. Section 1 has a mixture of large and small particles also, but it is predominately larger particles. Sections 3 and 4 contain mostly small particles of similar sizes.

Figure 4 contains a graph of the average TMA data collected. The data, examined at each temperature, is linear when volume % vs. CTE is plotted. The instantaneous CTE data has been tabulated in graphical form with error bars representing one standard deviation at each of the temperatures chosen. Figure 5 shows the ZrW_2O_8 volume % versus CTE at 25 °C. The data here is essentially linear. Data plotted as weight % zirconium tungstate is not linear because the density of the ZrW_2O_8 is 5.072 g/cm^3, whereas the density of the CTD-403 resin is only 1.239 g/cm^3.[3,24] Because of this difference in densities, the ratio of weight % to volume % decreases as particle loading increases.

Thus, the weight % is not increasing as quickly as the volume % and at the higher loadings the weight % numbers are closer together on the x-axis, decreasing the linearity of the graph. Because the volume % data is relatively linear, this form of the data was chosen for subsequent analysis.

There is some support for evaluating the CTE data with a linear fit. Holliday and Robinson evaluated the thermal expansion behavior of polymer-based composites by comparing the slope of the linear rule of mixtures line with the slope of a line run through experimental data.[25] In addition, several models of composite CTE behavior predict a linear relationship between CTE and volume %.[25-28]

Figures 6 and 7 show the volume % versus CTE data at 2 different temperatures: 100°C, and −100°C, respectively, complementing Figure 5. In Figures 5 and 6, the data

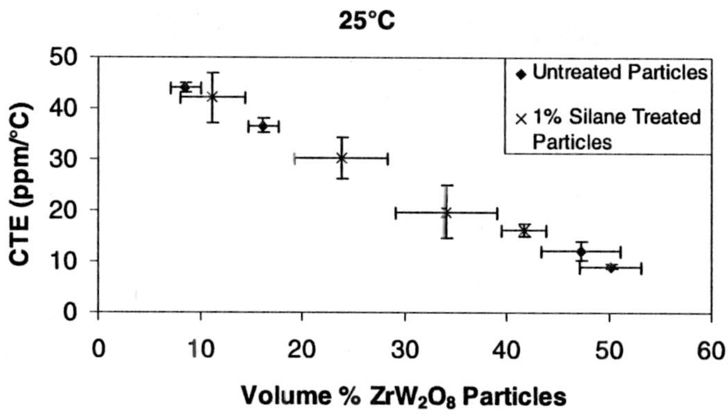

FIGURE 5 Particle volume % versus instantaneous CTE at 25°C

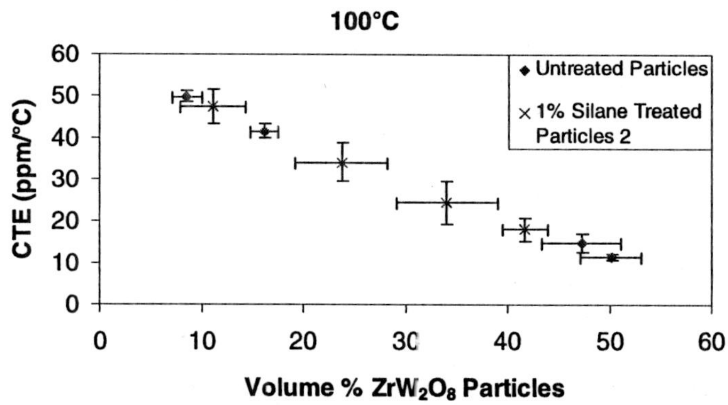

FIGURE 6 Particle volume % versus instantaneous CTE at 100°C

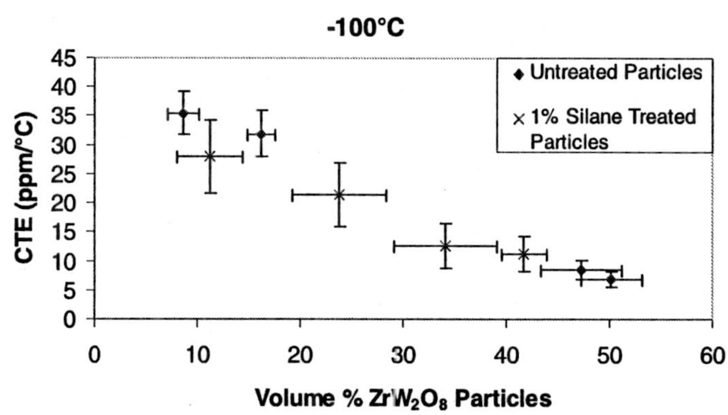

FIGURE 7 Particle volume % versus instantaneous CTE at –100°C

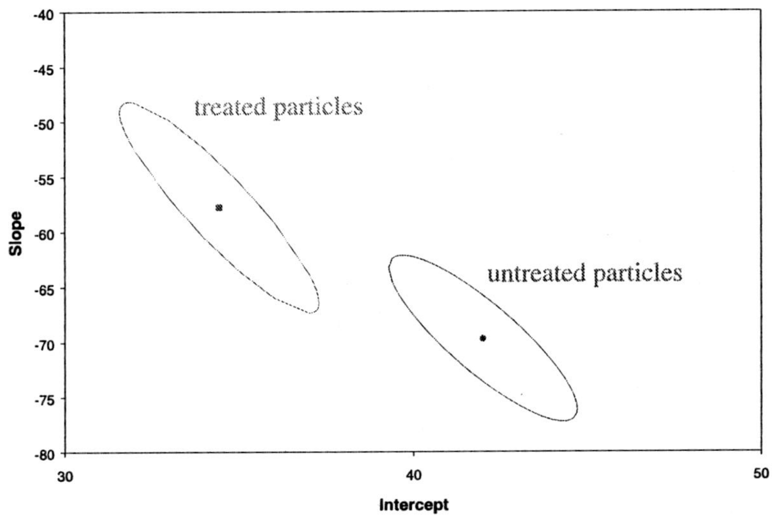

FIGURE 8 95% confidence ellipse for the linear regression coefficients found for composite CTE vs. volume % filler data at –100°C

for the samples with untreated particles and silane treated particles match closely, especially when sample variation is considered. A straight line drawn through the data will connect both the treated and untreated data points.

As seen in Figure 3, there is more variation in the CTE data as the temperature decreases. The instantaneous CTE data is more varied from –150°C to 0°C and smooths out at higher temperatures. Such perturbations are believed to be due to the variability in the instrumentation at low temperatures.

Composite samples made with silane treated particles appear to have a lower CTE at –100°C than those made with untreated particles (see Figure 7), however, this does not appear to be the case at 100°C and 25°C (Figures 5 and 6). Figures 5–7 also show this effect. At –100°C (Figure 7) the y-error bars are larger than the corresponding y-error bars at 25°C (Figure 5) and at 100°C (Figure 6). To examine whether or not the silane treated and untreated data sets at each temperature were different statistical analysis was performed. Linear regression analysis was performed on the individual data sets and the slope, intercept, and coefficient of determination were recorded. At each temperature studied, 95% confidence ellipses for the slope and intercept were constructed.[29–31] The comparison of these confidence ellipses are shown in Figures 8–10 for –100, 25 and 100°C, respectively. In Figure 8, the confidence ellipses for the treated and untreated filler composites do not overlap, indicating that the linear relationship between the CTE and volume % filler are not the same at –100°C. At the other two temperatures, the confidence ellipses do overlap showing no difference between the treated and untreated filler composites. Also, examination of Figures 8–10 shows that the intercept values at each temperature are less than the CTE of the pure CTD-403 resin at the temperature studied. This indicates that the initial compounding of the zirconium tungstate with the polymer must lead to a greater than expected decrease in the CTE.

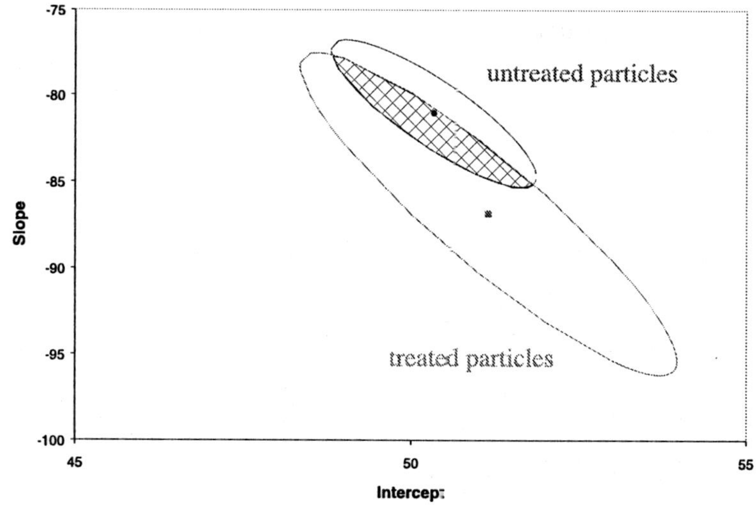

FIGURE 9 95% confidence ellipse for the linear regression coefficients found for composite CTE vs. volume % filler data at 25°C

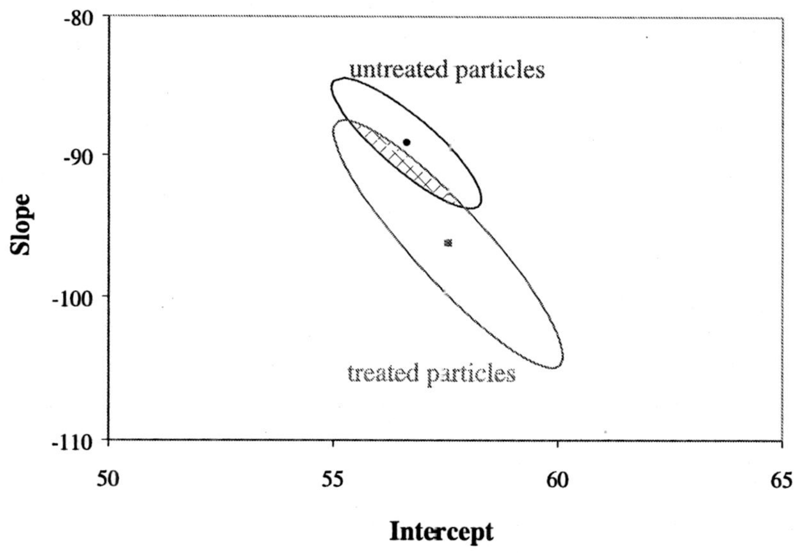

FIGURE 10 95% confidence ellipse for the linear regression coefficients found for composite CTE vs. volume % filler data at 100°C

SUMMARY

A definite relationship between volume % ZrW_2O_8 particles in a polymer matrix and CTE was established. As more ZrW_2O_8 particles were incorporated into the samples, the CTE of the composite decreased. This was expected and demonstrates that ZrW_2O_8 particles are very good at decreasing the overall CTE of polymer matrix composites. In fact, with the addition of 50.2 ± 3.0% particles by volume, the CTE of the neat CTD-403 resin was decreased by approximately 84%.

For comparison of samples made with untreated and silane treated ZrW_2O_8 particles, the instantaneous CTE was examined at 3 different temperatures: −100°C, 25°C, and 100°C. At each of these temperatures, the relationship between volume % particles vs. CTE was examined. In each case, the relationship exhibited a linear trend with volume % filler. At 100°C and 25°C, the CTEs for composite samples prepared with untreated particles and silane treated particles followed statistically identical lines. At −100°C, the CTE for silane treated particle samples were statistically different from the untreated samples at a 95% confidence level. The primary difference, at −100°C, was found in the intercept value as the silane-treated particulate composites. This difference may be due to physical factors in the testing. However, the silane treatment had little effect on the overall composite CTE.

A very high loading (~60 volume %) of ZrW_2O_8 particles would be needed to achieve dimensional stability. In this regard, work will need to be done to examine the mechanical properties of the resulting composite material. A composite material containing a large quantity of ZrW_2O_8 particles is expected to be very brittle and chip or fracture easily, although that did not occur during handling of the highly loaded samples analyzed in this research. Composites loaded with lower amounts of zirconium tungstate (up to ~12 volume%) have been studied,[32] but investigations of more highly loaded samples have not been reported in the literature. Also, zirconium tungstate filled composites lose utility for dimensional stability applications at temperatures in excess of the α-β phase transition at 155°C or if internal stresses build up in excess of 200 MPa causing the α-γ phase transformation of the zirconium tungstate.

REFERENCES

[1] Sleight, A.W., Mary, T.A., Evans, J.S.O., and Vogt, T., "Negative Thermal Expansion from 0.3 to 1050 Kelvin in ZrW_2O_8," 5 April 1996 *Science* **272** 90–92.

[2] ISO 11359-2:1999(E).

[3] Balch, D.K., and Dunand, D.C., "Copper-Zirconium Tungstate Composites Exhibiting Low and Negative Thermal Expansion Influenced by Reinforcement Phase Transformations," 2004 *Metallurgical and Materials Transactions A* **35A** 1159–1165.

[4] Sleight, A.W., "Isotropic Negative Thermal Expansion," 1998 *Annu. Rev. Mat. Sci.* **28** 29–43.

[5] David, W.I.F., Evans, J.S.O., and Sleight, A.W., "Zirconium Tungstate: The Incredible Shrinking Material," 1997 *ISIS Laboratory Scientific Highlights*. Retrieved January 18, 2005 from: http://www.isis.rl.ac.uk/ISIS97/feature1.pdf

[6] Li, W.H., Wu, S.Y., Lai, S.K., and Lee, K.C., "Thermal Contraction of Au Nanoparticles," 23 Sept. 2002 *Physical Review Letters* 135504.

[7] Yamamura, Y., Nakajima, N., and Tsuji, T., "Heat Capacity Anomaly Due to the α-to-β Structural Phase Transition in ZrW_2O_8," 2000 *Solid State Communications* **114** 453–455.

[8] Holzer, H., and Dunand, D.C., "Phase Transformation and Thermal Expansion of Cu/ZrW_2O_8 Metal Matrix Composites," 1999 *Journal of Materials Research*, **14**, 780–789.

[9] Pereira, A.S., Perottoni, C.A., and Da Jornada, J.A.H., "Raman Spectroscopy as a Probe for *In Situ* Studies of Pressure-Induced Amorphization: Some Illustrative Examples," 2003 *Journal of Raman Spectroscopy*, **34**, 578–586.

[10] Sleight, A.W., Evans, J.S.O., and David, W.I.F., "Structural Investigation of the Negative-Thermal-Expansion Material ZrW_2O_8," 1999 *Acta Crysallographica*, **B55**, 333–340.

[11] Ravindran, T.R., Arora, A.K., and Mary, T.A., "High-Pressure Raman Spectroscopic Study of Zirconium Tungstate," 2001 *J. Phys.: Condens. Matter* **13** 11573–11588.

[12] Evans, J.S.O., Jorgensen, J.D., Short, S., David, W.I.F., Ibberson. R.M., and Sleight, A.W., "Thermal Expansion in the Orthorhombic γ Phase of ZrW_2O_8," 1999 *Physical Review B*, **60**, 14643–14648.

[13] Yilmaz, S., and Dunand, D.C., "Finite-Element Analysis of Thermal Expansion and Thermal Mismatch Stresses in a Cu-60vol%ZrW_2O_8 Composite," 2004 *Composites Science and Technology*, **64**, 1895–1898.

[14] Composite Technology Development Staff, 2004 **CTD-403 Fact Sheet**, Lafayette, CO: CTD Inc.

[15] Summers, E., Thom, A.J., Cook, B., Akinc, M., "Extrusion and Selected Engineering Properties: Mo-Si-B Intermetallics," 2000 *Interrmetallics*, **8**, 1169–1174.

[16] TA Instruments Staff, "Thermal Analysis Application Brief: Use of TGA to Distinguish Flame-Retarded Polymers from Standard Polymers," TA 135 Application Brief, retrieved: May 20, 2005 from: <http://www.tainst.com/main.asp?n=2&id=46>.

[17] Callister, Jr., W.D. 2003 ***Materials Science and Engineering An Introduction, 6th ed.***, New York: John Wiley & Sons, pg. 659–662.

[18] BYU Optics lab. Retrieved: July 12, 2005 from: http://www.ee.byu.edu/photonics/fbg.phtml.

[19] Buryak, A.V., Kolossovski, K.Y., and Stepanov, D., "Optimization of Refractive Index Sampling for Multi-channel Fiber Bragg Gratings," retrieved: July 12, 2005 from: http://www.redferncomponents.com/dcg/downloads/IEEEQE.pdf.

[20] Sheppard, L.M. (1999 June) "ZrW2O8 Could Help Fiber Optics," *Photonics Spectra*, Retrieved: August 4, 2003 from: http://www.photonics.com/Spectra/tech/read.asp?techid=593.

[21] Futron.com "Space Transportation Costs: Trends in Price Per Pound to Orbit 1990–2000" 2002 Futron Corporation pg. 4, Retrieved: July 6, 2005, http://www.futron.com/pdf/FutronLaunchCostWP.pdf.

[22] Henderson, B.H., "Approaching Dimensional Stability Using the Functional Filler Zirconium Tungstate," Masters of Science Thesis, South Dakota School of Mines and Technology, 2005.

[23] Weyer, W.C., Cross, W.M., Henderson, B., Kellar, J.J., Kjerengtroen, L., Welsh, J.C., and Starkovich, J., "Achieving Dimensional Stability Using Functional Fillers," Published in: Proceedings of the 46th AIAA/ASME/ASCE/AHS/ASC Structures, Structural Dynamics, and Materials Conference, Austin TX, April 2005.

[24] Voss, G., Fabian, P., Feucht, S., Harte, M., and Terry, C., "Performance Evaluation of a Cyanate Resin," accessed: 10-20-04 from: http://www.fusion.org.uk/techdocs/sofe03_voss.pdf.

[25] Holliday, L., and Robinson, J., "Review: The Thermal Expansion of Composites Based on Polymers," 1973 *Journal of Materials Science* **8** 301–311.

[26] Wang, T.T., and Kwei, T.K., "Effect of Induced Thermal Stresses on the Coefficients of Thermal Expansion and Densities of Filled Polymers," 1969 *Journal of Polymer Science* **7** 889–896.

[27] Tummala, R.R., and Friedberg, A.L., "Composites, Carbides: Thermal Expansion of Composite Materials," *Journal of Applied Physics*, **41**, 5104–5107, 1970.

[28] Schapery, R.A., "Thermal Expansion Coefficients of Composite Materials Based on Energy Principles," *Journal of Composite Materials*, **2**, 380–404, 1968.

[29] Neter, J., and Wasserman, W., **Applied Linear Statistical Models**, Irwin-Dorsey Limited, Homewood IL, 1974, pp. 140–151.

[30] Montgomery, D.C., Peck, E.A., and Vining, G.G., **Introduction to Linear Regression Analysis**, 3rd Ed., Wiley-Interscience, New York, 2001, pp. 102–112.

[31] Draper, N.R., and Smith, H., **Applied Regression Analysis**, 3rd Ed., Wiley-Interscience, New York, 1998, pp. 142–146.

[32] Swanson, E., Cross, W.M., Starkovich, J., and Kellar, J.J., "Surface Modification of a Filler Having a Negative Coefficient of Thermal Expansion," in **Functional Fillers and Nanoscale Minerals**, J.J. Kellar, M.A. Herpfer and B.M. Moudgil, eds., Society for Mining, Metallurgy and Exploration, Littleton, CO, 2003, 269.

. . . .

Adsorption Behavior of Talc as a Filler Functionalized with Polysaccharides

Jing Wang* **and P. Somasundaran***

One of the main technical functions of talc is as fillers in strengthening products such as resins, rubber, ceramics and compacter. Toward its effective use, it is useful to modify the surface properties of talc and this can be achieved by adsorption of polymers. In this work, the mechanisms of interactions of talc with these polymers of different charge, galactose to mannose ratio and hydrophobic modification were investigated. The adsorption mechanisms of polysaccharides on talc were explored via electrokinetic, adsorption, fluorescence and wettability techniques. Adsorption of these polymers on talc was not found to be affected by changes in pH or ionic strength, ruling out electrostatic force as the controlling factor. Fluorescence spectroscopic studies conducted to investigate the role of hydrophobic bonding showed no evidence of hydrophobic domain formation at talc-aqueous interface. Urea, a hydrogen bond breaker, reduced the adsorption of polymers on talc significantly. Wettability measurements were first conducted to study the change of talc surface hydrophobicity after polymer adsorption. Upon comparison, it was found that nonionic and anionic polysaccharides (guar gum, locust bean gum (LBG) and carboxymethyl cellulose (CMC)) reduce the talc hydrophobicity more effectively than hydrophobically modified and unmodified ethylhydroxyethyl cellulose (EHEC). All the above results suggest the main driving force for polysaccharide adsorption on talc to be hydrogen bonding rather than electrostatic or hydrophobic force.

INTRODUCTION

The mineral talc is a hydrous magnesium silicate with the chemical formula $Mg_3Si_4O_{10}(OH)_2$.[1] Its elementary sheet is composed of a layer of magnesium-oxygen/hydroxyl octahedra sandwiched between two layers of silicon-oxygen tetrahedra. The

* NSF Industry/University Cooperative Research Center for Advanced Studies in Novel Surfactants, Columbia University, New York, New York

main surfaces of this sheet do not contain hydroxyl groups or active ions and hence the hydrophobicity. The edges of the sheets are, however, composed of hydroxyl, silicon, magnesium and oxygen ions and this composition contributes to its hydrophilicity. Inorganic filler, talc, is widely used in resins, rubber, ceramics and compacter to improve their mechanical and thermal properties[2–6]. Polysaccharides are often used as wetting agents to modify the surface of talc.[7] In spite of a number of studies conducted in the past, the mechanism of adsorption of polysaccharides on talc is not well understood. While Balajee,[8] Afenya[9] and Jucker[10] have attributed the adsorption of polysaccharides to hydrogen bonding, Liu and Laskowski[11] observed that the surface metallic sites were responsible for polysaccharide adsorption. Other workers suggested polysaccharide adsorption to involve physical interaction via hydrophobic bonding.[12–14] Steenberg and Harris[15] have proposed a combination of hydrophobic and hydrophilic interactions, with polysaccharides adsorbing firstly on the faces of talc via hydrophobic force, and then on the hydrophilic edges via hydrogen bonding.

The objective of this work was to clarify the mechanistic aspects of the interactions between polysaccharides and talc using a combination of spectroscopic and adsorption techniques. The hydrophobic or hydrophilic nature of the solids is determined by the wettability of the solid. Therefore, wettability measurements were conducted in this work to study the change of surface hydrophobicity of solids due to polymer adsorption.

EXPERIMENTAL METHOD

Materials

Talc sample obtained from Cytec Industries was of 40 microns size and BET surface area of 19.75m^2/g. The composition analysis was done by Galbraith Laboratory Inc. The samples were burned in a Schoniger flask and then the composition was read using ionic electrode. The result indicates that it is a very pure sample with only 0.7% tremolite ($Ca_2Mg_5Si_8O_{22}(OH)_2$).

Five polysaccharides with different functional groups and molecular weights were used in this work to study the mechanism of interactions between polymers and solids. These polymers were chosen so as to include hydrophilic modification (carboxymethyl cellulose (CMC)), hydrophobic modification (HM(C_{14})-EHEC), EHEC (ethylhydroxyethyl cellulose) without hydrophobic chains and different ratios of galactose to mannose units (locust bean gum and guar gum). The above polysaccharides have well-defined structures with different linkage and allow study of the effect of most parameters related to polymer structures. The repeat units of those polymers are schematically shown in Figure 1. Guar gums (142C and IMP4) and carboxymethyl cellulose (CMC) were obtained from Cytec Industries. The weight average molecular weights (Mw) of 142C and IMP4 are 1,450,000 and 242, 000 respectively. Their polydispersities were 1.5 and 25.9 separately. The CMC from Cytec has Mw of 714,900 and polydispersity of 4.59. Locust bean gum of Mw of 1,000,000 was supplied by AEP colloids. EHEC and Hydrophobically modified ethyl (hydroxyethyl) cellulose (HM(C_{14})-EHEC) were obtained from Akzo-Nobel. They had approximately the same weight average molecular weight of 100,000. HM(C_{14})-EHEC was washed by acetone and stirred for 30 minutes. The washed sample was vacuum filtered and dried. Then all the stock solutions were prepared by quickly adding the proper amount of polymer powder into vigorously stirred

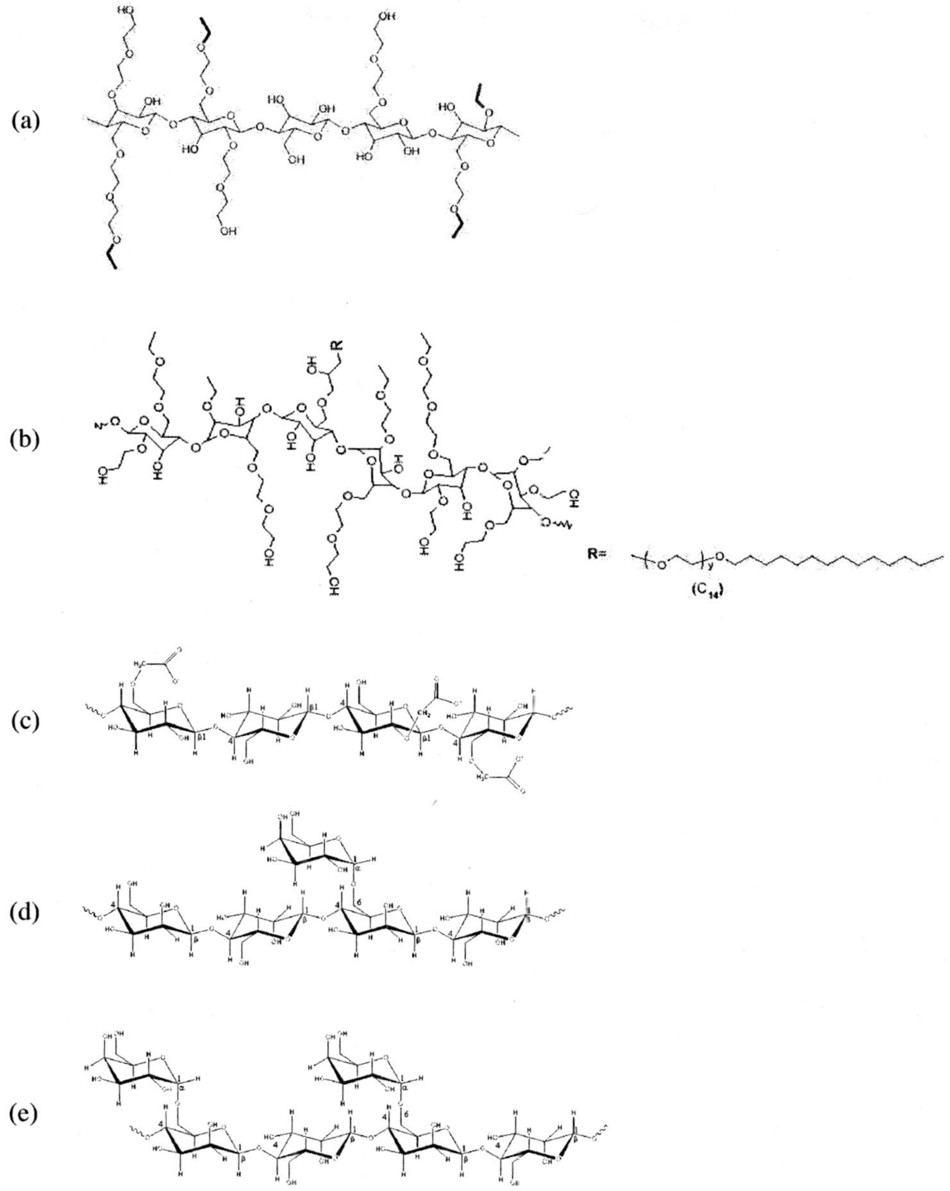

FIGURE 1 Repeat units of (a) EHEC, (b) HM-EHEC, (c) CMC, (d) LBG (G/M: 1/3.5) and (e) Guar gum (G/M: 1/2)

triple distilled water (TDW) and further stirring for 30 minutes. The solution was refrigerated overnight to ensure complete hydration or dissolution and then filtered through filter paper (Whatman #4) to remove any undissolved impurities.

The water used was triply distilled. pH of the solutions and suspensions were adjusted by Fisher standard hydrochloric acid and sodium hydroxide solutions.

Experiments

Adsorption Measurements. The suspensions of solids, with ionic strength adjusted to the desired level using KCl were ultrasonicated at 10% solid loading, for 30 minutes, and then the pH was adjusted to the desired value using HCl or KOH and stirred magnetically for 2 hours. Polymer stock solution was then added to the solid suspension and left for overnight conditioning with stirring. The suspensions were then centrifuged and the supernatants analyzed for polymer concentration by a Total Organic Carbon Analyzer. The adsorption density of the polymer on solid was calculated from the difference between the initial and residual polymer concentrations.

Wettability Measurements

The samples for hydrophobicity tests were prepared in the same way as these for the adsorption experiments. After 16 hours of equilibration, 20 ml of slurry was transferred to a separatory funnel to which 15 ml of toluene was added. The talc–polymer–toluene dispersion was shaken for 1 minute manually and then allowed to settle for 10 minutes. The bulk of the aqueous phase with hydrophilic solids, as well as the toluene phase with hydrophobic solids, was then allowed to flow out of the funnel separately. The two phases containing the solids were evaporated and the weight of talc was recorded. The relative percentage of the hydrophobicity was determined as: (weight of solid in toluene phase)/(weight of solid in aqueous phase + weight of solid in toluene phase).

Fluorescence Spectroscopy. Suspensions of talc at 10% solid loading were ultrasonicated for 30 minutes and mixed for 2 hours with a magnetic stirrer. Stock solution of pyrene-labeled polymer was then added to the talc suspension. The vials containing the suspensions were quickly wrapped with aluminum foil to keep the light away and left to condition overnight. Finally, the hydrophobicity of the supernatant and the solid were analyzed separately using a LS-1 fluorescence spectrophotometer (Photon Technology International).

RESULTS AND DISCUSSION

Influence of pH on Adsorption of Anionic Polysaccharide and Hydrophobicity of Talc

To study the effect of polymer charge on modifying talc hydrophobicity, wettability tests were conducted for CMC/talc system at different pH values and the results obtained show the talc to become less hydrophobic with CMC adsorption more sharply at pH 3 than at pH 9 (Figure 2). Since the polymer chain is negatively charged at pH 9, it should be more hydrophilic at this pH than at pH 3. But this does not correspond to the wettability results. From adsorption results given in Fig. 9, it can be seen that the adsorption density of CMC is reduced with increase in pH and this is attributed to the increase in negative charge of talc surface and the CMC chain and the resultant increase in electrostatic repulsion

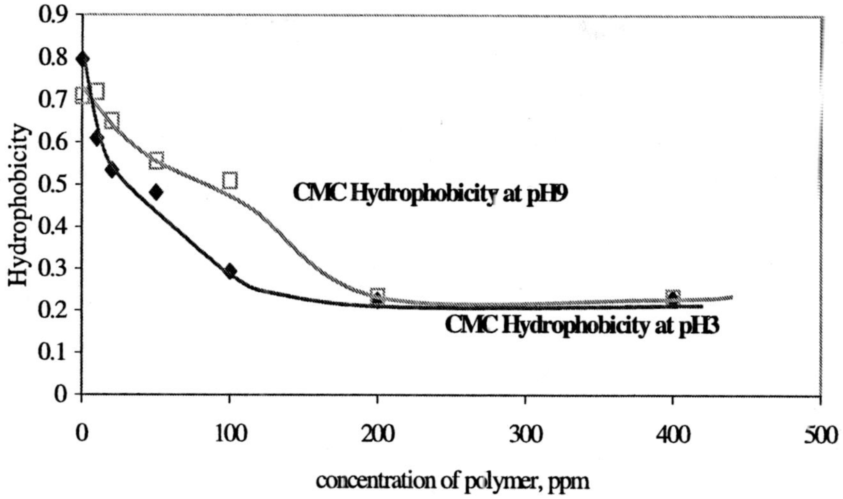

FIGURE 2 Surface hydrophobicity of talc with adsorption of CMC at different pH (no salt, S.L.: 10%)

between them. Therefore, the adsorbed amount of polymer can be proposed to play an important role in controlling the surface hydrophobicity of solids within the low polymer concentration range. At concentrations higher than 50ppm, the talc hydrophobicity does not decrease further at both pH values according to the wettability results.

Influence of Mannose/Galactose Ratio of Polysaccharide on Surface Hydrophobicity of Talc

To further study the structural effect on polysaccharide wettability modification performance, two polysaccharides, guar gum and LBG with different ratio of galactose and mannose units, were studied. Figure 3 shows wettability of talc along with adsorption of LBG (G/M: 1/3.5) and guar gum (G/M: 1/2). It is evident that at polymer concentrations below 25ppm, the adsorption of guar gum decreases the talc hydrophobicity more effectively than LBG. However, from adsorption measurements, it is seen that LBG adsorbs more than guar gum. This result suggests that adsorbed amount is not the only factor governing surface wettability modification. It is proposed that the conformation of adsorbed polymer also plays an important role.

Influence of Hydrophobic Modification of Polysaccharide on Talc Hydrophobicity

Tests were done with unmodified EHEC and hydrophobically modified EHEC (C_{14}-EHEC) to explore the effect of hydrophobic modification of polysaccharides on the wettability. As shown in Figure 4, the adsorption of C_{14}-EHEC does not reduce talc hydrophobicity at all. It is expected that C_{14}-EHEC may interact with talc surface via hydrogen bonding, leaving the hydrophobic tetradecyl chains protruding into the solution to make the surface hydrophobic.

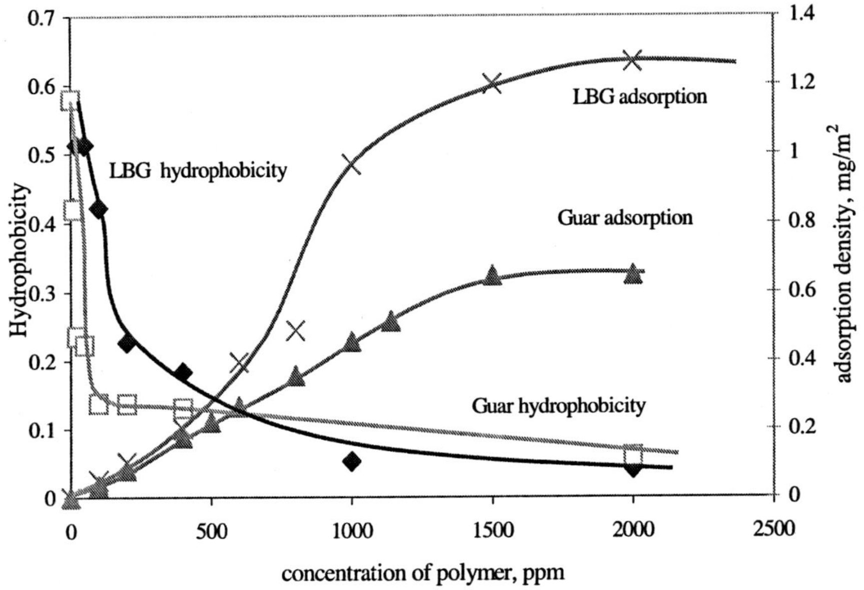

FIGURE 3 Surface hydrophobicity of talc with adsorption of guar gum and LBG (no salt, S.L.: 10%)

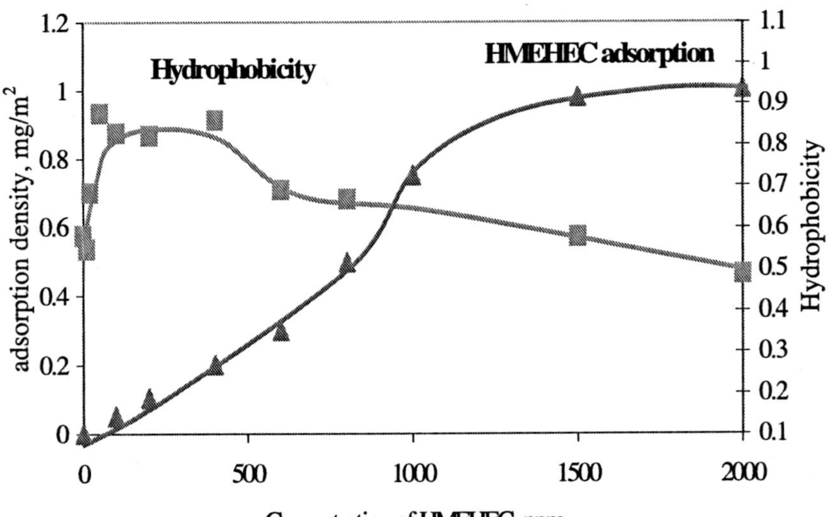

FIGURE 4 Surface hydrophobicity of talc with adsorption of HMEHEC (no salt, S.L.: 10%)

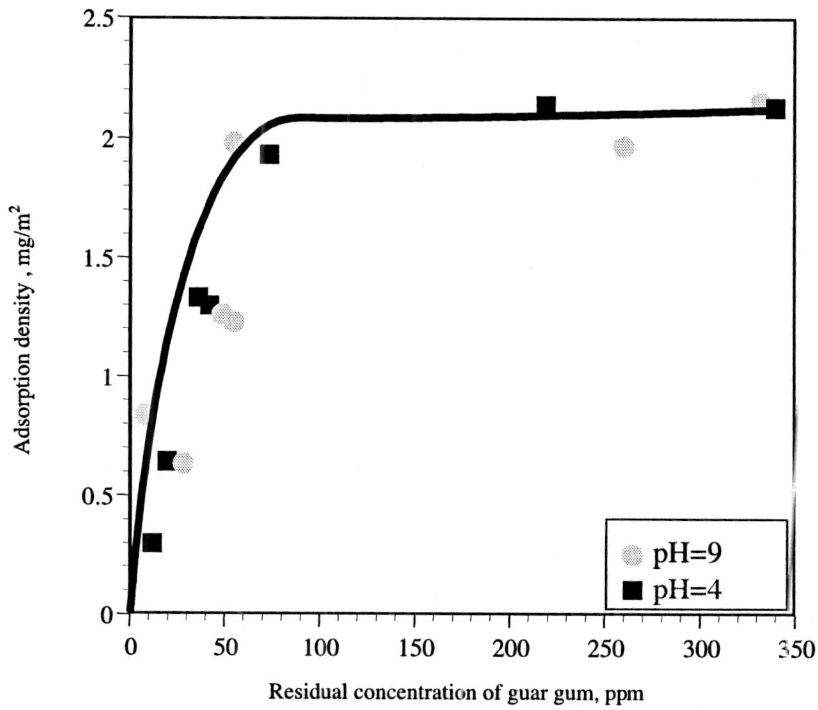

FIGURE 5 Adsorption isotherm of guar gum on talc at different pH (no salt, S.L.: 10%)

Adsorption at Varying pH[18, 19, 20]

Adsorption isotherms of guar gum, LBG, EHEC and C_{14}-EHEC on talc at different pH are presented in Figures 5–8. It can be seen that the adsorption density does not vary measurably between high and low pH. If electrostatic interaction plays an important role in the adsorption process, the adsorption density of polysaccharide on talc should have been affected significantly since the surface charge of the talc does change with change in pH from 2 to 9. Such is not the case here, and it can be concluded that electrostatic interaction is not the dominant force for the adsorption of nonionic guar gum, LBG, EHEC and C_{14}-EHEC on talc.

While for the adsorption of CMC on talc, the result (Figure 9) shows that the adsorption density of CMC is reduced with the increase of pH from 3 to 9. It is apparent that electrostatic repulsion plays an important role in this system.

Influence of Ionic Strength on Adsorption[18, 19, 20]

Adsorption isotherms of guar gum and LBG on talc at pH 9 are shown in Figure 10 and 11 at different ionic strengths. It is obvious that the influence of ionic strength is also negligible. If hydrophobic force makes a significant contribution to adsorption of a polymer on a mineral, the adsorption density would increase with the addition of salt due to "salting-out" effect. This is not in agreement with the experimental results. Thus it would appear that the hydrophobic force is not the main driving force for the adsorption of

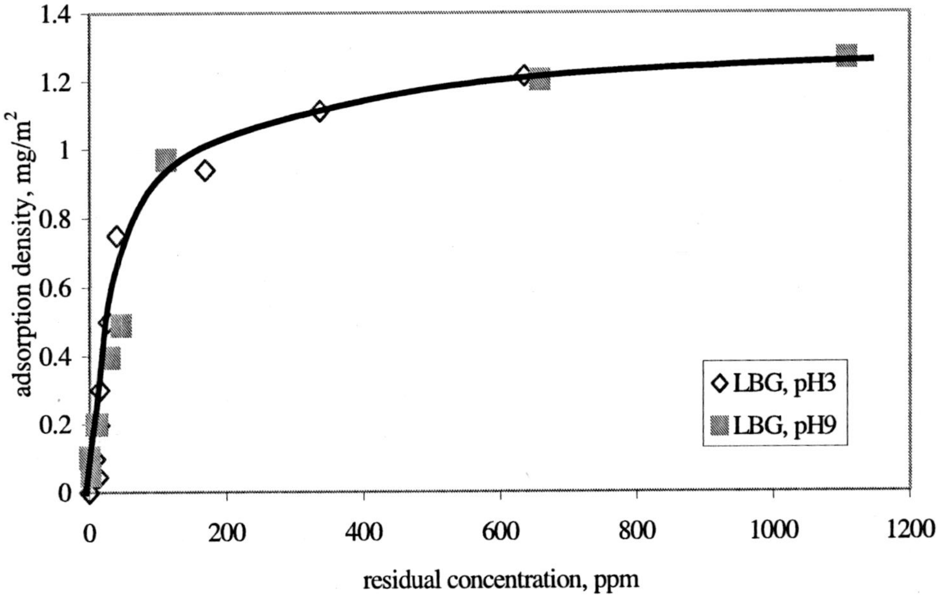

FIGURE 6 Adsorption isotherm of LBG on talc at different pH (no salt, S.L.: 10%)

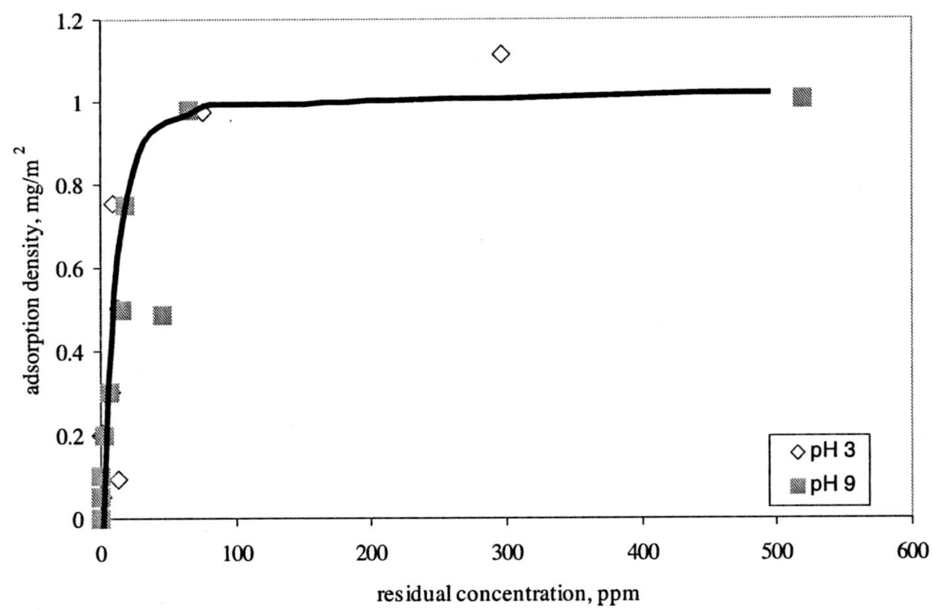

FIGURE 7 Adsorption isotherm of HMEHEC on talc at different pH (no salt, S.L.: 10%)

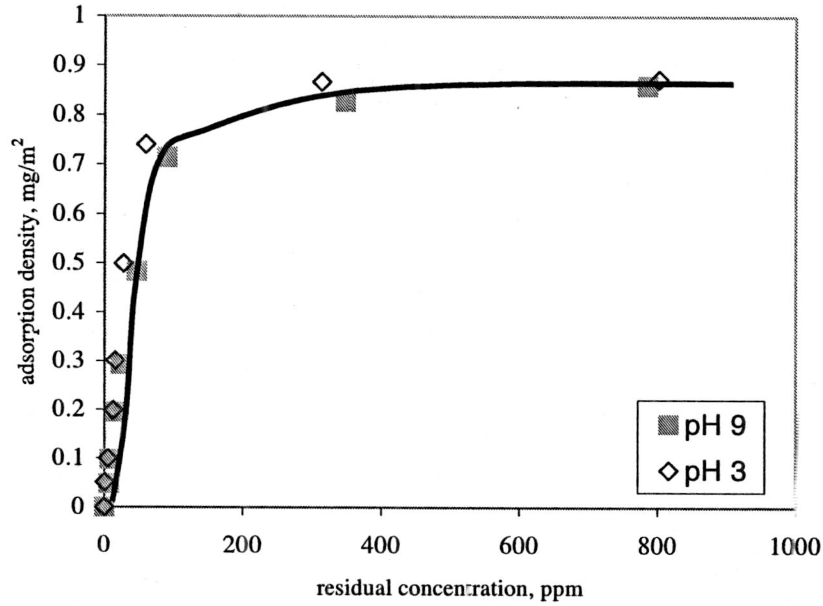

FIGURE 8 Adsorption isotherm of EHEC on talc at different pH (no salt, S.L.: 10%)

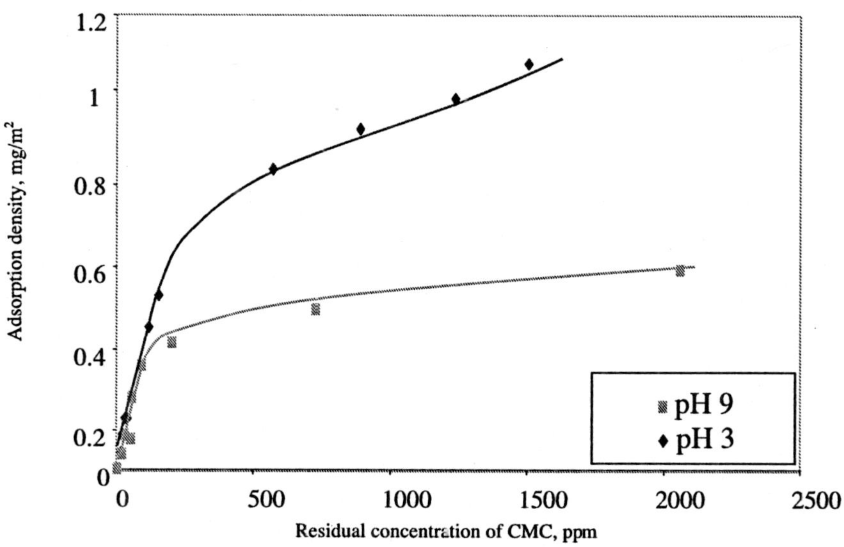

FIGURE 9 Adsorption isotherm of CMC on talc at different pH (no salt, S.L.: 10%)

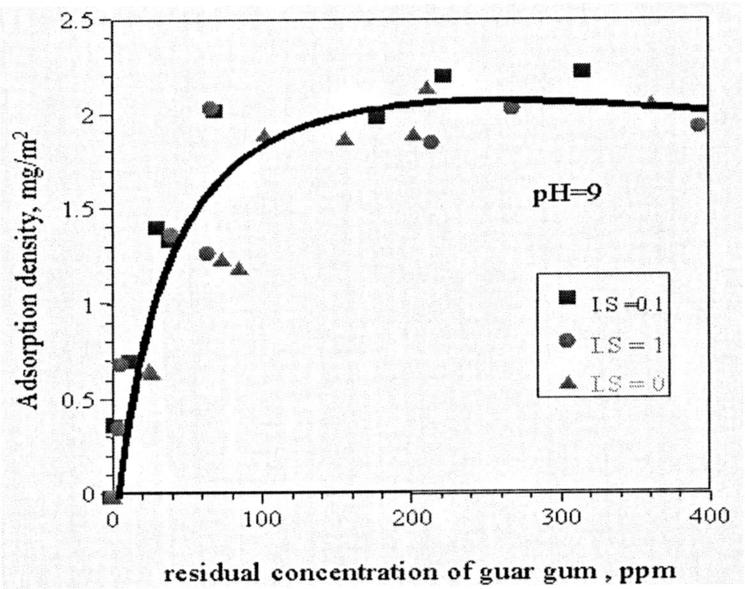

FIGURE 10 Adsorption isotherm of guar gum on talc at different ionic strength (pH: 9, S.L.: 10%)

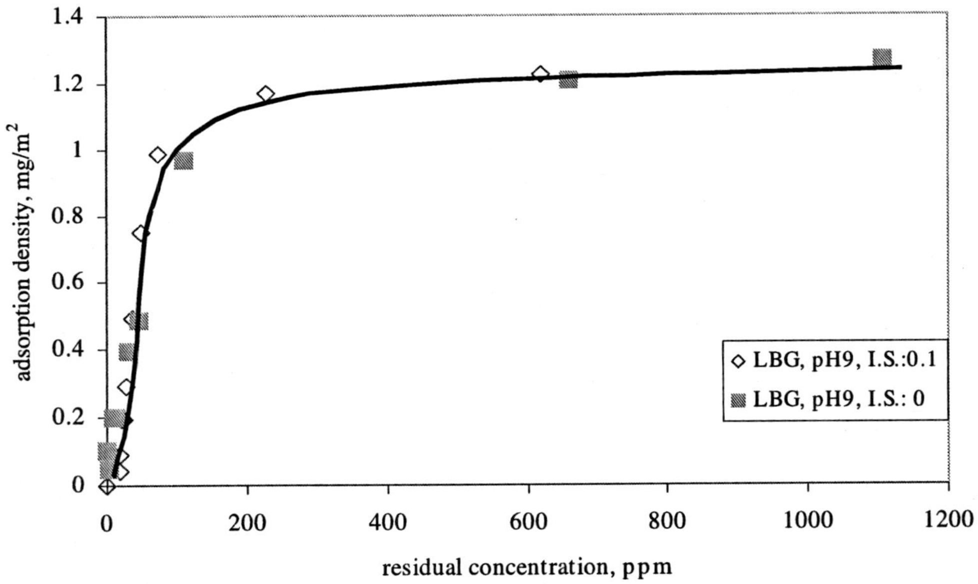

FIGURE 11 Adsorption isotherm of LBG on talc at different ionic strength (pH: 9, S.L.: 10%)

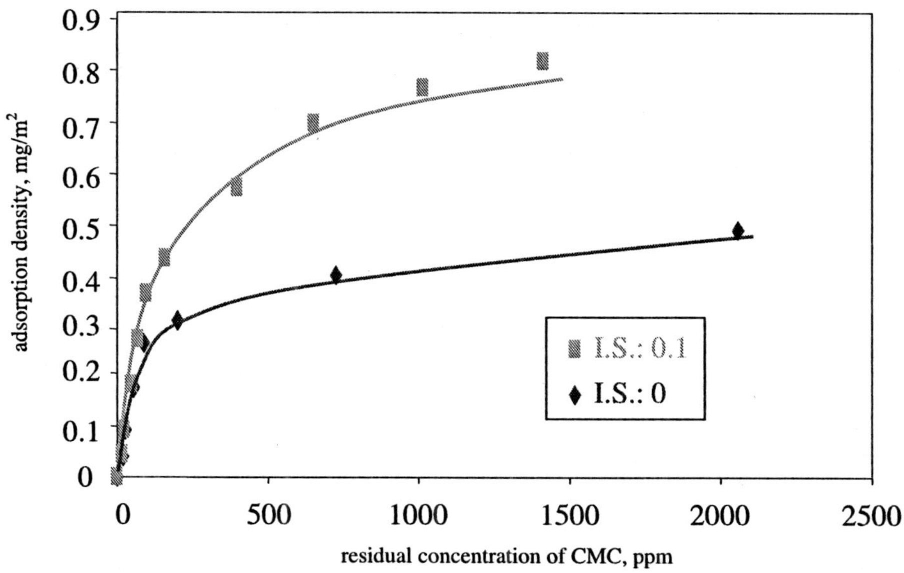

FIGURE 12 Adsorption isotherm of CMC on talc at different ionic strength (pH: 9, S.L.: 10%)

guar gum on talc. However, in contrast to guar gum and LBG, for CMC (Figure 12), the adsorption density increased with the increase of ionic strength, which may be due to the "screening" effect of salt. Adsorption isotherms of EHEC and C_{14}-EHEC on talc at pH 9 are shown in Figures 13 and 14 at different ionic strengths. If hydrophobic force makes a significant contribution to adsorption of a polymer on a mineral, the adsorption density would increase with the addition of salt due to "salting-out" effect. For EHEC, there is no significant change of the adsorption density as ionic strength increased from 0 to 0.1 and 1. In the case of C_{14}-EHEC, almost no effect can be observed with the increase of ionic strength from 0 to 0.1. However, as ionic strength is increased further from 0.1 to 1, adsorption density increased markedly. Hydrophobic force may play a role in the adsorption of HMEHEC on talc.

ADSORPTION ON SOLIDS WITH AND WITHOUT UREA[18, 19, 20]

To explore the role of hydrogen bonding, adsorption tests were carried out in the presence of urea, a hydrogen bond breaker. Tests were also done with alumina which is known to adsorb guar gum by hydrogen bonding.[16] Adsorption isotherms for guar with different molecular weight on talc and alumina in the presence and absence of urea at pH 8.5 are shown in Figures 15–17 respectively. A comparison of the results presented in these figures shows that urea, a hydrogen bond breaker, reduces the adsorption of guar on talc to the same extent as that on alumina. Similar results were also found for LBG, HMEHEC and CMC adsorption on talc (Figures 18–20). Since as mentioned above, the driving force for the adsorption of guar gum on alumina is hydrogen bonding, it is suggested that hydrogen bonding plays a similar important role in the adsorption of polysaccharide on alumina and talc.

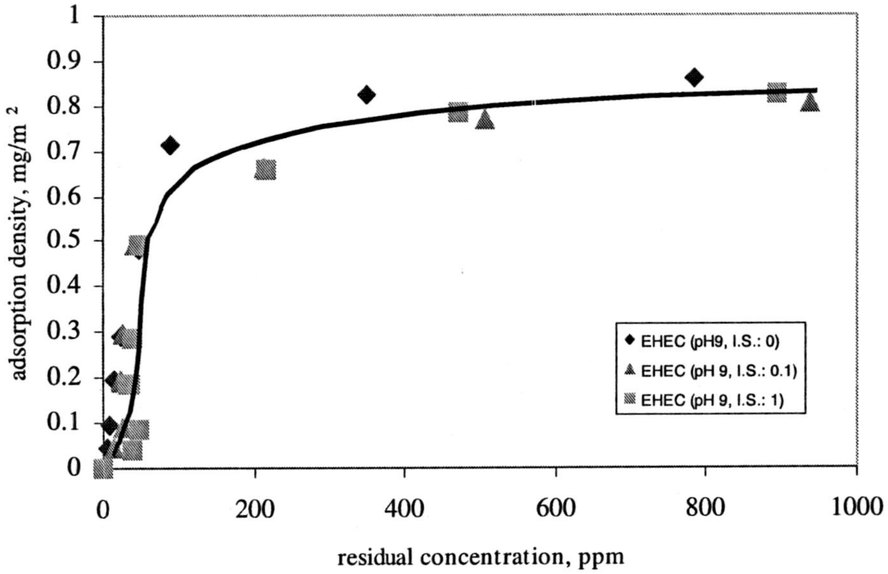

FIGURE 13 Adsorption isotherm of EHEC on talc at different ionic strength (pH: 9, S.L.: 10%)

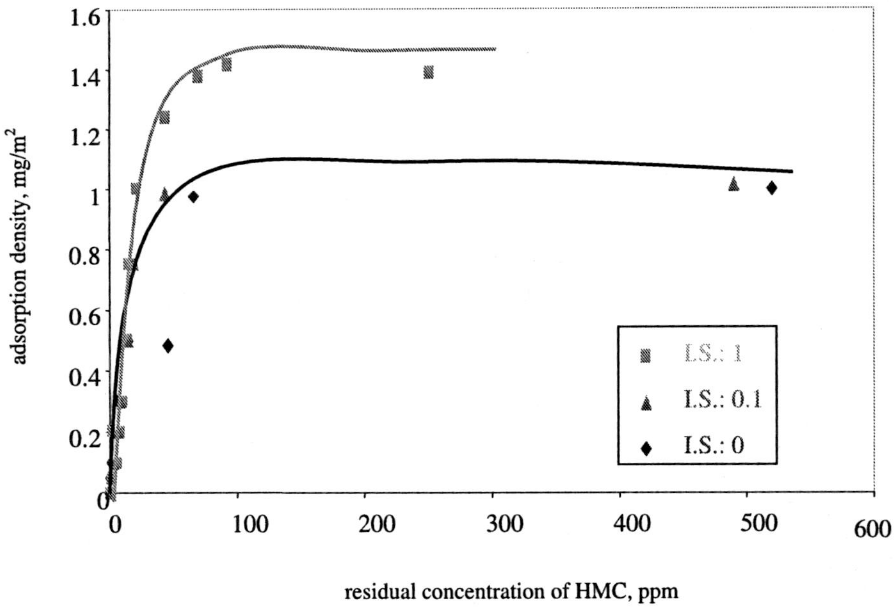

FIGURE 14 Adsorption isotherm of HMEHEC on talc at different ionic strength (pH: 9, S.L.: 10%)

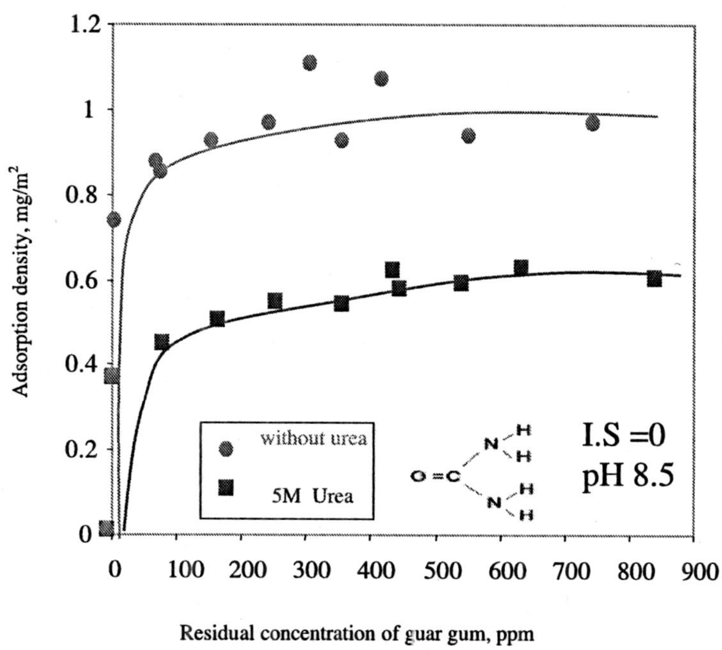

FIGURE 15 Adsorption isotherms of guar gum (142C) on talc with and without urea (I=0)

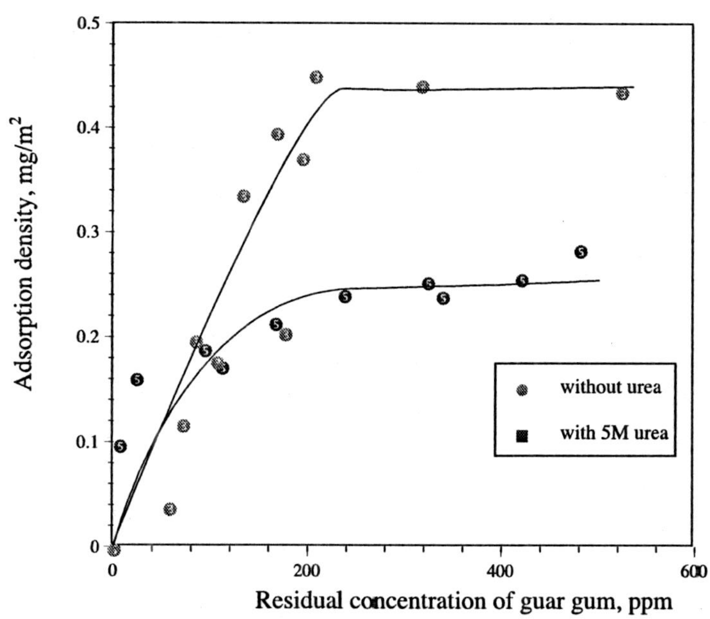

FIGURE 16 Adsorption isotherms of guar gum (142C) on alumina with and without urea (I=0)

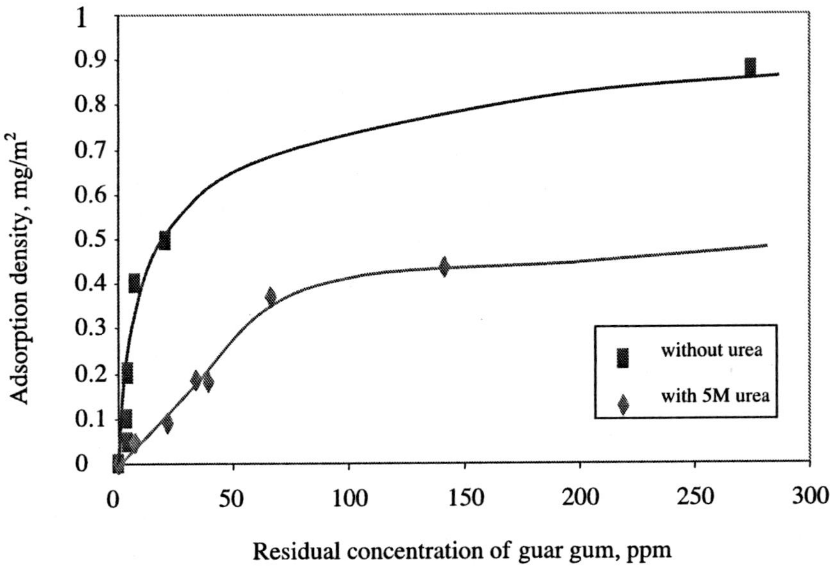

FIGURE 17 Adsorption isotherms of guar gum (IMP4) on talc with and without urea (I=0)

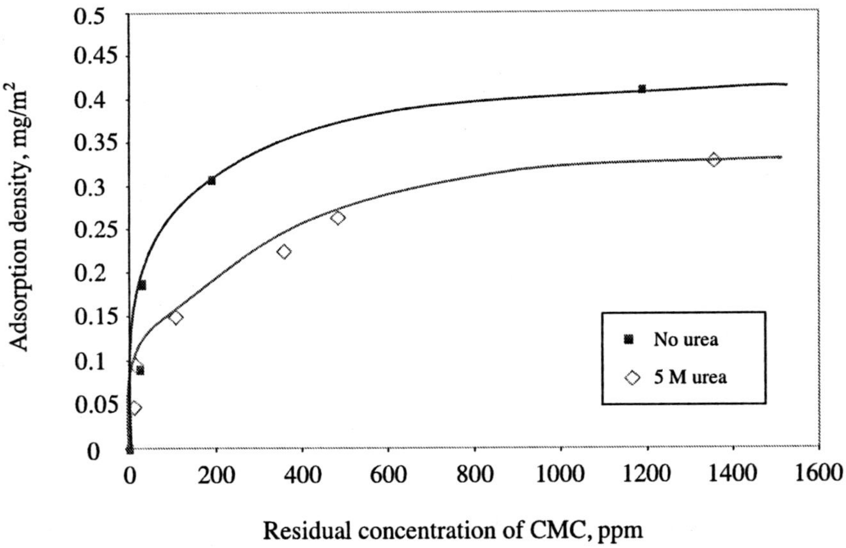

FIGURE 18 Adsorption isotherms of CMC on talc with and without urea (I=0, pH9)

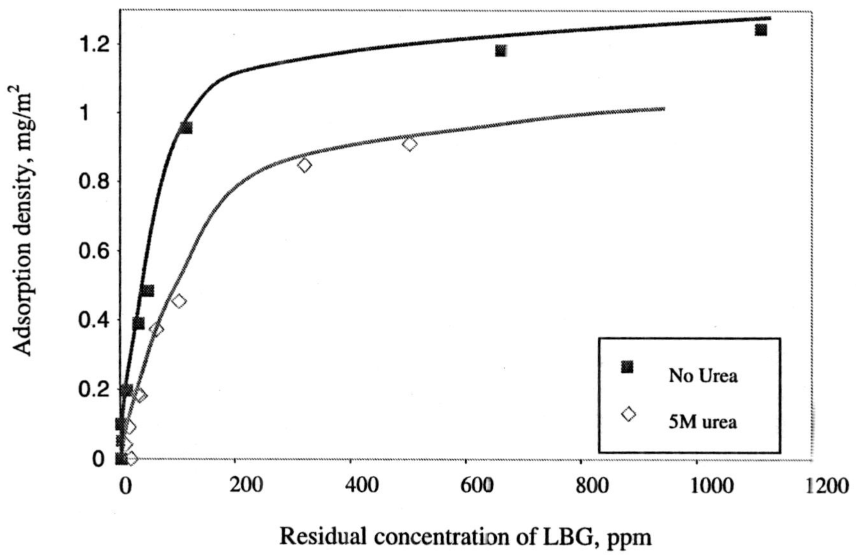

FIGURE 19 Adsorption isotherms of LBG on talc with and without urea (I=0, pH9)

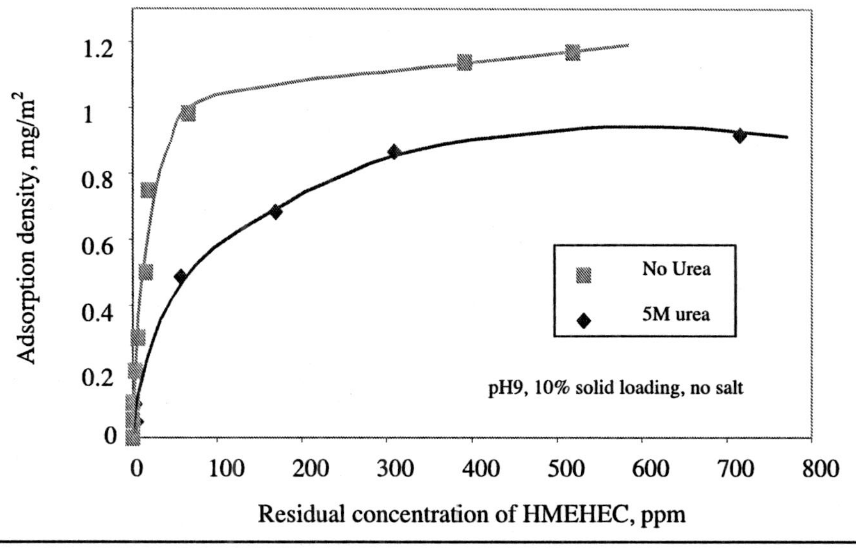

FIGURE 20 Adsorption isotherms of HMEHEC on talc with and without urea

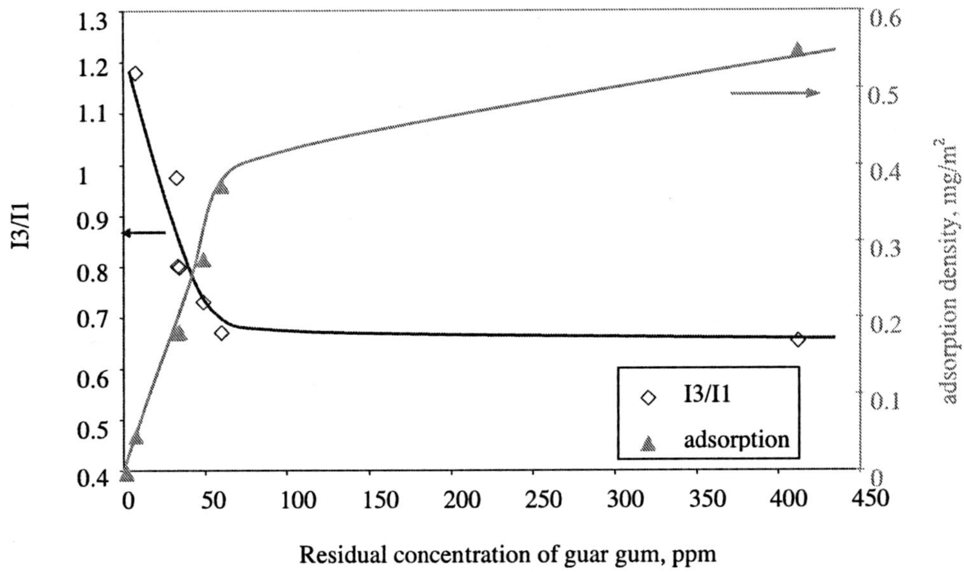

FIGURE 21 **Pyrene fluorescence for guar gum and talc interface**

FLUORESCENCE STUDIES[18, 19, 20]

To detect the role of hydrophobic force in polysaccharide adsorption, the nanostructure of the adsorbed layer was probed using fluorescence spectroscopy. Fluorescence spectroscopy is a well-developed method[17] for monitoring hydrophobic domain formation during adsorption. Fluorescence of pyrene is sensitive to the medium in which it resides. The ratio of the intensities of the first and third peaks (I3/I1) on a pyrene emission spectrum will increase as the environment becomes more hydrophobic. Since this ratio can be used to characterize the polarity of environments, it is termed the polarity parameter. Fluorescence data of pyrene probe at the talc-solution interface (after guar adsorption) and in the supernatants are shown in Figure 21. Pyrene fluorescence showed no sign for the existence of hydrophobic domains at the talc-aqueous interface because the polarity parameter for the guar adsorbed talc interface kept decreasing instead of increasing with increase in concentration of guar gum below the maximum adsorption. Above that concentration, the polarity parameter kept almost constant, which indicates that the hydrophobicity of the interface does not change any more. As shown in Figures 22–24, similar results were also found for CMC, EHEC and HMEHEC adsorption. Therefore, it is evident that the interface between them is more hydrophilic due to polysaccharide adsorption as shown in the figures. On the basis of these results, it can be proposed that more hydrogen bonding forms between polymer and the solid surface due to adsorption. Hydrophilic interaction rather than a hydrophobic one is therefore proposed to be the dominant force for polysaccharide adsorption process.

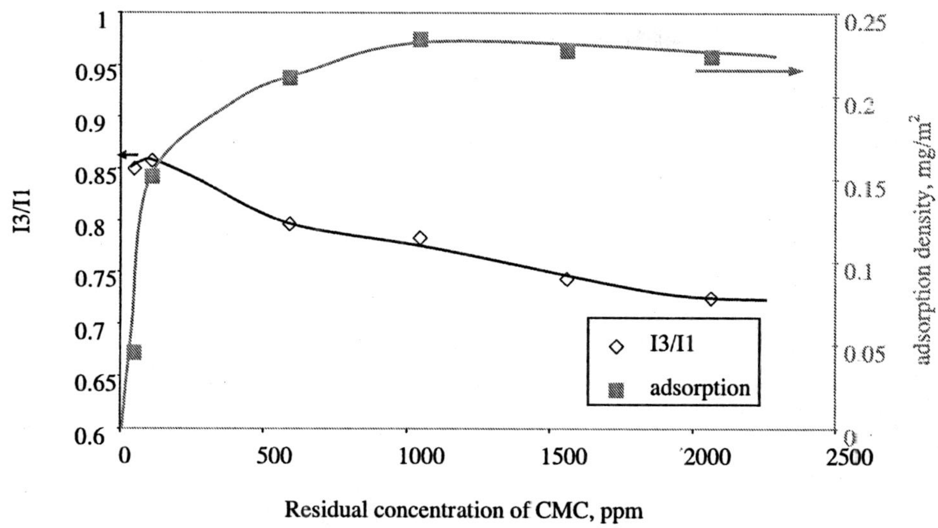

FIGURE 22 Pyrene fluorescence for CMC and talc interface

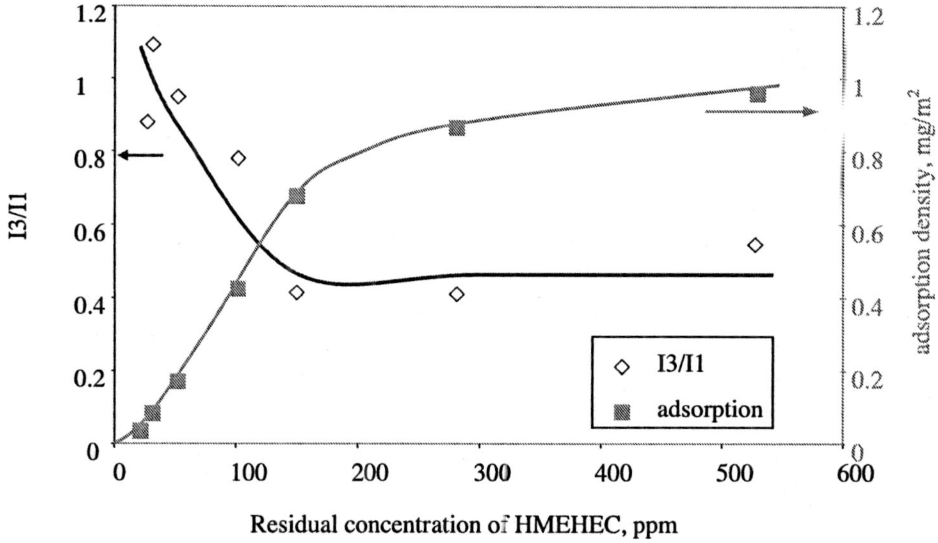

FIGURE 23 Pyrene fluorescence for HMEHEC and talc interface

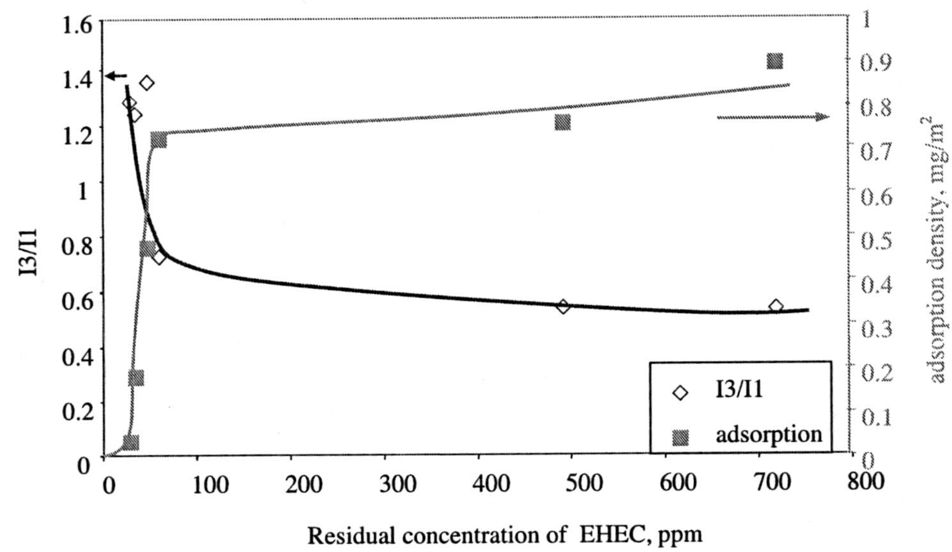

FIGURE 24 Pyrene fluorescence for EHEC and talc interface

SUMMARY

1. Adsorption studies showed that the adsorption of these polymers on talc was not affected by changes in pH or ionic strength level. These results suggest that the electrostatic force is not the dominant force for these adsorption processes. However, adsorption of carboxymethyl cellulose (CMC) on talc decreased the negative zeta potential of talc with a shift of the i.e.p. (isoelectric point). In addition, the adsorption of CMC on talc was affected significantly by changes in pH and ionic strength. These results suggest that the electrostatic force plays an important role in this adsorption process. Moreover, hydrophobic force may play a role in adsorption of HMEHEC on talc since the adsorption of HMEHEC on talc was found to be affected significantly by the ionic strength level.
2. Pyrene fluorescence showed no hydrophobic domain formation upon the adsorption of polysaccharides at talc-aqueous interfaces. Measurements with a new fluorescence probe, dansyl, support the governing role of hydrophilic force. Urea, a hydrogen bond breaker, reduced the adsorption of polymers on talc markedly. This result also supports a major mechanism involving hydrogen bonding rather than a hydrophobic one.
3. Nonionic and anionic polysaccharides (CMC, guar gum and LBG) modify the talc surface to be hydrophilic. Adsorption of hydrophobically modified and unmodified EHEC has almost no effect on wettability of talc surface. It is expected that HM(C_{14})-EHEC and EHEC may interact with talc surface via hydrogen bonding, leaving the hydrophobic chain or the ethylene oxide side chains protruding into the solution to make the surface much less hydrophilic than guar gum.

ACKNOWLEDGMENTS

The authors acknowledge the support of the National Science Foundation (Grant# CTS-00-89530, EEC-03-2861) and Cytec Industries.

REFERENCES

1. R.J. Piniazkiewicz, E.F. McCarthy, N.A. Genco, Talc. Industrial minerals and rocks, 6th edn., 1990.
2. Krysztafkiewicz, Andrzej; Domka, Ludwik. Surface-modified microporous talcs as fillers of polymers and pigments. Journal of Materials Chemistry (1997), 7(8), 1,655–1,659.
3. Arroyo-Ramos, M.; Lopez-Manchado, Miguel Angel. Impact behavior of modified talc filled PP/LDPE blends. Journal of Polymer Engineering (1995), 14(4), 237–52.
4. Bigg, D.M. Mechanical properties of particulate filled polymers. Polymer Composites (1987), 8(2), 115–22.
5. Varga, Jozsef. Filler dispersion in polypropylene/elastomer blends. Journal of Polymer Engineering (1991), 10(1–3), 231–51.
6. Krysztafkiewicz, A.; Domka, L.; Wieczorek, W. The modified talcs as semi-active fillers of rubbers. Colloid and Polymer Science (1985), 263(10), 804–11.
7. Krysztafkiewicz, Andrzej; Rager, Bozena; Wieczorek, Wojciech. Surface modification methods for mineral fillers used in plastics, Physicochemical Problems of Mineral Processing, 30, 107–117 (1996), 30, 107–117.
8. S.R. Balajee, I. Iwasaki. Adsorption mechanism of starches in flotation and flocculation of iron ores. *Trans. AIME*, 1969, 244, pp401–406.
9. P.M. Afenya. Adsorption of xanthate and starch on synthetic graphite. *Int. J. Miner. Process.*, 1982, 9, pp303–319.
10. B.A. Jucker, H. Harms, S.J. Hug, A.J.B. Zehnder. Adsorption of bacterial surface polysaccharides on mineral oxides is mediated by hydrogen bonds, *Colloids Surf. (B)*, 1997, 9, pp331–343.
11. Q. Liu, J.S. Laskowski. The role of metal hydroxides at mineral surfaces in dextrin adsorption, *Int. J. Miner. Process.*, 1989, 26, pp297–316.
12. R.J. Pugh, Macromolecular organic depressants in sulfide flotation—a review, 1. principles, types and applications. *Int. J. Miner. Process.*, 1989, vol. 25, pp101–130.
13. H.H. Haung, J.V. Calara, D.L. Bauer, J.D. Miller. Adsorption reactions in the depression of coal by organic colloids. In: N.N. Li (ed.), *Recent Developments in Separation Science*. CRC Press, Chicago, IL, 1978, 4, pp115–133.
14. S.R. Grano, L.F. Griffin, N.W. Johnson, R.St.C. Smart, J. Ralston. Treatment of naturally hydrophobic gangue minerals at the copper concentrator of Mt Isa mines limited, In *Proceedings of the Fourth Mill Operator's Conference*, The Australia Institute of Mining and Metallurgy, 1991, pp121–128.
15. E. Steenberg, J.P. Harris. A study of the adsorption, zeta potential and dispersion characteristics of certain polymers in relation to their effect on the floatability of talc, *S. Afr. J. Chem.*, 1984, 37, pp 85.
16. B.A. Jucker, H. Harms, S.J. Hug, A.J.B. Zehnder (1997). Adsorption of bacterial surface polysaccharides on mineral oxides is mediated by hydrogen bonds, *Colloids Surf. (B)*, 9, pp331–343.
17. J.K. Thomas (1984). The chemistry of excitation at interfaces, ACS, Washington, DC.
18. Jing Wang, P. Somasundaran and D.R. Nagaraj, "Adsorption mechanism of guar gum at solid-liquid interfaces," Minerals Engineering (2005), 18(1), 77–81.

19. Jing Wang and P. Somasundaran, "Adsorption and Conformation of Carboxymethyl Cellulose at Solid-Liquid Interfaces using Spectroscopic, AFM and Allied Techniques," Journal of colloid and interface science, 2005, 291, 75–83.
20. Jing Wang and P. Somasundaran, "Mechanisms of Ethylhydroxyethyl Cellulose-Solid Interaction: Influence of Hydrophobic Modification," Journal of colloid and interface science, 2005, In press.

PART 4

Formation and Characterization of Nanoparticles/Surfaces

Preparation of Nanoparticles: The Microemulsion Nanoreactor Approach

K. Osseo-Asare*

The availability of nanosize colloidal particles (1–100nm) with tailored properties is critical to the development of high performance materials in many areas of modern technology. A promising approach to the synthesis of such materials is to constrain the reactor size to match the required nanodimensionality. Microemulsions, compartmentalized fluids characterized by surfactant-stabilized nanosize domains of the type water-in-oil (w/o), oil-in-water (o/w), or bicontinuous, represent convenient vehicles for exploring the nanoreactor concept. In this presentation, metal oxides (including silica) are used as model materials to examine the effects of microemulsion properties and synthesis protocols on particle evolution and characteristics.

INTRODUCTION

Liquid-phase synthesis of nanoparticles relies primarily on two techniques, based respectively on use of growth-modifiers and nanoreactors (Burda et al. 2005; Pillai et al. 1995; Schmidt 2001; Osseo-Asare and Arriagada 1990a; Osseo-Asare 1999; Osseo-Asare and Arriagada 2000). In the first case, chemical additives (e.g., organic solvents, organic polymers) are used to constrain the basic steps of particle formation (e.g., nucleation, growth, and aggregation) in bulk liquid media. With the nanoreactor approach (e.g., liquid atomization) the reactor size itself is controlled to match the required nanodimensionality of the product particles. Water-in-oil microemulsions, compartmentalized liquid systems characterized by aqueous nanodrops (2–30 nm) dispersed in a continuous organic phase, have the unique advantage that they offer both growth-modifier and nanoreactor functionalities. A number of different inorganic materials have been

* Penn State University, University Park, Pennsylvania

TABLE 1 Critical issues surrounding synthesis of nanometer scale materials in microemulsions

Scientific Issue	Synthesis Issue
1. What quantitative differences exist in chemical equilibria between bulk aqueous solution and microemulsion water pools?	Can synthesis routes be accurately predicted from bulk solution chemistry?
2. How does the mode and sequence of reactant mixing affect the growth kinetics and characteristics of nanostructures?	Can predictive methods be developed for selecting optimum methods of mixing reactants?
3. How are the equilibrium properties of a microemulsion affected by the presence of reactants and subsequent products?	Can general methods always be found to sustain microemulsion stability during synthesis?
4. How are the dynamics and mechanisms of intermicellar transport critical factors in the nucleation/growth process?	How accurately can we predict growth rates and size dispersions from a knowledge of intermicellar dynamics?

prepared via microemulsion-mediated synthesis, including metals, metal sulfides, selenides, halides, carbonates, oxides, and hydroxides; synthesis of organic polymers and organic-inorganic nanocomposites has also been reported (Burda et al. 2005; Pillai et al. 1995; Schmidt 2001; Osseo-Asare and Arriagada 1990a; Osseo-Asare 1999; Osseo-Asare and Arriagada 2000).

A central problem in the microemulsion-mediated synthesis of nanosize materials is the relationship between the properties of the microemulsion fluid phase and those of the resulting particles. Table 1 presents a partial list of critical scientific issues which flow from this central problem. To address these questions systematically, it is necessary to consider how the properties of the microemulsion influence each stage in the solid formation process, i.e., nucleation, growth of primary particles, ripening, and coagulation of primary particles, as summarized in Figure 1 for the case of microemulsion-mediated synthesis of silica via alkoxide ($Si(OR)_4$) hydrolysis. Among the key microemulsion properties of interest are: the domain of the one-phase microemulsion region in the pseudo-ternary phase diagram, the size of the water pools (radius = r_w), the surfactant aggregation number (N), the nature of the surfactant polar groups, the nature of the solubilized water, and the rate of intermicellar communication (rate constant, k_{ex}). These properties, in turn, may be altered by varying the water-to-surfactant molar ratio (R), the nature of the continuous oil phase, the type of surfactant and/or cosurfactant, the concentration of electrolyte in the water cores, and the temperature (Luisi and Straub 1984; Friberg and Bothorel 1987; Kumar and Mittal 1999).

CONTROL OF MATERIAL COMPOSITION, SIZE, AND STRUCTURE VIA CHEMICAL REACTION PATHS (CRITICAL ISSUES #1 AND #2)

A major issue (#1 in Table 1) concerns the feasibility of transferring to nanoheterogeneous fluid media the well proven synthetic chemistries and quantitative thermochemistry of conventional bulk solutions. For precipitation from a metal ion, a solid product of elemental metal, or a metallic compound such as an oxide, a sulfide, or a salt (e.g., carbonate, chloride, sulfate, oxalate, phosphate, etc.) may result, depending upon the reactant. The latter can be added directly or may be generated *in situ* (Salutsky and Willard 1959), e.g., decomposition of thioacetamide releases sulfide ions, while the decomposition of urea

(I) TEOS Partition :

$$Si(OR)_4(b) \rightleftharpoons Si(OR)_4(m)$$

$$Si(OR)_4(m) + H_2O(m) \overset{OH^-}{\rightleftharpoons} Si(OR)_3OH(m) + ROH(m)$$

$$Si(OR)_4(b) + H_2O(m) \overset{K_p}{\rightleftharpoons} Si(OR)_3OH(m) + ROH(m)$$

(II) Hydrolysis:

$$Si(OR)_3OH(m) + xH_2O(m) \xrightarrow{OH^-} Si(OR)_{3-x}(OH)_x(m) + xROH(m)$$

$x = 1,2,3$

$$Si(OR)_{3-x}(OH)_{x+1}(m) = \text{"monomer" } (m)$$

$x = 0,1,2,3$

(III) Nucleation:

n.monomer(m) $\longrightarrow$ nucleus(m)

monomer(m) + monomer(m') $\longrightarrow$ nucleus(m,or m')

(IV) Particle: Growth

monomer(m) + nucleus(m) $\longrightarrow$ growth (monomer addition)

monomer(m') + nucleus(m) $\nearrow$

nucleus(m) + nucleus(m') $\longrightarrow$ growth (nuclei aggregation)

FIGURE 1 General reaction scheme for the formation of SiO_2 particles in reverse micellar systems (after Arriagada and Osseo-Asare 1999a)

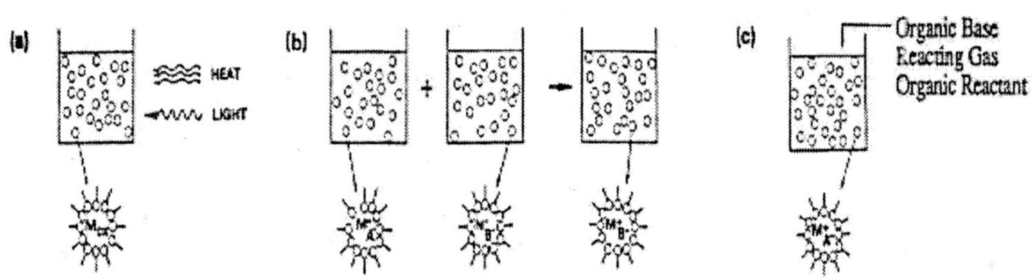

FIGURE 2 Micro-emulsion-mediated materials synthesis methods: (a) microemulsion-plus-trigger (MPT), (b) microemulsion-plus-microemulsion (MPM), and (c) microemulsion-plus reactant (MPR) (Osseo-Asare and Arriagada 1990a; Osseo-Asare 1999; Osseo-Asare and Arriagada 2000)

gives hydroxyl ions. When a sample of a bulk aqueous solution is solubilized in the polar core of a reverse micellar microemulsion, the local pH, ion concentration, as well as complex stereochemistries may change significantly from the corresponding bulk solution values (Luisi and Straub 1984; Friberg and Bothorel 1987). Taking advantage of these factors in the rational design of reaction pathways remains a major challenge.

Figure 2 shows a classification of microemulsion-based synthesis methods: (a) the microemulsion-plus-trigger (MPT) method, (b) the microemulsion-plus-microemulsion (MPM) method, and (c) the microemulsion-plus-reactant (MPR) method (Osseo-Asare

TABLE 2 Synthesis routes for selected oxide materials

Material	Polar Phase Reactants	Synthesis Route
$Al(OH)_3$	Al(III), NO_3^-	MPR (NH_3 addition)
$Al(OH)_3$	Na^+, $Al(OH)_4^-$	MPT (Heating)
α-FeOOH	Fe(II), NH_4^+, SO_4^{2-}	MPR(O_2 (air addition)
β-FeOOH	Fe(III), Cl^-, H^+	MPT (Heating)
α-Fe_2O_3	Fe(III), NO_3^-	MPR (NH_3OH addition)
Fe_3O_4	Fe(II), Fe(III), Cl^- (μE 1); NH_4OH (μE 2)	MPM (μE 1 + μE 2)
SiO_2	H_2O (+ HCl, pH 2)	MPR (Alkoxide addition)
SiO_2	Na^+, $SiO(OH)_3^-$	MPR (Acid addition)
SiO_2	H_2O (+ NH_4OH)	MPR (Alkoxide addition)
TiO_2	H_2O	MPR (Alkoxide addition)
TiO_2	$TiCl_4$ (μE 1); NH_4OH (μE 2)	MPM (μE 1) + (μE 1)
V_2O_5	H_2O	MPR (Alkoxide addition)
ZrO_2	H_2O	MPR (Alkoxide addition)

and Arriagada 1990a; Osseo-Asare 1999; Osseo-Asare and Arriagada 2000). Table 2 presents a list of selected oxide materials that have been prepared by these techniques. When a given material can be synthesized by more than one protocol (e.g., $Al(OH)_3$: MPR vs. MPT; SiO_2: MPR–alkoxide addition vs. acid addition), a relevant question is whether the choice of synthesis protocol makes any difference in terms of the resulting materials characteristics. The issue (#2 in Table 1) has been little addressed. It is likely that intermicellar communication is a hidden variable here. For example, the MPM method requires intermicellar matter exchange before precipitation of a solid can occur. In contrast, the MPT route does not require intermicellar exchange.

CONTROL OF MICROEMULSION STABILITY DURING SYNTHESIS (CRITICAL ISSUE #3)

The stability and extent of the one phase microemulsion region is dependent on a number of factors foremost among which are the nature of the surfactant molecules, the type of oil, the nature of the solubilized aqueous phase, and temperature (Luisi and Straub 1984; Friberg and Bothorel 1987; Kumar and Mittal 1999). The isothermal ternary diagrams for the AOT/isooctane/water, Triton X-100/toluene/water, and DDAB/decane/water systems, shown in Figure 3, illustrate how different surfactant/oil combinations give rise to reverse micellar regions of different size and location within the phase diagram. Not all reactant-microemulsion combinations can be exploited in materials synthesis, because of solubility and stability constraints. In the presence of a given reactant, phase instability may arise as a consequence of complexation (which draws surfactant monomers from the surfactant aggregates or chemically decomposes surfactant molecules), preferential interaction of reactants with water molecules (which decreases the amount of water available to the surfactant polar groups), or consumption of water molecules (in cases where water is an active participant in the synthesis reaction).

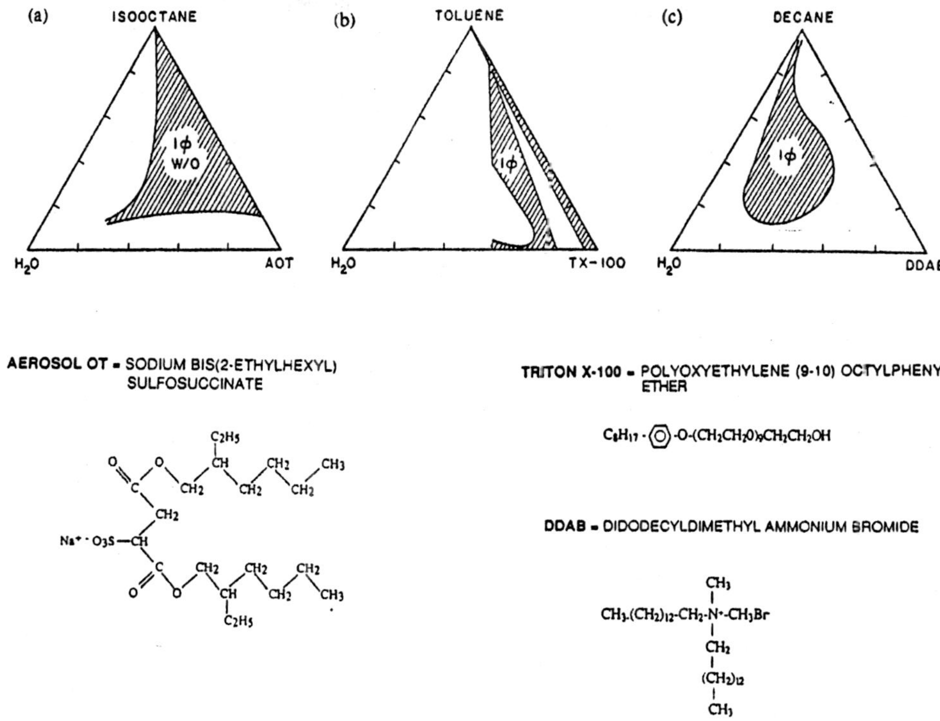

FIGURE 3 One phase water-in-oil microemulsion regions in ternary phase diagrams: (a) Aerosol OT/isooctane/water system (Kunieda and Shinoda 1979); (b) Triton X-100/toluene/water system (Almgren et al. 1986); (c) DDAB/decane/water system (Puig et al. 1988)

Reaction-induced phase instability is an especially important issue in systems where water plays the dual role of reaction medium (i.e., water pool) and reactant, as in the case of metal oxide synthesis via acid- or base-catalyzed hydrolysis of an alkoxide ($M(OR)_n$):

$$M(OR)_n + nH_2O \rightarrow M(OH)_n\ (s) + nROH \qquad \textbf{(EQ 1)}$$

The polyoxyethylene (5) nonylphenyl ether (NP-5)/cyclohexane/NH_4OH and the AOT/decane/NH_4OH water-in-oil microemulsions have been used as model systems for preparation of silica nanoparticles via the MPR technique (Osseo-Asare and Arriagada 1990a; Osseo-Asare 1999; Osseo-Asare and Arriagada 2000; Arriagada and Osseo-Asare 1999a; Osseo-Asare and Arriagada 1990b; Arriagada and Osseo-Asare 1992; Arriagada and Osseo-Asare 1994; Arriagada and Osseo-Asare 1995; Arriagada and Osseo-Asare 1999b; Osseo-Asare and Arriagada 1999). Tetraethoxysilane (TEOS) served as the alkoxide precursor and silica nanoparticles in the 50–70 nm size range were obtained. A representative TEM micrograph of the resulting silica particles is presented in Figure 4 (Osseo-Asare and Arriagada 1990b). In experiments (Arriagada and Osseo-Asare 1992) where phase separation occurred during the reaction, a bimodal size distribution was obtained. This phase behavior is attributable to water loss from

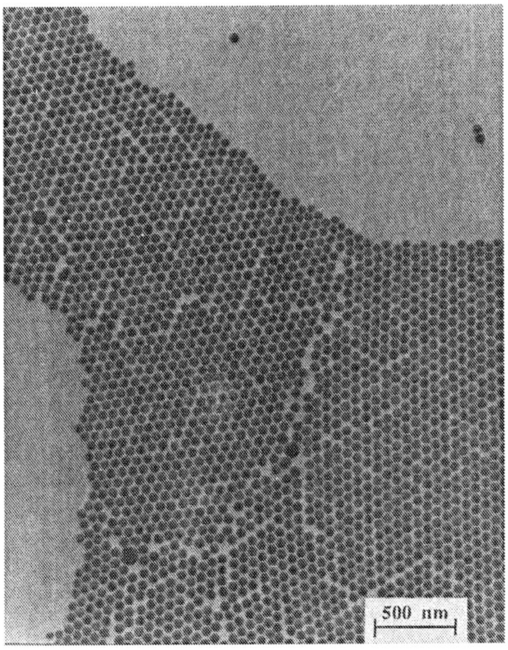

FIGURE 4 EM micrograph of SiO2 nanoparticles prepared by TEOS hydrolysis in the NP-5/cyclohexane/water ammonia microemulsion system (Osseo-Asare and Arriagada 1990b)

the system and the ability of the ethanol product (Equation 1) to destabilize the microemulsion (Arriagada and Osseo-Asare 1992).

CONTROL OF PARTICLE FORMATION VIA CONTROL OF INTERMICELLAR DYNAMICS (CRITICAL ISSUE #4)

For cluster formation to proceed to completion in the nucleation process, the monomers which represent potential candidates for a given critical nucleus must be retained for a sufficiently long period of time within a volume of molecular dimensions. Thus, intermicellar transport becomes a critical issue. If intermicellar matter exchange proceeds extremely rapidly, then the candidate monomers will be redistributed before nucleation can occur. Accordingly it may be deduced that an increase in the intermicellar communication rate will decrease the probability of nucleus formation. On the other hand, since the final particle sizes are frequently greater than the microemulsion droplet sizes, it is obvious that intermicellar matter exchange is essential for particle growth. Thus, one would expect an increase in communication rate to favor particle growth. It appears then that the manner in which the rate of intermicellar matter exchange (k_{ex}) influences particle formation and, therefore, particle characteristics such as size and polydispersity, may be complex.

Figure 5 shows the effect of the water-to-surfactant molar ratio (R) on the size and size distribution of silica nanoparticles prepared in the AOT/decane/water/ammonia system (Arriagada and Osseo-Asare 1995). It can be seen that for both the 18-h and

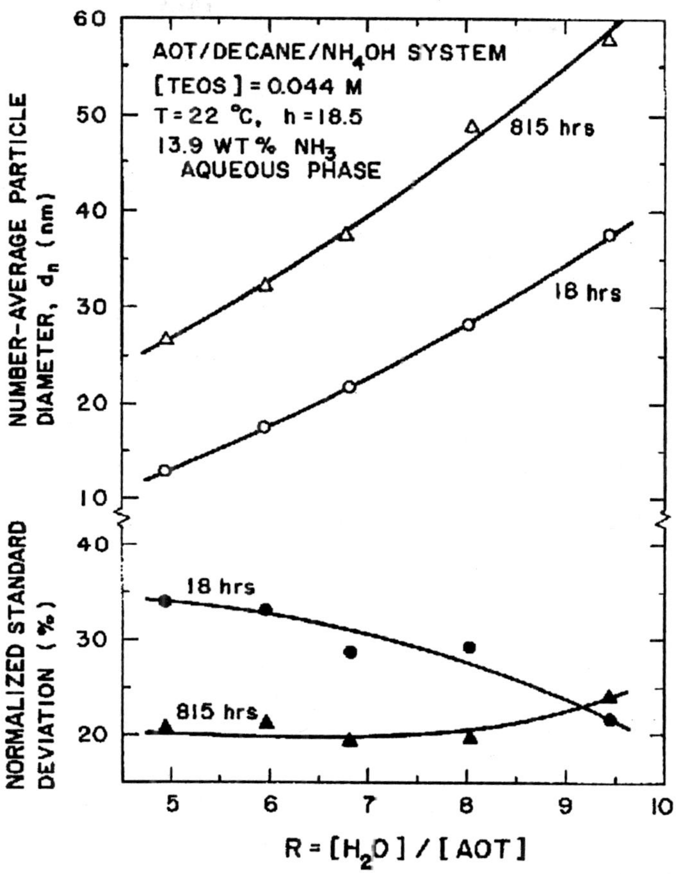

FIGURE 5 Effect of the water-to-surfactant molar ratio (R) on the size and size distribution of SiO_2 particles: h = water-to-TEOS molar ratio (Arriagada and Osseo-Asare 1995)

815-h samples, particle size increased monotonically with R. Analysis of the aggregate statistics for the AOT/decane/water/ammonia system (Table 3) indicates that the AOT reverse micellar population (N_m) decreases by about one-fourth as R increases from 5.0 to 9.5. At the same time the average number of solubilized TEOS molecules per reverse micelle (N_t) increases from about 8 to 24 (Arriagada and Osseo-Asare 1995). A plot of the average 18-h particle size as a function of the average TEOS occupancy number (N_t) gives a linear relationship (Arriagada and Osseo-Asare 1995). The implication is that particle nucleation is primarily an intramicellar process and that the water core plays the role of a nanoreactor. That is, the surfactant film is sufficiently protective, so that once a nucleus is formed in a given reverse micelle, further growth is initially determined largely by the number of monomers available to the confined water core. This behavior is consistent with the relatively low intermicellar exchange rates characteristic of AOT microemulsions (Luisi and Straub 1984; Friberg and Bothorel 1987; Kumar and Mittal 1999).

TABLE 3 Aggregate statistics in the AOT/decane/water/ammonia system (Arriagada and Osseo-Asare 1995)

[AOT](M)	R	$2r_w$ (nm)	N_m	N_t	d_n (nm)
0.16	5.0	2.5	9.0×10^{18}	8.2	13.1
0.14	6.0	2.9	6.1×10^{18}	11.1	17.6
0.12	6.8	3.2	4.5×10^{18}	13.9	21.6
0.10	8.0	3.5	3.3×10^{18}	17.4	28.4
0.09	9.5	4.0	2.2×10^{18}	23.8	37.5

Further examination of Table 3 reveals however, that the 18-h particle sizes (d_n) are about 5–10 times larger than the pool sizes ($2r_w$). It is reasonable to suppose that additional particle growth beyond the pool size must rely on mass exchange between water pools; this requires intermicellar collision and transient merging of the colliding reverse micelles.

The dependence of the rate of intermicellar exchange (k_{ex}) on temperature and microemulsion parameters such as composition and nature of components is known (Luisi and Straub 1984; Friberg and Bothorel 1987; Kumar and Mittal 1999). For example, in the above AOT microemulsion system, benzyl alcohol (BA) acts as a cosurfactant at the water/oil interface, thereby increasing intermicellar exchange rates (Arriagada and Osseo-Asare 1994). When BA was introduced into the AOT/decane/water/ammonia system, it was found that the silica particle size increased and polydispersity decreased as the BA/AOT ratio increased. This result may be rationalized as follows (Arriagada and Osseo-Asare 1994): (a) aggregation of the nuclei initially formed occurs, resulting in fewer growing particles, a process favored by rapid intermicellar exchange, or (b) nucleation of silica is inhibited in fast-exchanging systems, possibly as the result of rapid rearrangement of hydrolyzed TEOS species. The result is a decrease in the number of nuclei formed and, therefore, an increase in particle size (in the presence of a large supply of monomeric growth units).

REFERENCES

Almgren, M., J. van Stam, S. Swarup, and J. E. Lofroth. 1986. Structure and transport in the microemulsion phase of the system Triton X-100-toluene-water. *Langmuir.* 2:432.

Arriagada, F. J., and K. Osseo-Asare. 1992. Phase and dispersion stability effects in the synthesis of nanosize silica in a nonionic reverse microemulsion. *Colloids and Surfaces.* 69:105.

Arriagada, F. J., and K. Osseo-Asare. 1994. Silica nanoparticles produced in aerosol OT reverse microemulsion: effect of benzyl alcohol on particle size and polydispersity. *Journal of Dispersion Science and Technology.* 15:59.

Arriagada, F. J., and K. Osseo-Asare. 1995. Synthesis of nanosize silica in aerosol OT reverse Microemulsion. *Journal of Colloid and Interface Science.* 170:8.

Arriagada, F. J., and K. Osseo-Asare. 1999a. Controlled hydrolysis of tetraethoxysilane in a nonionic water-in-oil microemulsion: a statistical model of silica nucleation. *Colloids and Surfaces.* 154:311.

Arriagada. F. J., and K. Osseo-Asare. 1999b. Synthesis of nanosize silica in a nonionic water-in-oil microemulsion: effects of the water/surfactant molar ratio and ammonia concentration. *Journal of Colloid and Interface Science.* 211:210.

Burda, C., X. Chen, R. Narayanan, and M. A. El-Sayed. 2005. Chemistry and properties of nanocrystals of different shapes. *Chemical Review*. 105:1025.

Friberg, S. E., and P. Bothorel. 1987. *Microemulsions: Structure and Dynamics*. Florida:CRC Press.

Kumar, P., and K. L. Mittal. 1999. *Handbook of Microemulsion Science and Technology*. New York:Marcel Dekker.

Kunieda, H., and K. Shinoda. 1979. Solution behavior of aerosol ot/water/oil system. *Journal of Colloid and Interface Science*.70:577.

Luisi, P. L., and B. E. Straub. 1984. *Reverse Micelles*. New York:Plenum.

Osseo-Asare, K., and F. J. Arriagada. 1990a. Synthesis of nanosize particles in reverse microemulsions. In *Ceramic Powder Science III*. Edited by G. L. Messing, S. Hirano, and H. Hausner. Westerville, OH:ACS. 3–16.

Osseo-Asare, K., and F. J. Arriagada. 1990b. Preparation of SiO_2 nanoparticles in a non-ionic reverse micellar system. *Colloids and Surfaces*. 50:321.

Osseo-Asare, K. 1999. Microemulsion-mediated synthesis of nanosize oxide materials. In *Handbook of Microemulsion Science and Technology*. Edited by P. Kumar and K. L. Mittal. New York:Marcel Dekker. 549–603.

Osseo-Asare, K., and F. J. Arriagada. 1999. Growth kinetics and nanosize silica in a nonionic water-in-oil microemulsion: a reverse micellar pseudophase reaction model. *Journal of Colloid and Interface Science*. 218:68.

Osseo-Asare, K., and F. J. Arriagada. 2000. Hydrolysis of silicon alkoxides in microemulsions. In *Fine Particles: Synthesis, Characterization, and Mechanisms of Growth*. Edited by T. Sugimoto. New York:Marcel Dekker. 147–188.

Pillai, V., P. Kumar, M. J. Hou, P. Ayyub, and D. O. Shah. 1995. Preparation of nanoparticles of silver halides, superconductors and magnetic materials using water-in-oil microemulsions as nano-reactors. *Advanced Colloid Interface Science*. 55:241.

Puig, J. E., I. Rodriguez-Siordia, L. I. Rangel-Zumudio, J. F. Billman, and E. W. Kaler. 1988. Properties and structures of three- and four-component microemulsions. *Langmuir*. 4:806.

Salutsky, M. L., and H. H. Willard. 1959. *Precipitation from Homogeneous Solution*. New York: John Wiley & Sons.

Schmidt, H. 2001. Nanoparticles by chemical synthesis, processing to materials and innovative application. *Applied Organometallurgical Chemistry*. 15:331.

Measurement and Prediction of Electrochemical Removal of Copper and Subsequent Sub-Micron Particle Removal Associated with Chemical Mechanical Planarization in the Presence of Surfactants

Michael L. Free*

Chemical mechanical planarization (cmp) is a critical process in the manufacture of integrated circuits. It requires the use of nano-scale mineral particles in an oxidizing solution environment. The particles are maintained in suspension by careful control of solution conditions and through the addition of surfactant, which also reduces surface tension and reduces the force needed to transfer wafers from polishing pads to cleaning stations. This paper discusses the effects of surfactant on the chemical etching process during cmp processing and the particle removal process that follows. The results show that measured etching and particle removal data can be reasonably predicted as a function of surfactant concentration, surfactant chain length, and dissolved salt concentrations using the equations presented herein.

INTRODUCTION

The production of integrated circuits is a very challenging process that requires a large number of steps and incredibly high precision. This process generally utilizes silicon wafers as the starting material onto which multiple integrated circuits are produced as depicted in Figure 1. The sets of devices and associated wiring inside each integrated circuit are coated during several stages of manufacturing. Because the coatings follow

* Department of Metallurgical Engineering, University of Utah, Salt Lake City, Utah

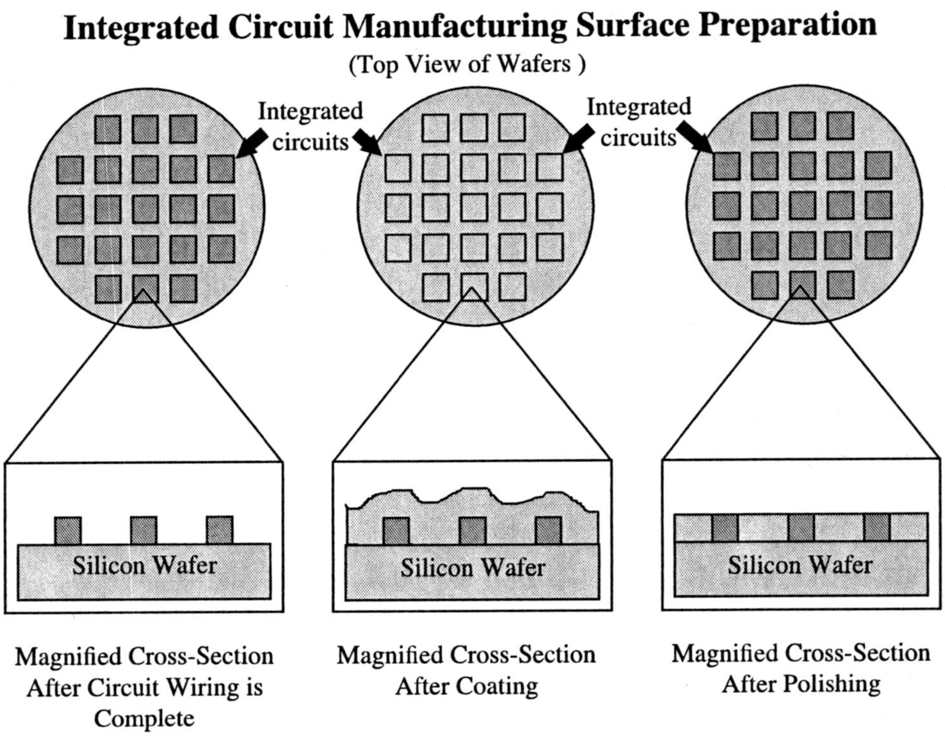

FIGURE 1 Schematic diagram illustrating the relationship between integrated circuit manufacturing coatings and the need for planarization

underlying topography as shown in Figure 1, excess portions of the coating are removed during a step called chemical mechanical planarization (cmp). The cmp process utilizes a polishing slurry that contains chemicals that etch the excess coating and polishing particles that abrade excess coating. The combination of chemical and mechanical action produces the desired planar surface. The need for proper control of the chemical etching and particle removal are clearly essential to the process that must result in precision at the 100 nanometer scale. The post-cmp rinsing/cleaning process must leave a surface that is nearly free of residual polishing particles. Creating the needed precision in material removal and subsequent particle removal can be improved through the use of surfactants. The importance of particle removal and material removal control during the cmp process is evident from published literature [Du, et al., 2004; Saxena, et al., 2004; Ein-Eli, et al., 2004; Meyers, et al., 1995; Martinez, 1994; Patrick, et al., 1991; Iqbal, et al., 1994; Hayashi, et al., 1995; Kaufman, et al., 1991; Itano, et al., 1993; Verhaverbeke, et al., 1995; Fusstetter, et al., 1995; Riley and Carbonell, 1993; Miyashita, et al., 2000; Babel and Mackay, 2000; Roy, et al., 1995; Free and Shah, 1999a; Free and Shah, 1999b].

EXPERIMENTAL PROCEDURES

The copper corrosion evaluations in this study were performed using an EG&G 273 potentiostat (operated using PowerCORR software made by Princeton Applied Research) and an AFASR rotator (Pine Instruments). The steel (AISI 1018) corrosion tests were performed using a Gamry Instruments P4 potentiostat with DC105 software. All corrosion tests were performed at 32°C with a Teflon-sealed and epoxy precoated (on unexposed sides) metal disk that was prepared by fine grinding using 600 grit abrasive followed by polishing with 1 μm gamma alumina polishing powder (Buehler) followed by ultrasonic cleaning in water. The metal disc was rotated at 1000 RPM. Corrosion rates were determined using a traditional polarization resistance approach with anodic Tafel slopes of 0.12 V/decade for copper and 0.04 V/decade for steel, and a cathodic Tafel slope of -0.12 V/decade for both copper and steel.

Particle removal evaluations involving alumina particles on quartz substrates utilized, optical-grade, single-crystal, quartz windows (approximately 2.5 cm in diameter) obtained from Harrick Scientific Co. and high purity alumina powder (0.7 μm average diameter, Sumitomo).

Particle deposition was performed by immersing vertically the quartz window into a suspension of water and alumina particles (0.06 wt %) at pH 4.4 for 20 seconds followed by rinsing for 15 seconds under a stream of deionized and distilled water (~2 m/s water velocity perpendicular to the surface).

All data for residual particle density involving copper substrates were obtained using copper blanks of 99.99 % purity (Kennecott Utah Copper Corporation) that were polished on Texmet® 1000 (Beuhler) polishing pads in a slurry composed of the specified solution and 3 wt. % alumina particles (0.3 μm–Beuhler Micropolish). All solutions were made using Type I water (18 MΩ·cm) and reagent grade chemicals. Copper substrates were prepared by polishing with 600 grit silicon carbide paper followed by polishing with 1 and 0.05 μm alumina particles (Beuhler Micropolish) that were subsequently removed by ultrasonic cleaning using a Bransonic ultrasonic cleaner, manual brushing with a polymer soft bristle brush in water, and water rinsing. All polishing related to data acquisition for this study was performed manually at approximately 30°C for 30 seconds.

Residual particle density measurements were made using a Zeiss Axioskop microscope, a Sony high-resolution color video camera, and Snappy image acquisition hardware and software. Images were analyzed to determine the number density of residual particles using Uthisca Image Tool version 1.28 software.

RESULTS AND DISCUSSION

Chemical Etching of Copper

The chemical etching of copper during chemical mechanical planarization is often performed using ferric ions or hydrogen peroxide as the added oxidant and oxygen as a secondary oxidant that is naturally present, but usually unaccounted for. Surfactant, which is added to cmp slurries to stabilize the suspension of particles and reduce surface tension inhibits etching by the oxidant. Figure 2 shows three sets of data that involve different levels of ferric nitrate and surfactant. The data in Figure 2 illustrate the increasing

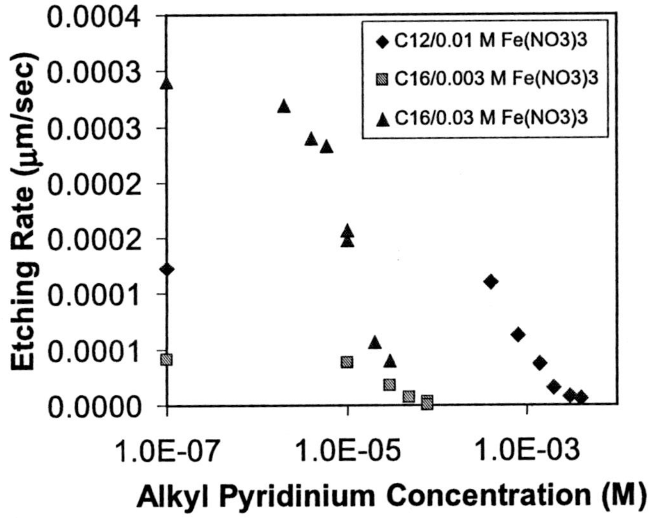

FIGURE 2 Comparison of copper etching rates and surfactant concentration. Rotating copper disk tests performed in aqueous media with alkyl trimethyl ammonium bromide compounds with the number of CH_2 units in the hydrocarbon chain length noted in the legend along with the concentration of ferric nitrate.

etching rate associated with increasing the ferric concentration as well as the decreasing etching rate that results from increasing surfactant concentration.

The chemical etching associated with the cmp process can be described using a combination of solution thermodynamics, mass transport, and electrochemical kinetics. The change in free energy for the reaction aA ↔ bB can be expressed by:

$$\Delta G_r = \Delta G_r^o + RT\ln\left(\frac{\gamma_B^b m_B^b}{\gamma_A^a m_A^a}\right) \qquad \text{(EQ 1)}$$

where ΔG_r is the change in free energy associated with the reaction, ΔG_r^o is the standard free energy change for the reaction, R is the gas constant, T is the absolute temperature, γ_A is the activity coefficient of "A," and m_A is the dimensionless concentration that is equivalent to the molality of "A." Equilibrium is achieved when the change in free energy for a given reaction is zero. Consequently, for the reaction of aA ↔ bB, equilibrium is established when $\Delta G_r = 0$. Therefore, at equilibrium:

$$\Delta G_r^o = -RT\ln\left(\frac{\gamma_B^b m_B^b}{\gamma_A^a m_A^a}\right) = -RT\ln K \qquad \text{(EQ 2)}$$

where K is the equilibrium constant for the reaction. The involvement of common ions in multiple reactions necessitates the solution of a set of free energy based equations to determine the equilibrium speciation for the system. The ionic activity coefficients in most practical solutions can be calculated using the Davies equation, which is expressed as [Butler, 1998]:

$$-\log(\gamma) = 0.5083z^2\left(\frac{\sqrt{I}}{1+\sqrt{I}} - 0.2I\right) \quad \textbf{(EQ 3)}$$

where z is the charge of the ion and I is the ionic strength, which is expressed as:

$$I = \frac{1}{2}\sum_{j=1}^{n} z^2 m_j \quad \textbf{(EQ 4)}$$

The use of the Davies equation to determine the activity coefficient values needed in the free energy expressions for the system of reactions requires the equilibrium molalities of all of the species in the system. Because the molalities are not known initially, an iterative approach is used to solve for the correct molalities based upon the combination of equilibrium free energy and ion activity coefficient equations.

The electrochemical aspects of thermodynamics are usually described using the Nernst Equation. One form of the Nernst Equation for a typical reaction of $M^{2+} + 2e^- = M$ is expressed as:

$$E = E^o - \frac{RT}{nF}\ln\left(\frac{\gamma_{M^{2+}} m_{M^{2+}}}{\gamma_M m_M}\right) \quad \textbf{(EQ 5)}$$

Note that because the activity, which is expressed by multiplying the activity coefficient and the molality, is equal to one for the solid metal, the expression shown can be further simplified.

The rate at which electrons are exchanged during electrochemical reactions is can be described mathematically using a form of the Butler-Volmer equation, which can be expressed as [Newman and Thomas-Alyea, 2004]:

$$i = i_o\left[\exp\left(\frac{\alpha_a F\eta}{RT}\right) - \exp\left(\frac{-\alpha_c F\eta}{RT}\right)\right] \quad \textbf{(EQ 6)}$$

in which i is the current density, i_o is the equilibrium exchange current density, α_a is the anodic charge transfer coefficient, F is the Faraday constant, η is the surface overpotential or the difference between the surface potential and the equilibrium reaction potential, R is the gas constant, T is the absolute temperature, and α_c is the cathodic charge transfer coefficient. A more useful form of the Butler-Volmer equation in which surface and bulk solution concentrations are shown is:

$$i = k'C_b\left(\frac{C_s}{C_b}\right)^{\lambda}\left[\exp\left(\frac{\alpha_a F\eta}{RT}\right) - \exp\left(\frac{\alpha_c F\eta}{RT}\right)\right] \quad \textbf{(EQ 7)}$$

where k' is constant that is directly related to the equilibrium exchange current density, C_b is the bulk concentration, C_s is the surface concentration, and λ is a factor that depends upon reaction mechanisms and related factors and it is usually between 0.25 and 1. The value of k' is determined by the equation:

$$k' = k_f nF\left[\exp\left(\frac{\alpha F E^o}{RT}\right)\right] \quad \textbf{(EQ 8)}$$

where k_f is the rate constant, n is the number of electrons in the half-cell reaction, α is the charge transfer coefficient, and E^o is the equilibrium potential for the half-cell reaction.

Each half-cell reaction in the system has a unique Butler-Volmer expression that describes its kinetics. In order for an electrochemical reaction to occur, the transfer of electrons and charge must balance. In other words for reactions occurring on a common surface the sum of current densities must be zero, which can be expressed mathematically as:

$$\sum i = 0 \tag{EQ 9}$$

Because there are nearly always more than two half-cell reactions occurring simultaneously, they must all be solved simultaneously to achieve a balance of current.

The value of the surface concentration is often determined by mass transport limitations. Under steady state conditions with uniform geometries such as semi-infinite plates, the maximum rate at which ions can be transported to a surface to react under steady-state conditions can be approximated using Fick's first law of diffusion, which can be rearranged for the rotating disk geometry to yield:

$$C_s = C_b - \frac{1.61\omega^{-1/2}\nu^{1/6}D^{1/3}i}{nFD} \tag{EQ 10}$$

where n is the number of moles, t is time, ω is the angular velocity, and v is the kinematic viscosity, and D is diffusivity.

The effect of surfactant inhibitor molecules can be described mathematically below the surface aggregation concentration (sac), which is the concentration at which surfactant molecules aggregate into micelle-like structures at the surface, using the equation [Free, et al., 2004]:

$$i_{etch} = \frac{i_{baseline}}{\left(1 + \dfrac{K'C}{\exp\left(\frac{1}{RT}[(L-x)\Delta G_{c.l.} + k(L-x)RT\ln(\gamma)]\right)}\right)} \tag{EQ 11}$$

where i_{etch} is the copper etching rate (current density), $i_{baseline}$ is the etching rate (current density) in the absence of surfactant, L is the total hydrocarbon chain length in terms of CH_2 units, x is the minimum number of CH_2 units needed for surface activity (generally close to 5), $\Delta G_{c.l.}$ is the free energy change associated with each CH_2 unit, and k is a solvent interaction factor, C is the surfactant concentration, and K' is adsorption equilibrium constant. Above the sac, but not more than an order of magnitude above the sac, a modified expression can be used:

$$i_{etch} = i_{baseine}\left(1 - \frac{K'}{1+K'} - \frac{qC}{\exp\left(\frac{1}{RT}[(L-x)\Delta G_{c.l.} + k(L-x)RT\ln(\gamma)]\right)}\right) \tag{EQ 12}$$

where q is a proportionality constant determined from experimental data.

Using the preceding formulas, solved simultaneously, the effects of ferric nitrate concentration, oxygen partial pressure, acidity, and surfactant concentration can be determined. Figures 3–6 show the various effects of parameters on copper etching. Note that in each figure various effects are only significant within a range of values. For example, oxygen transport is rate controlling when the concentration of ferric ions is low. At

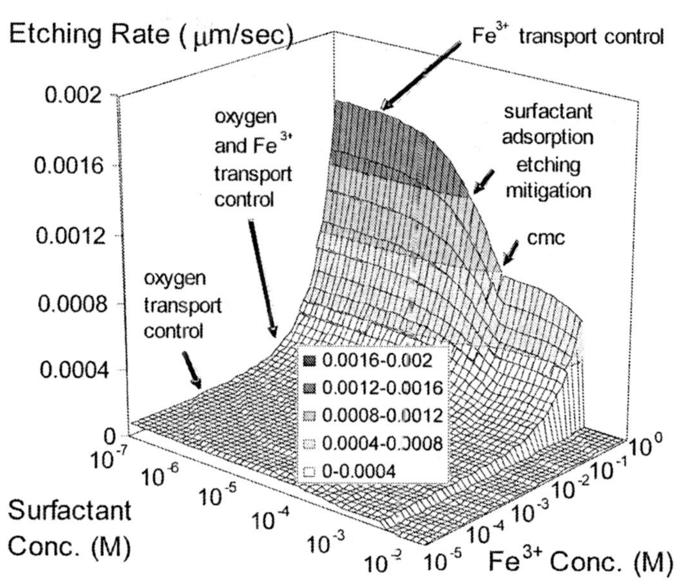

FIGURE 3 Copper etching rate in acidic aqueous media (0.0001 M HCl) as a function of surfactant concentration (cetyl trimethyl ammonium bromide) and ferric ion (added as $FeCl_3$) concentration as determined using METER software (University of Utah). The constants used in the model included: $\alpha_c = 0.5$, $\alpha_a = 0.5$, $k_f = 0.2$ for $Fe^{3+} + e^- = Fe$, ; $\alpha_c = 0.5$, $\alpha_a = 0.5$, $k_f = 0.2$ for $2H^+ + 2e^- = H_2$; $\alpha_c = 1.5$, $\alpha_a = 0.5$, $k_f = 0.072$ for $Cu^{2+} + 2e^- = Cu$; $\alpha_c = 1.5$, $\alpha_a = 0.5$, $k_f = 0.4$ for $4H^+ + 4e^- + O_2 = 2H_2O$; $\alpha_c = 1.5$, $\alpha_a = 0.5$, $k_f = 0.00000015$ for $Fe^{2+} + 2e^- = Fe$; $\Delta G_{c.l.} = -1800$ J/mole, $k = 0.89$, $K' = 1$, $x = 5$, $q = 0.005$, $v = 0.01$ cm²/sec, $\lambda = 1$, $\omega = 104.7$ sec⁻¹, diffusivities of ions were assumed to be 1×10^{-5} cm²/sec. Standard thermodynamic data for species were used.

moderate ferric ion concentrations, the rate of etching is controlled by ferric ion mass transport. At high surfactant concentrations, the rate of etching is significantly reduced in all the tests evaluated. The rate of etching is reduced most rapidly as the concentration of surfactant approaches the critical micelle concentration as shown in Figures 3–6. Acidity does not dramatically alter the rate of etching as expected based on the standard potential of copper (0.34 V) relative to that of the hydrogen ion/hydrogen gas standard potential (0 V). However, the addition of surfactant to acidic solutions reduces the rate of etching at lower surfactant concentrations as the acid concentration increases. An overall comparison of etching rate values for measured data and data predicted using the equations previously discussed is presented in Figure 7.

Enhanced Particle Removal

The removal of particles from surfaces following the cmp process is determined by the relative ratio of the removal force, which is usually fluid flow combined with megasonic vibrations, and the adhesion force, which is determined by the interaction forces between the particles and the surface. The interaction forces between the particles and the surface have been altered in this study using surfactants, whereas the removal forces

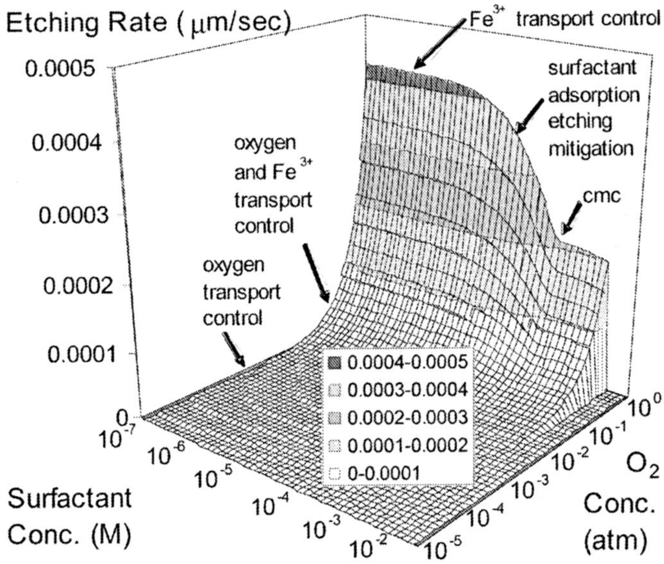

FIGURE 4 Copper etching rate as determined using METER software (University of Utah) for acidic aqueous media (0.0001 M HCl) as a function of surfactant concentration (cetyl trimethyl ammonium bromide) and oxygen partial pressure with 0.001 M $FeCl_3$ using the same parameters given in the caption for Figure 3

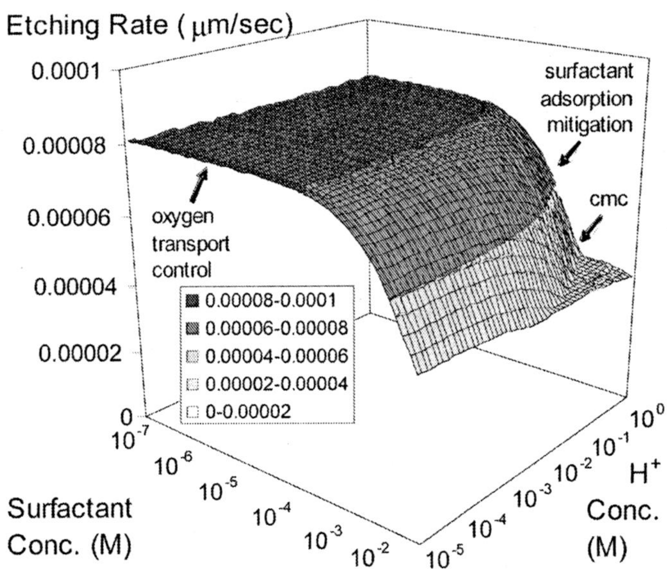

FIGURE 5 Copper etching rate as determined using METER software (University of Utah) as a function of surfactant concentration (dodecyl trimethyl ammonium bromide) and acidity with 0.001 M $FeCl_3$ using the same parameters given in the caption for Figure 3

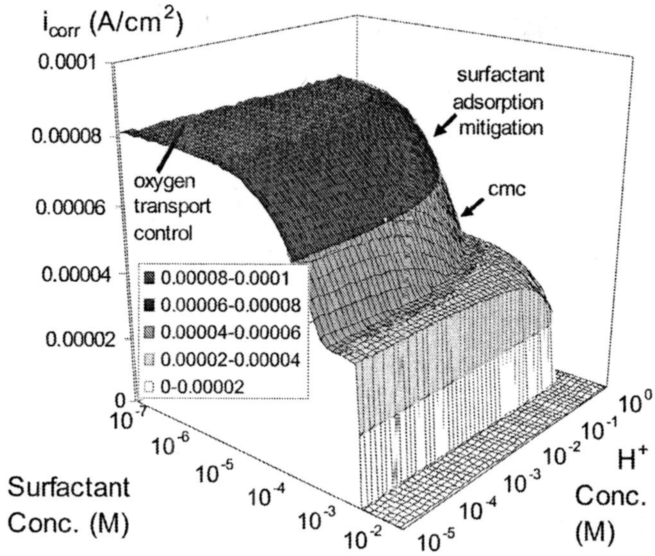

FIGURE 6 Copper etching rate as determined using METER software (University of Utah) as a function of surfactant concentration (cetyl trimethyl ammonium bromide) and acidity with 0.001 M $FeCl_3$ using the same parameters given in the caption for Figure 3

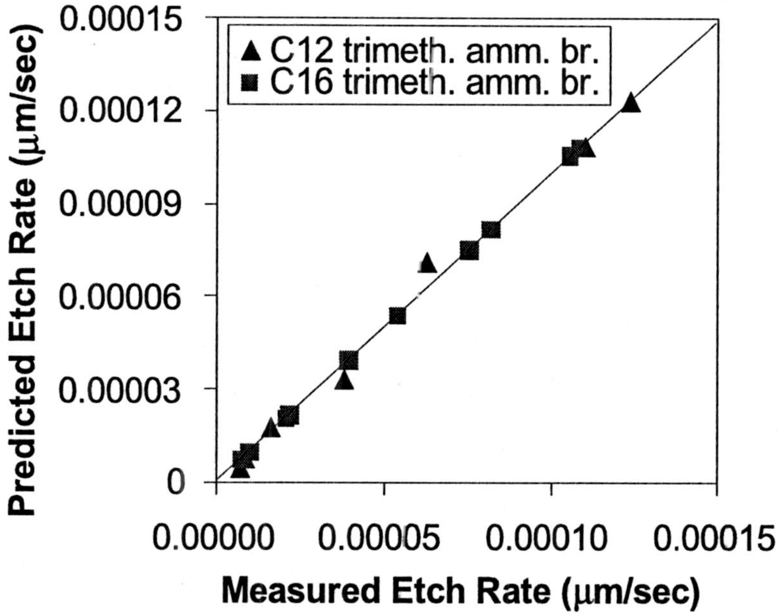

FIGURE 7 Comparison of most of the measured copper etching rate values previously presented in Figure 2 to predicted values presented in Figures 3–6

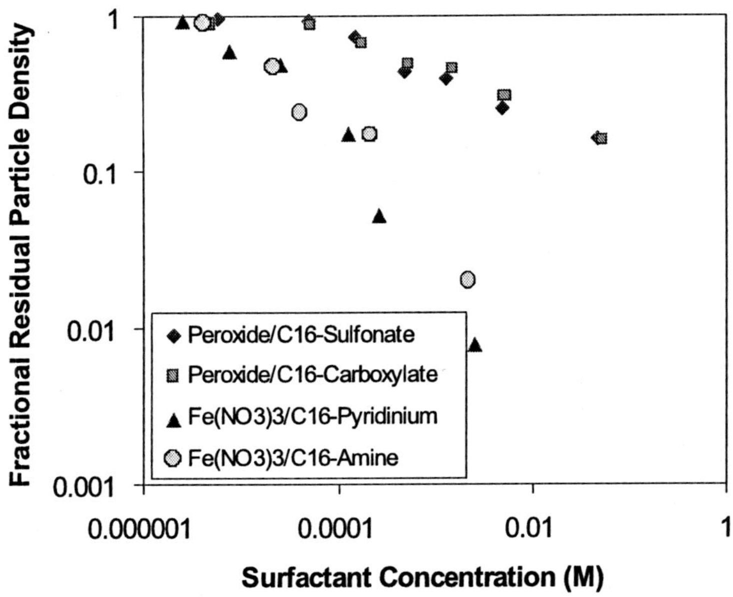

FIGURE 8 Comparison of residual particle fraction and surfactant concentration in ferric nitrate (1 %) and peroxide (0.1 % sodium peroxide and 1 % ammonium hydroxide) media as noted in the legend. The legend specification of C16 indicates 16 CH_2 hydrocarbon units in the respective surfactant.

have been maintained at a constant level. A comparison between the residual particle fraction and surfactant concentration is presented in Figure 8. As surfactant adsorbs on the surface and creates two or more layers of surfactant at the surface, the adsorbed surfactant creates a hydrophilic, steric barrier that reduces adhesion between particles and substrates that facilitates particle removal [Free, 2005]. Consequently, as the surfactant concentration increases above the level needed for the hydrophilic film, the removal of particles improves as shown in Figure 8. The basis for modeling the effect of surfactant in enhanced particle removal is described in subsequent paragraphs.

The sum of the interaction forces between particles and surfaces includes contributions made by specific forces that are often described for aqueous systems as van der Waals, electrostatic, and in some cases structural and hydrophobic forces [Horn, 1990]. The sum of the van der Waals and electrostatic contributions are often referred to as DLVO forces which were described and evaluated by Derjaguin, Landau, Verwey, and Overbeek. Their work and use of surface force theory has become widely known as the DLVO theory [Horn, 1990]. The additional structural and hydrophobic forces are not as well understood [Horn, 1990; Vinogradova, 1998; Yoon and Ravishankar, 1994; Ducker, et al., 1994; Yaminsky, et al., 1996; Ducker, et al., 1999], although the structural forces are attributed to water structuring effects near the surface and hydrophobic forces are also attributed to water structuring and/or gas entrainment/cavitation effects that are associated with hydrophobic surfaces [Yaminsky, et al., 1996; Ducker, et al., 1999].

Van der Waals force between a spherical particle and a flat substrate can be expressed as [Horn, 1990]:

$$F(h) = -\frac{Ar}{6h^2} \quad \textbf{(EQ 13)}$$

where A is the Hamaker constant, r is the radius of curvature of the particle, and h is the separation distance between the particle and the plate.

The electrostatic force between a sphere and a plate is given as [Horn, 1990]:

$$F(h) = \frac{128\pi R\rho_{\infty}kT\gamma_1\gamma_2}{\chi}\exp(-\chi h) \quad \textbf{(EQ 14)}$$

where the surface charge for each surface is defined as [Horn, 1990]:

$$\gamma = \tanh\left(\frac{ze\Psi_o}{4kT}\right) \quad \textbf{(EQ 15)}$$

and the Debye length, χ, is given as [Horn, 1990]:

$$\chi = \left(\frac{2000e^2IN_A}{\varepsilon_o\varepsilon_m kT}\right)^{1/2} \quad \textbf{(EQ 16)}$$

in which z is the ion valence, e is the electrical charge, ρ_{∞} is the bulk dissolved ion density, Ψ_o is the surface potential for the particle or substrate surface of interest. I is the ionic strength ($0.5\sum z^2\rho_{\infty}$ for all positive and negative ions), N_A is Avogadro's number, ε_o is the dielectric permittivity of vacuum, and ε_m is the dielectric constant of the medium between the two surfaces.

The aforementioned equations reasonably describe the adhesion of hard particles to hard substrates such as mineral particles and metals but not for soft media such as plastics. Surfactant reduces adhesion primarily by altering surface charge and by creating a steric barrier that effectively increases the separation distance between particles and surfaces, thereby reducing the effect of the strongly attractive van der Waals force.

The removal of particles using a constant flow of rinsing water is related to the quantity of adsorbed surfactant as well as the surface hydrophilicity. A useful equation that describes the relationship between residual particle density and surfactant concentration is [Free, 2005]:

$$P = P_0\exp\left[w\left(\frac{K\frac{C}{sac}}{1+K\frac{C}{sac}}\right)^n\right] \quad \textbf{(EQ 17)}$$

where P is the residual particle density, P_o is the baseline residual particle density in the absence of surfactant, K is the surfactant adsorption constant, w is a proportionality constant, n is a constant, and sac is the surface aggregation concentration, which can be determined from the equation [Free, et al., 2004]:

$$sac \cong \exp\left(\frac{1}{RT}[(L-x)\Delta G_{c.l.} + k(L-x)RT\ln(\gamma)]\right) \quad \textbf{(EQ 18)}$$

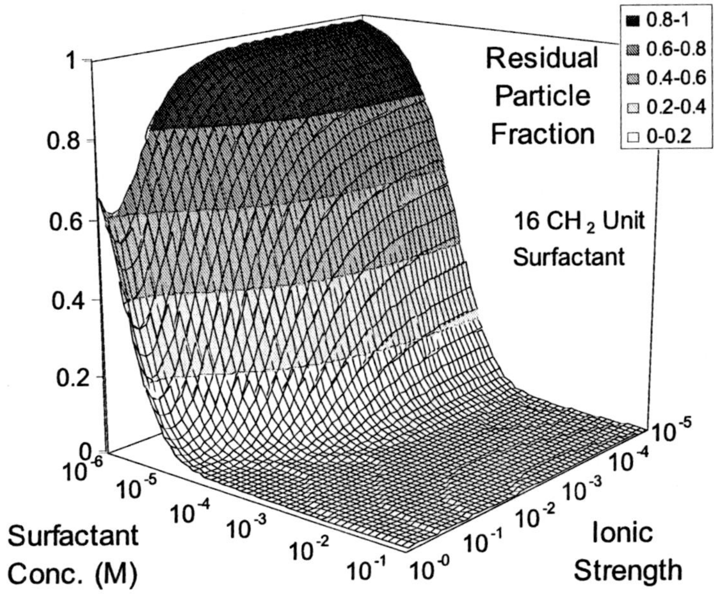

FIGURE 9 Residual particle fraction as a function of surfactant concentration (cetyl trimethyl ammonium bromide) and ionic strength. The constants used in Equations 17 and 18 included: $\Delta G_{c.l.}$ = –1650 J/mole, k = 0.89, K = 0.15, x = 5, w = –5, n = 1, z = 11688. The ion activity coefficient was determined using Equation 3 assuming a monovalent ion.

where R is the gas constant, T is temperature, L is the total hydrocarbon chain length in terms of CH_2 units, x is the minimum number of hydrocarbon chain units needed for surface activity, ΔG_{cl} is the free energy change per CH_2 unit, k is the solvent polarity factor, and γ is the characteristic ionic activity coefficient for a monovalent ion.

Equations 17 and 18 allow for the prediction of the residual particle fraction as a function of surfactant concentration and ionic strength as shown in Figure 9. The data in Figure 9 show that as the surfactant concentration increases fewer particles are predicted to remain after the rinsing process. The predictions also indicate that increasing ionic strength lowers the concentration at which the surfactant enhances particle removal. A comparison of the predicted residual particle fraction, which is shown in Figure 8, and the measured residual particle fraction, which is presented in Figure 7, is shown in Figure 10. The correlation between the measured and predicted residual particle fraction is reasonable as evidenced by correlation coefficients that are greater than 0.93 for each data set.

CONCLUSIONS

Chemical etching rates during the cmp process in the presence of surfactant can be reasonably predicted using a combination of thermodynamic equilibria, electrochemical kinetics, and mass transport equations. The agreement between theoretical and measured copper etching rates resulted in a correlation coefficient greater than 0.995 over

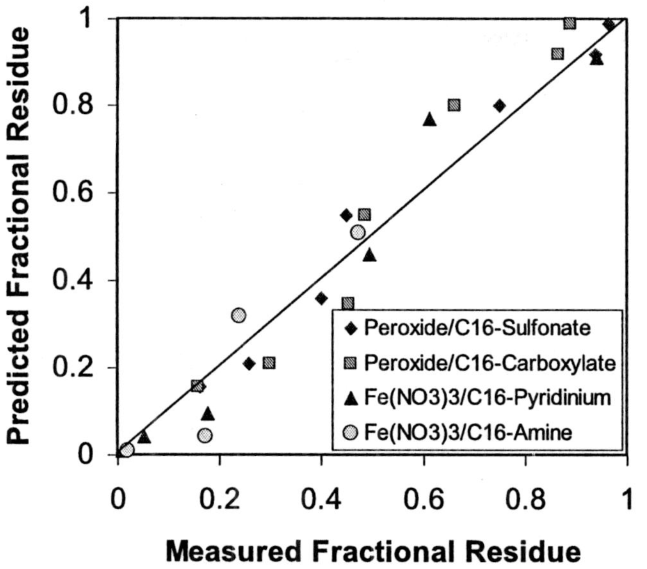

FIGURE 10 Comparison of measured data presented in Figure 8 and predicted data presented in Figure 9

the range of conditions evaluated. Particle removal effects were also predicted effectively using a combination of adsorption and removal equations. The comparison of predicted and measured particle removal revealed correlation coefficients for each data set that exceeded 0.93 over the range of conditions evaluated.

ACKNOWLEDGMENTS

Financial support for the development of the understanding of surfactant adsorption was provided by the National Science Foundation (DMR-9983945) for a related project. Laboratory assistance provided by Eric Riddle and Justin Fuller is also gratefully acknowledged.

REFERENCES

Babel, A. K., and R. A. Mackay, 2000. Mater. Res. Soc. Symp. Proc., 566:135.

Butler, J. N., 1998. "Ionic Equilibrium: Solubility and pH Calculations," New York, NY: John Wiley & Sons, Inc., p. 49.

Du, Tianbao, Ying Luo, Vimal Desai, 2004. Microelectronic Engineering, v 71, n 1, p 90–97.

Ducker, W. A., Z. Xu, and J. N. Israelachvili, 1994. Langmuir, 10, 3279.

Ducker, W. A., Z. Xu, D. R. Clarke, and J. N. Israelachvili, 1999. J. Am. Ceramic Soc., 77:437.

Ein-Eli, Y., E. Abelev, D. Starosvetsky, 2004. Electrochimica Acta, v 49, n 9–10, p 1499–1503.

Free, M. L., and D. O. Shah, 1999a. Micro, 16(5):29.

Free, M. L., and D. O. Shah, 1999b. in *Particles on Surfaces 5 & 6: Detection, Adhesion, and Removal*, K. L. Mittal, (Ed.), p. 95, VSP, Utrecht.

Free, M. L., W. Wang, D. Y. Ryu, 2004. Corrosion, 60(9), pp. 837–844.

Free, M. L., 2005, invited paper in International Journal of Particulate Science and Technology.
Fusstetter, H., A. Schnegg, D. Graf, H. Kirschner, M. Brohl, and P. Wagner, 1995. Mater. Res. Soc. Symp. Proc., 386:97.
Hayashi, Y., M. Sakurai, T. Nakajima, and K. Hayashi, 1995. Jpn. J. Appl. Phys. 34:1037.
Horn, R. G., 1990. J. Am. Ceram. Soc., 73:1117.
Iqbal, A., S. R. Roy, G. B. Shinn, S. Raghavan, R. Shah, and S. Peterman, 1994. Microcontamination, No. 10, 45.
Itano, M., F. W. Kern Jr., M. Miyashita, and T. Ohmi, 1993. IEEE Trans. Semiconductor Manuf., 6:258.
Kaufman, F. B., D. B. Thompson, R. E. Boroadie, M. A. Jaso, W. L. Guthrie, D. J. Pearson, and M. B. Small, 1991. J. Electrochem. Soc., 138:3460.
Martinez, M. A., 1994. Solid State Technol., No. 5, 26.
Meyers, T. L., M. A. Fury, and W. C. Krusell, 1995. Solid State Technol., No. 10, 59.
Miyashita, N., Y. Mase, J. Takayasu, Y. Minami, M. Kodera, M. Abe, and T. Izumi, 2000. Mater. Res. Soc. Symp. Proc., 566:253.
Newman, J., K. E. Thomas-Alyea, 2004. "Electrochemical Systems," 3rd Edition, (New York, NY: John Wiley & Sons, Inc., p. 7.
Patrick, W. J., W. L. Guthrie, C. L. Standley, and P. M. Schiable, 1991. J. Electrochem. Soc., 138, 1778.
Riley, D. J., and R. G. Carbonell, 1993. J. Colloid Interface Sci. 158:259.
Roy, S. R., I. Ali, G. Shinn, N. Furusawa, R. Shah, S. Peterman, K. Witt, S. Eastman, and P. Kumar, 1995. J. Electrochem. Soc., 142:216.
Saxena, Ravi, Dipto G. Thakurta, Ronald J. Gutmann, and William N. Gill, 2004. Thin Solid Films, v 449, n 1–2, p 192–206.
Verhaverbeke, S., R. Messoussi, H. Morinaga, and T. Ohmi, 1995. Mater. Res. Soc. Symp. Proc., 386:3.
Vinogradova, O. I., 1998. Langmuir, 14:2827.
Yaminsky, V. V., B. W. Ninham, H. K. Christenson, and R. M. Pashley, 1996. Langmuir, 12:1936.
Yoon R-H., and S. A. Ravishankar, 1994. J. Colloid Interface Sci., 166:215.

Atomic Force Microscopy Investigation of Interaction Forces Between Polyethylene and Asphaltene Surfaces in Nonaqueous Systems

Robert Aranowski,* Jakub Nalaskowski,† Jan Hupka,* and Jan D. Miller†

Many petroleum-derived asphalts used for road construction exhibit poor rheological properties. These properties of asphalt can be improved by addition of various polymer fillers to the asphalt blend. Recycled polyolefines, especially polyethylene and polypropylene have been used as the asphalt modifying fillers. Unfortunately the polyolefine-asphalt mixtures are not stable and have a tendency to separate because of the chemical incompatibility of the components. In our research, surfactants were used to improve polymer-asphalt compatibility. Interaction forces between a polyethylene particle and a polyethylene surface and between a polyethylene particle and an asphaltene-surface were measured in surfactant solutions in n-tetradecane and pyridine using the atomic force microscopy colloidal probe technique. Dodecylamine, lauric acid and tetraethoxylated lauric alcohol ($C_{12}E_4$) were used as surfactants. The most pronounced change in the interactions was observed for polyethylene-polyethylene surfaces in the solution of lauric acid. At the same conditions polyethylene-asphaltene interactions were decreased. The most pronounced increase was the increase in the repulsion range (to 26 nm) between the spherical PE probe and flat polyethylene surface for 10^{-3} M n-tetradecane solution of ethoxylated lauryl alcohol ($C_{12}E_4$). In the case of lauric acid the interaction range extended to 24 nm and for dodecylamine to 21 nm. The interaction between the polyethylene sphere and flat asphaltene surface extended to 16 nm for dodecylamine, to 18 nm for lauryl alcohol ethoxylated ($C_{12}E_4$) and to 14 nm for lauric acid.

* Gdansk University of Technology, Department of Chemical Technology, Gdansk, Poland

† University of Utah, Department of Metallurgical Engineering, Salt Lake City, Utah

INTRODUCTION

A wide spectrum of modifying polymers has been used for asphalt fillers. In order for a polymer to be effective, it must blend with the bitumen and improve the flow resistance at higher temperature without making the bitumen too viscous at the mixing temperature or too brittle at low temperature. If used in road asphalt, the polymer should be capable of being processed by conventional equipment. In any case it must be easily available and inexpensive. Recycled polyolefines satisfy all the mentioned concerns. The addition of dispersed polyolefines to the asphalt blends helps to retard the low-temperature thermal stress cracking as well as rutting of the pavement at elevated temperatures. Other benefits include: greater fatigue resistance, reduced temperature susceptibility, increased tire traction and greater adhesion to the aggregate. Polymono-olefine polymers (e.g., polyethylene) do not contain double bounds and therefore have good thermal and aging resistance [1]. The major obstacles to widespread application of polyolefines for asphalt modification is low stability of the polymer dispersion in asphalt at elevated temperature, which results in separation of the polymer phase from the asphalt blend [2, 3]. The understanding of interaction forces existing between the polyolefine particles in the asphalt matrix and between the polyolefine particles and asphalt itself may help in further modification of these forces in order to improve the stability of the polymer-asphalt blends.

THEORETICAL CONSIDERATIONS

Hydrocarbons and other apolar solvents have low dielectric constants, and a very small surface-charge density is usually developed at the surface in apolar solution. Thus the thickness of Debye's layer is very high. These facts account for a relatively weak electric double layer interaction force. For this reason, the van der Waals attractions in pure solvents cause coagulation. To prevent this phenomenon from occurring, surfactants can be used for steric stabilization. The van der Waals attractions are reduced on account of steric interaction between adsorbed polymer and surfactants chains [4, 5].

The relationship for attractive potentials, originating from the London-van der Waals dispersion forces for two equally sized spheres interacting in vacuum as derived by Hamaker, is as follow [4, 6]:

$$\Delta G_S^{att.} = -\frac{A_{11}}{6}\left[\frac{2}{s^2-4}+\frac{2}{s^2}+\ln\frac{s^2-4}{s^2}\right] \quad \textbf{(EQ 1)}$$

where:

$s = a/R$

a = the distance between the particle centers

R = the particle radius

A_{11} = the Hamaker constant for the particles.

For small interparticle distances compared to the particle diameters, this relationship reduces to [4, 7]:

$$\Delta G_S^{att.} = -A_{11}\frac{R}{12h} \quad \text{(EQ 2)}$$

where:

h = the surface to surface distance between the particles

The above relationships are only valid for two spheres interacting in vacuum. In dispersion medium, which attenuates the interaction, an effective Hamaker constant should be used:

$$A^* = A_{11} + A_{22} - A_{12} \approx \left(A_{11}^{\frac{1}{2}} - A_{22}^{\frac{1}{2}}\right)^2 \quad \text{(EQ 3)}$$

where:

A_{11} = the Hamaker constant for the particles,

A_{22} = the Hamaker constant for the solvent,

A_{12} = the Hamaker constant for particle-solvent interaction

The repulsive interaction energy can be calculated from following equation:

$$V_R = 2\pi\varepsilon R\xi^2 \ln[1 + \exp(-\kappa h)] \quad \text{(EQ 4)}$$

where:

ε = the medium permittivity

ξ = zeta potential of the emulsion

κ = the Debye-Hückel parameter

The total interacting energy is:

$$V_T = V_A + V_R = -A^*\frac{a}{12h} + 2\pi\varepsilon a\xi^2 \ln[1 + \exp(-\kappa h)] \quad \text{(EQ 5)}$$

The classical DLVO theory cannot predict exactly the stability of the systems where the molecular structure and heterogeneity of the medium in which the particles are dispersed must be considered [8, 9, 10]. This is the case of the asphalt dispersions where asphalt matrix has a complicated structure.

The repulsive forces in sterically stabilized colloidal systems are often divided into those with enthalpic origin and those origin with an entropic or volume restriction origin. The domination of the enthalpic or entropic force determines parameters like: the interparticle distance in relation to the steric layer thickness, the particle size, the concentration of steric stabilizer and the interaction between the stabilizer and the continuous phase.

Fischer [11] first derived equation (6) for the interaction free energy due to enthalpic or mixing effects when interparticle distances are less then two times the steric layer thickness. This relation gives the enthalpic or mixing contribution to the interaction energy as function of the interpartcle distance.

$$\Delta G_S^{mix} = \frac{4\pi}{3V_1}\varphi_2^2\left(\frac{1}{2}-\chi_1\right)\left(L-\frac{h}{2}\right)^2\left(6R+2L+\frac{h}{2}\right) \quad \text{(EQ 6)}$$

where:

V_1 = the molar volume of the continuous phase

φ_2 = the polymer volume fraction in the steric layer

χ_1 = the Flory-Huggins polymer-solvent interaction parameter

L = the steric layer thickness

R = the particle radius

h = the surface-to-surface interparticle distance

The elastic or entropic contribution to the steric stabilization mechanism for two spherical particles covered with polymer or surfactants tails is obtained from equation (7), as reported in the literature [12].

$$\Delta G_S^{el} = 4L\nu\pi RKT\left(\frac{2}{3}-\frac{\delta_o}{2}-\frac{\delta_o^3}{6}+\delta_o \ln\delta_o\right) \quad \text{(EQ 7)}$$

where:

ν = the total number of chains per unit area

$\delta_o = h/L$

L = the steric layer thickness

With these equations we can estimate the total interaction free energy in an elastically stabilized system:

$$\Delta G_S^{tot} = \Delta G_S^{att} + \Delta G_S^{el} \quad \text{(EQ 8)}$$

It is also possible to obtain a better understanding of how the surface coverage (stabilizer concentration) and steric layer thickness influence the colloidal stability.

EXPERIMENTAL

Materials

The polyethylene (PE) powder was a low-density PE (Scientific Polymer Product, Inc.) with a molecular weight (MW) of 1,800 and a melting point (mp) of 117°C. The spherical particles of PE were prepared using the method described by Nalaskowski et al. [13]. Powdered polymer was mixed with glycerol to form a 0.1% wt suspension. The suspension was agitated using an Ultra-Turrax type homogenizer (Silverson L4R) at 2,000 rpm and heated above the melting temperature of the suspended material, i.e., 130°C for PE (mp 117°C). The temperature and stirring were maintained for about 10 min and then the homogenizer's head was removed and the suspension was left to cool for about 15 min until the temperature reached 50°C. The suspension was then diluted about ten times with cold water and left for another 5 min. The spherical particles were filtered from the suspension using 0.2 μm pore size filter unit with a cellulose acetate membrane (Sigma Chemical Co.). The filter with separated particles was then rinsed with water several times to remove any residual traces of glycerol. Finally, the filter unit was dried at room temperature in a dust-free environment. The membrane with spherical particles was removed from the filtration unit. The spherical particles were eventually dried again in an oven at 75°C for 24 h in nitrogen at ambient pressure and subsequently stored in an evacuated (66.5 Pa) desiccator for 1 h.

The spherical particles of asphaltene were prepared in similar way. Because of the very high melting point (above 280°C) of asphaltenes and decomposition of polycondensed aromatic rings at high temperature, melting of asphaltene in glycerol (boiling point about 200°C) is impossible. Therefore, asphaltene powder was dissolved in toluene to form 30% wt. mixture and next mixed with glycerol to form a 0.1% wt suspension. The suspension was agitated at 2,000 rpm and heated above 150°C. The temperature and stirring were maintained for about 15 min and then the homogenizer's head was removed and the suspension was cooled to 50°C. The suspension was then diluted about ten times with cold water and the particles were filtered from the suspension using 0.2 μm pore size filter. The filter with separated particles was then rinsed with water several times to remove any residual traces of glycerol. Finally, the asphaltene particles were dried at 130°C temperature in an evacuated (60 Pa) desiccator to a constant weight. The asphaltene particles had regular spherical shape and roughness was measured using atomic force microscope (see Figure 1).

The asphaltene was extracted from an asphalt sample (D50 Rafineria Gdanska Ltd.). Small samples of asphalt (about 10 g) were dissolved in toluene, centrifuged and the mixture was filtered for removal of mineral precipitates. After that, n-heptane was added, the asphaltene was filtered from the suspension, and then rinsed 3 times by n-heptane. Finally, the filter unit was dried at elevated temperature (40°C) and stored in an evacuated desiccator.

For preparing a flat surface of asphaltene the spin-coating method was used. Small amounts of asphaltene (0.2 g) were dissolved in 3 cm^3 volume of toluene and several drops of the mixture were deposited at the surface of a glass disk rotating at 300 rpm. After evaporation of toluene, the disk with the film of asphaltene was dried in a desiccator to a constant weight.

The disc of PE was prepared by melting powdered PE in a vacuum oven. The melt was casted onto a freshly cleaved mica surface. After cooling the PE disc was separated from mica under water. The obtained PE disc had very smooth surface as was determined by AFM imaging (see Figure 2).

Methods

Topography features of the prepared asphaltene disc and PE spherical particles were measured using a Veeco Nanoscope III E AFM. Measurements were done in the contact mode (in air) by means of triangular oxygen-sharpened silicon nitride cantilevers (spring constant 0.12 N/m). Topographic images were obtained from the "height"

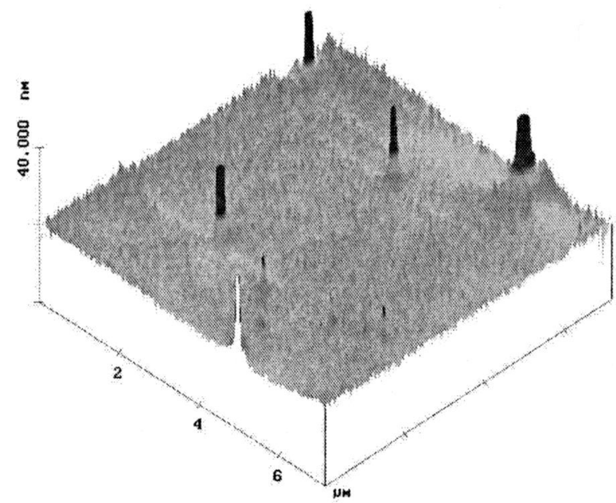

FIGURE 1 Surface of asphaltene sphere prepared using the technique described in materials section

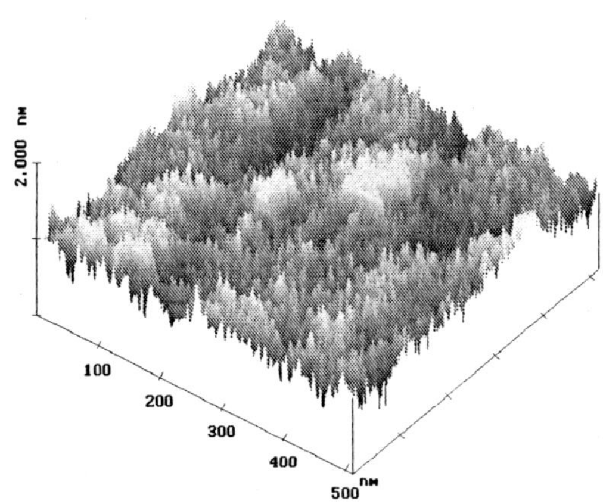

FIGURE 2 Surface of the PE disc as measured by the AFM method

channel with relatively high values of integral and proportional gain. Images were planarized using a third-order planarization algorithm. Roughness measurements were performed using the Veeco Nanoscope III E software for 2.5 μm^2 areas of the sample.

Force measurements for the different systems were done using silicone nitride tipless cantilevers obtained from Veeco. A triangular cantilever (205 μm width, 195 μm length, 36 μm thickness) with a nominal spring constant 0.12 N/m were used. Before measurements, the spring constant was measured by Cleveland's method [14]. A small sphere of PE or asphaltene was glued to a cantilever using a two-component epoxy glue. The radii of the spheres were measured from the SEM photos [15].

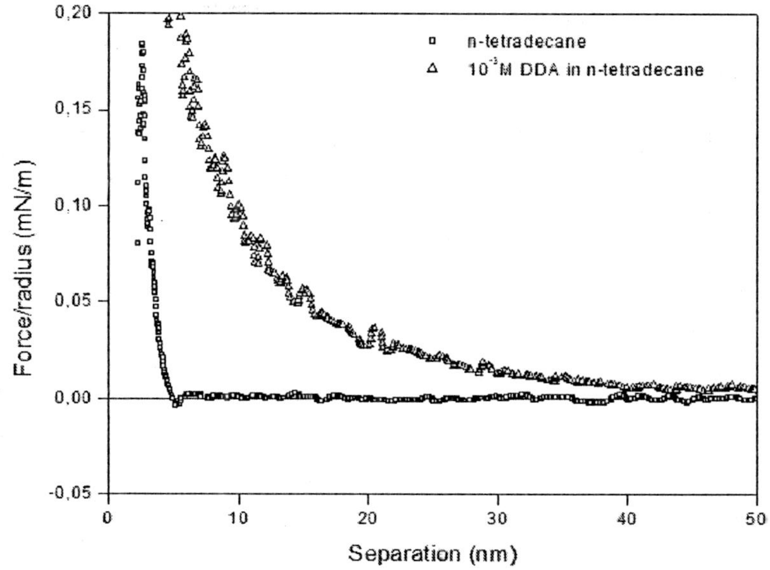

FIGURE 3 Force of interaction between polyethylene particle and smooth polyethylene surface in n-tetradecane and 10^{-3} M dodecyloamine solution in n-tetradecane

RESULTS AND DISCUSSION

Interactions between polyethylene particles and the smooth polyethylene surface in n-tetradecane and dodecylamine solution at 23°C are shown in Figure 3. It can be seen from the curve for the polyethylene-polyethylene interaction in n-tetradecane a small minimum at a distance of 5 nm. In such a case the polyethylene particles may flocculate or coalesce. In the case of the dodecylamine solution the attraction force between polyethylene surfaces is not observed and the repulsive force has a range almost ten times greater than in pure n-tetradecane.

In the case of polyoxyethylene-4-lauryl ether and lauric acid solution in n-tetradecane we have observed an increase of the repulsive interaction (Figure 4). In both cases the force range increases about six times when compared to pure n-tetradecane. On the curve for the lauric acid solution a step at the 5–10 nm range can be seen which is most probably due to the interpenetration of absorbed surfactant chains.

The interaction between polyethylene surfaces in polar (pyridine) and apolar (n-tetradecane) solvents has also been studied (see Figure 5). In polar solution (pyridine) the interaction range is about six times greater than in n-tetradecane. The local short-range attractive minimum, which was visible in n-tetradecane case, was not detected. These differences can be explained by swelling of the polyethylene chains as well as a stronger electric double layer contribution in pyridine. In the case of polyethylene interactions in pyridine/surfactant solutions the increase in the repulsive force is not as evident as it was in the case of n-tetradecane solutions.

The steric stabilization in pyridine by surfactants is less effective than in n-tetradecane. The largest steric layer thickness was in 10^{-3}M polyoxyethylene-4-lauryl ether solution and

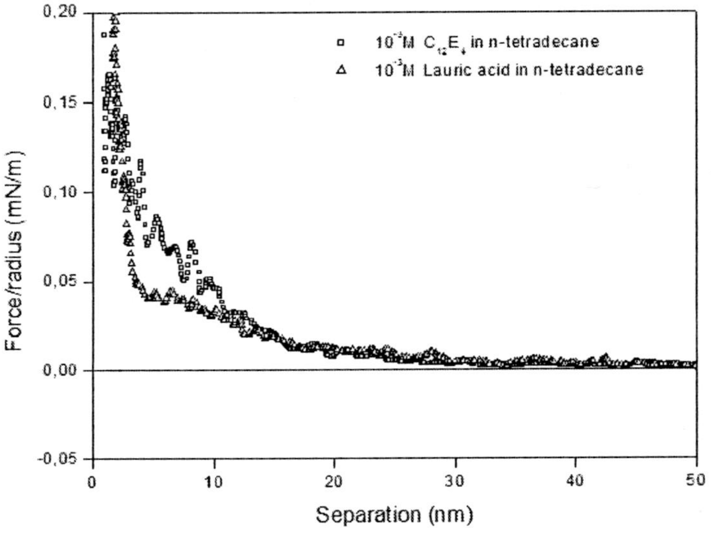

FIGURE 4 Force of interaction between a polyethylene particle and a smooth polyethylene surface in 10^{-3} M polyoxyethylene-4-lauryl ether and 10^{-3} M lauric acid solution in n-tetradecane

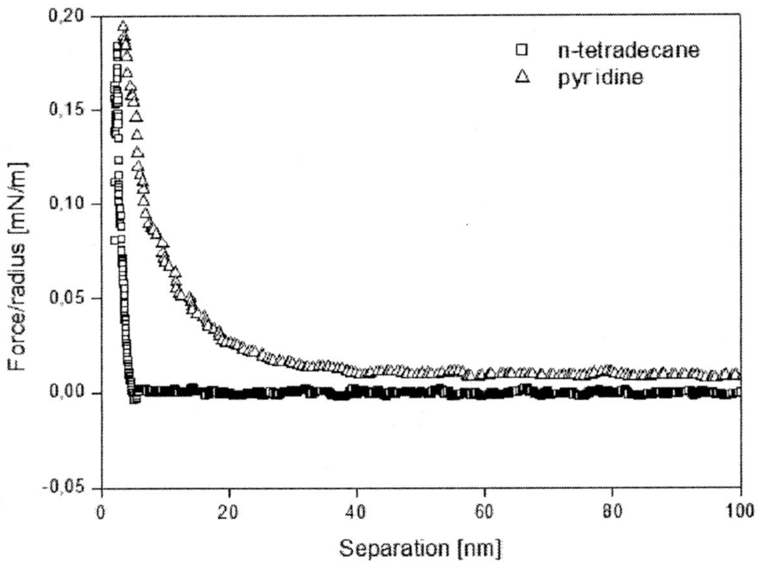

FIGURE 5 Force of interaction between a polyethylene particle and a polyethylene surface in n-tetradecane and pyridine

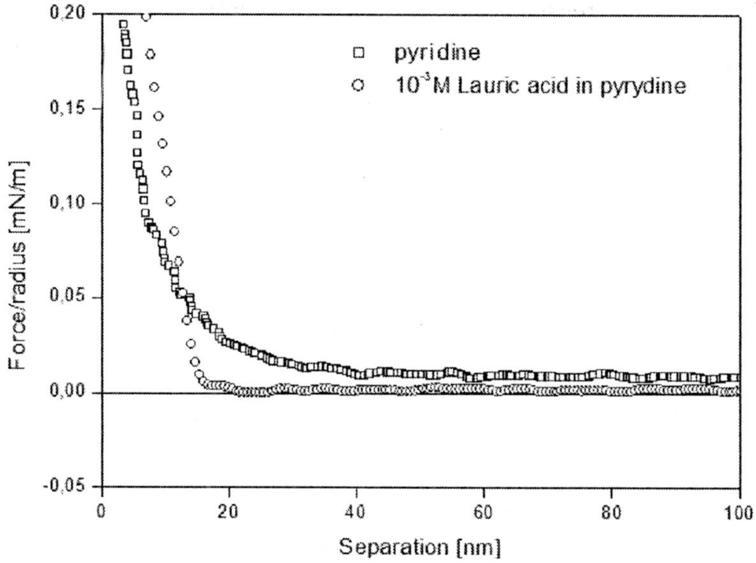

FIGURE 6 Force of interaction between a polyethylene particle and polyethylene surface in pyridine and 10^{-3}M lauric acid in pyridine solution

almost the same range for dodecyloamine and lauric acid (about 20 nm). The repulsive interaction in pure pyridine increases gradually but in the case of dodecylamine and lauric acid (in spite of a smaller range) increases suddenly (Figure 6 and Figure 7).

The interaction force between polyethylene particles and an asphaltene surface are presented on Figure 8 and Figure 9 in various solutions. In all cases the interaction is repulsive. Because of very complicated structure of asphaltene molecules, that have many polar groups, the adsorption of the surfactants at the asphaltene surface is difficult to describe.

The repulsive force for these systems was found to increase with increasing temperature. At high temperature the adsorbed surfactant chains can move more easily, so the steric layer thickness increases (Figure 10 and Figure 11). Additional swelling of the polymer or aphaltene is also possible.

CONCLUSIONS

Surfactant molecules in nonaqueous systems can significantly increase the repulsion between PE filler particles. Despite the small size of surfactant molecules the repulsion can be sufficient to provide for an increase in the stability of PE filler suspended in asphalt blends. At the same time the interaction between PE and asphaltene surfaces was found to be much less affected by the presence of surfactants.

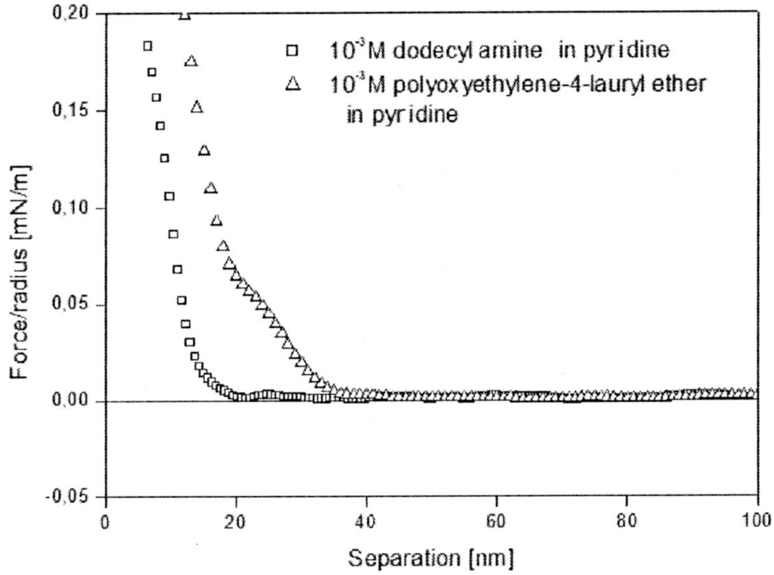

FIGURE 7 **Force of interaction between a polyethylene particle and a polyethylene surface in 10^{-3}M dodecylamine and 10^{-3}M polyoxyethylene-4-lauryl ether in pyridine solution**

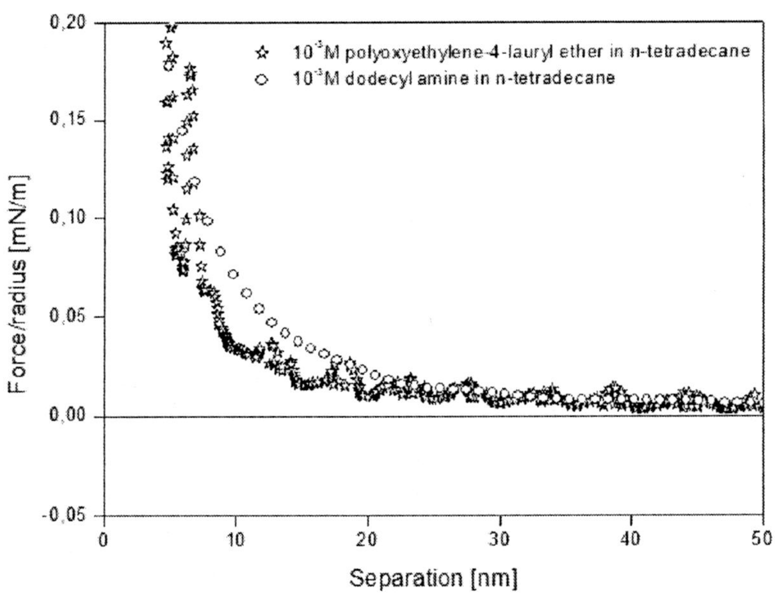

FIGURE 8 **Force of interaction between a polyethylene particle and a flat surface of asphaltene in 10^{-3} M dodecylamine and 10^{-3}M polyoxyethylene-4-lauryl ether in n-tetradecane solution**

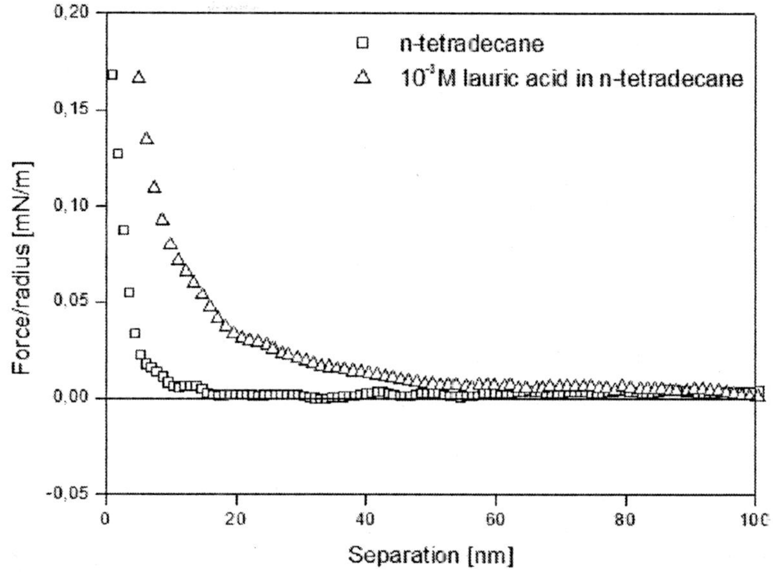

FIGURE 9 Force of interaction between a polyethylene particle and an asphaltene surface in n-tetradecane and 10^{-3} M lauric acid n-tetradecane solution

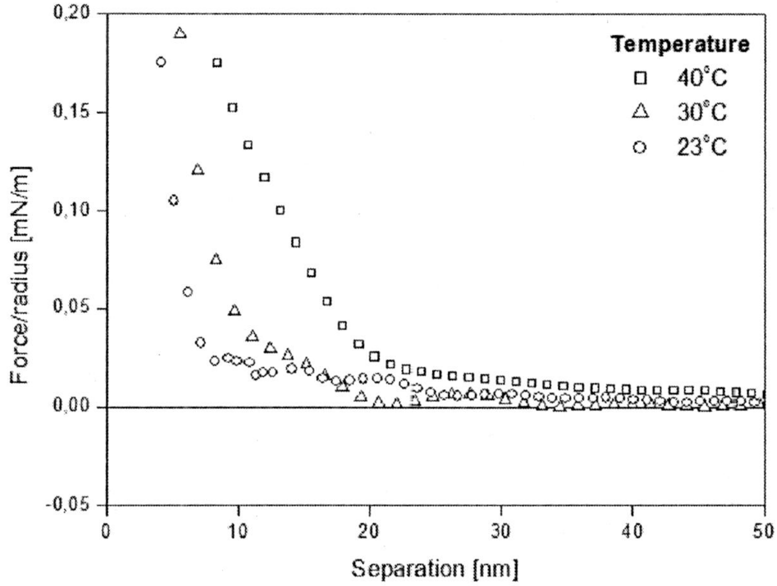

FIGURE 10 Force of interaction between a polyethylene particle and a polyethylene surface in 10^{-3} M lauric acid in n-tetradecane solution at different temperatures

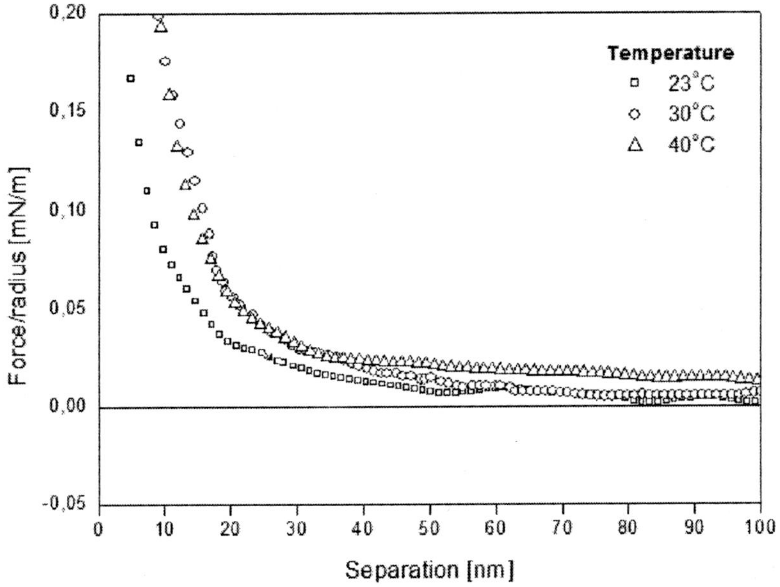

FIGURE 11 Force of interaction between an asphaltene particle and an asphaltene surface in n-tetradecane at different temperatures

REFERENCES

[1] C. Giavarini, P. De Filippis, M.L. Santarelli, M. Scarsella, Fuel, 6 (75), (1996), 681.
[2] S.A.M. Hesp, R.T. Woodhames, Colloid Polym Sci., 269, (1991), 825.
[3] T. Zeiler, S. Kellermann, H. Münstedt, J. Adhesion Sci. Technol., 14 (5), (2000), 619.
[4] G.R. Morrison, M. Hedmark, S.A.M. Hesp, Colloid Polymer Sci., 272, (1994), 375.
[5] Z. Liang, S.A.M. Hesp, Colloids and Surfaces A: Physicochemical and Engineering Aspects, 81, (1993), 239.
[6] S.R. Raghavan, J. Hou, G.L. Baker, S.A. Khan, Langmuir, 16, (2000), 1066.
[7] X. Wu, T. Dabros, J. Czarnecki, Langmuir, 15, (1999), 8706.
[8] J. Israelachvili, Intermolecular and Surface Forces, Academic Press, London, 1992.
[9] H. Yotsumoto, R.H. Yoon, J. Colloid Interface Sci., 157, (1993), 426.
[10] B.P. Binks, W-G. Cho, P.D.I. Fletcher, D.N. Petsev, Langmuir, 16, (2000), 1025.
[11] Stanley I. Sandler, Johann Fischer, Frank Reschke, Fluid Phase Equilibria, 45, (1989), 251.
[12] R. Evans, J.B. Smitham, D.H. Napper, Colloid Polym Sci., 255, (1977), 161.
[13] J. Nalaskowski, J. Drelich, J. Hupka, J.D. Miller, J. Adhesion Sci. Technol., 13 (1), (1999), 1.
[14] J.P. Cleveland, S. Manne, Rev. Sci. Instrum., 64 (2), (1993), 403.
[15] M. Salou, B. Siffert, A. Jada, Colloids and Surfaces A: Physicochemical and Engineering Aspects, 142, (1998), 9.

Biomimetic Formation of Nano-crystalline Calcium-Phosphate Films on Self-Assembled Monolayers of Alkanethiols: Effects of End Functionality and Deposition Methodology

Kevin C. Baker* and Jaroslaw Drelich*

The influence of chemical functionality on the coating morphology and substrate coverage of biomimetic calcium phosphate was investigated by the use of self-assembled monolayers of alkanethiols. Monolayers possessing carboxyl, hydroxyl, and methyl terminal functional groups or their mixture were constructed on gold films that were sputter coated on either glass slides or polystyrene sections. Self-assembled monolayers were immersed in a supersaturated calcium phosphate solution for three days with daily solution refreshments to grow calcium phosphate coatings on their surfaces. The resulting calcium phosphate coatings were characterized by scanning electron microscopy to compare coating morphology and substrate coverage. Energy dispersive x-ray spectrometry was used to determine the calcium-to-phosphorus ratio. The experiments revealed that both surface functionality and deposition methodology affect deposition, growth and morphology of biomimetic calcium phosphate porous coatings.

INTRODUCTION

Calcium phosphate (CaP) coatings have been employed in several orthopedic applications to provide an environment conducive to bone growth at the surface of implants.[1-10] Plasma sprayed hydroxyapatite (HA) is the most common form of bioactive coating used to enhance hard tissue integration with orthopedic implants. Since the plasma spray

---

* Department of Materials Science and Engineering, Michigan Technological University, Houghton, Michigan

deposition method is a line of sight process, coatings produced on implants with complex geometries (screws, etc.) are often non-uniform in terms of substrate coverage and coating thickness.[1,2] The plasma spray deposition process also entails the use of high temperatures which can lead to heterogeneous coating properties and can be damaging to polymers.[2] To solve these problems, research has focused on biomimetic processes, which have simple chemical immersion techniques as their basis. In a biomimetic deposition process, the substrate of appropriate surface characteristic is immersed in either a simulated body fluid or solutions saturated or saturated with calcium and phosphate ions at room or physiological temperature. Calcium phosphate films grow on the substrate surface through heterogeneous nucleation process.[2] A biomimetic process that produces uniform, homogenous coatings on metallic surfaces has also been developed at Michigan Tech and is described elsewhere.[1]

Coating-substrate adhesion is a major problem facing the use of biomimetic CaP films with polymeric materials. Chemical etching, flame treatment, ultraviolet (UV) irradiation, corona treatment, and plasma treating are all methods currently employed in various non-medical industries to improve the adhesion of polymers with other materials. Over the years, similar treatments of polymers have been tested in laboratories for biomedical application. Many reports also suggest that surface polar groups act as nucleation sites for formation of CaP structures and increase the rate of CaP deposition.[2,3,5,6] For example, exposing the substrate to UV radiation prior to immersion in a simulated body fluid solution has been shown to enhance substrate-apatite adhesion properties.[5] The exposure of a substrate to UV radiation increased the density of polar groups at the substrate surface, which interact favorably with CaP nuclei.[5] Alkali pretreatments and glow-discharge (GD) processing of substrates prior to biomimetic formation of CaP coatings also increased deposition rate and substrate-coating adhesion through a similar mechanism.[3,6] It is interesting to note, however, that GD processing was the only method that reduced apatite formation induction time and enhanced substrate adhesion on polymers such as polyethylene terephthalate (PET), poly(methyl methacrylate) (PMMA), polyamide 6 (PA6), polyethersulfone (PESF), polyethylene (PE) and polytetrafluoroethylene (PTFE), while other treatment methods were only effective for certain polymers.[3,5,6] It is therefore, rational to assume that the surface chemistry plays a key role in the formation and growth of biomimetic CaP films. Understanding the effect of surface chemistry on nucleation, growth, and adhesion of CaP films on polymers could stimulate further progress in the development of orthopedic devices with improved biocompatibility.

In this study, self-assembled monolayers (SAMs) of alkanethiols possessing differing surface functionalities were prepared on gold-coated glass and polystyrene substrates. The monolayers served as model organic substrates in the biomimetic deposition of CaP coatings from supersaturated calcium phosphate solutions. Methyl (CH_3) terminated SAMs were meant to serve as a plausible model of a saturated hydrocarbon surface, such as polyethylene (common biomaterial). SAMs with oxygen-based functionality, either homogeneous (COOH, OH) or heterogeneous (OH+COOH, CH_3+COOH, OH+CH_3, OH+COOH+CH_3), were selected to simulate organic polymer surfaces that had been modified by a generic oxidation pretreatment. Tanahashi and Matsuda also used SAMs in studying the formation of apatite structures from a simulated body fluid.[11] They found that the growth rate of apatite coatings on SAMs with PO_4H_2 and COOH functional groups was substantially higher than for SAMs with $CONH_2$, OH, and NH_2 groups. There was practically no growing of the apatite structures on SAMs with methyl as the end functionality. In other studies on the formation of HA on Langmuir-Blodgett monolayers, Sato et al. reached a similar conclusion.[12] The authors found nucleation and growth of HA on monolayers with carboxyl groups and practically no nucleation on the monolayers with amino groups.

In contrast to previous studies, we investigate the influence of chemical functionality on the CaP coating morphology and substrate coverage when the biomimetic deposition process is carried out using supersaturated calcium phosphate solutions (SCPS) instead of simulated body fluids. Our previous research showed that the deposition rate for CaP coatings is accelerated from supersaturated solutions.[1] In this paper, we report that both substrate surface chemistry and immersion technique influence the formation and morphology of growing CaP coatings.

EXPERIMENTAL PROCEDURE

Substrates

High impact polystyrene (Dow Chemical, PS) tensile bars were sectioned into 17 × 17 mm squares using a silicon carbide cutting wheel and smoothed using a surgical scalpel. Glass slides (Eerie Scientific, GS) were cut into 37 × 25 mm rectangles using a diamond scalpel. Both PS and GS samples were subjected to ultrasonic cleaning in a diluted surfactant solution (Micro 90, International Products Inc.) for five minutes. The samples were then rinsed with de-ionized water and ultrasonically cleaned in methanol (histological grade, Fisher Scientific) for an additional five minutes. After ultrasonic cleaning, the GS samples were rinsed once more with methanol and acetone, placed in a covered Pyrex dish and allowed to dry in an L-C oven for at least 1 hour at 100°C. PS samples were rinsed with methanol following the ultrasonic cleaning and dried in the L-C oven at 70°C for 15–20 minutes.

Both sides of the GS and PS samples were sputter-coated with a 2.5 nm thick gold film (Hummer 6.2, Anatech LTD.). The average pressure of the vacuum was 0.075 Torr with a plasma discharge current of 0.015 A during deposition.

Preparation of Self-Assembled Monolayers

Formation of SAMs on the sputtered gold surfaces took place from 1–10 mM solutions of alkanethiols in histological grade methanol at 22–25°C. Hexadecanethiol ($HS(CH_2)_{15}CH_3$, MW = 258.51 g/mol, 95.0 %, Fluka Chemical), 16-mercaptohexadecanoic acid ($HS(CH_2)_{15}COOH$, MW=288.50 g/mol, 90.0 %, Aldrich) and 11-mercapto-1-undecanol ($HS(CH_2)_{11}OH$, MW = 204.24 g/mol, 97.0 %, Aldrich) were used without further purification.

SAMs with CH_3, OH, or COOH end functionality were prepared on gold-coated GS and PS substrates by immersion in 50 mL of respective thiol solutions in covered beakers for approximately 36 hours. Heterogeneous SAMs (CH_3+COOH, OH+COOH, OH+CH_3, or OH+COOH+CH_3) were also prepared according to the same protocol using a mixture (1:1, 2:1, 1:2, 1:4 or 1:10) of respective solutions. Substrates with SAMs were rinsed with methanol to remove organic debris and physically adsorbed thiols. They were then placed in covered Pyrex dishes and dried in an L-C oven at 100° C for at least 15 minutes. The substrates were stored in clean, covered Pyrex dishes for a maximum of three hours before they were immersed into supersaturated CaP solutions.

Characterization of Self-Assembled Monolayers

The self-assembled monolayers were characterized by measurements of advancing and receding (static) water contact angles using the sessile drop technique.[13] A syringe equipped with a 0.5 mm Luer-tipped needle was filled with deionized water (18MΩ). A small drop was placed on the solid surface of interest and adding more water from the syringe increased its volume. When the size of the drop was increased, the three-phase contact line advanced over the dry solid surface creating conditions for the measurement of advancing contact angle. Measurements of advancing contact angles were carried out 10–15 seconds after enlargement of the drop size to allow the three phase contact line to stabilize its shape. Then, the needle tip was carefully lowered into the sessile drop and water was removed to reinforce the reduction in the length of the three-phase contact line and create conditions for receding contact angle measurements. The receding contact angles were measured 10–15 seconds after change in the drop volume was made. A Kruss-G10 goniometer with drop shape analysis software was used for measuring seven advancing and seven receding contact angles. Average values and their standard deviations are reported in this paper. Additional surface energy characterization of self-assembled monolayers will be provided in a forthcoming publication.

Surfaces of selected substrates were imaged with the Dimension 3000 atomic force microscope (AFM, Digital Instruments). A Si_3N_4 cantilever tip (300kHz, 40N/m) in intermittent contact mode was used. Scan sizes from $2 \times 2\ \mu m^2$ to $20 \times 20\ \mu m^2$ were used for characterization of surface roughness and asperity geometries.

Formation of Calcium Phosphate Coating

A supersaturated calcium-phosphate solution (SCPS) was prepared by combining 5.6 mM $CaCl_2.2H_2O$ (MW=147.02 g/mol, Fisher Scientific) and 3.34 mM $NaH_2PO_4.H_2O$ (MW=137.99 g/mol, 99.4 %, Fisher Scientific) in 1:1 volumetric ratio. The resting pH of the solution was approximately 3.2 and a tris-buffer (tris(hydroxymethyl)-aminomethane, $C_4H_{11}NO_3$, MW=121.14 g/mol, 99+ %, Aldrich) was used to elevate and maintain the solution pH to approximately 7.35. Solution pH was monitored during preparation (pH-Meter, Metrohm Inc.).

Chemical immersion in the SCPS took place in one of three ways listed below.

Type I Immersion. This is a combined immersion technique, in which all samples were immersed in an 800 mL beaker containing 700 mL of SCPS. Beakers were covered during immersion by a sheet of paper to reduce dust and foreign objects in the solution. The samples were positioned by placing them in plastic microscope slide holders. Although orientation and position were held constant during the experiment, the samples were not perfectly vertical throughout immersion and faced the bottom of the beaker at an angle of 65–70 degrees. Using this technique sample surfaces were between 5 and 8 mm from each other. Surfaces facing downwards (not predisposed to sedimentation) were analyzed during characterization.

Type II Immersion. In the first set of experiments using the Type I immersion technique, it was found that the orientation of the sample has a profound effect on formation of CaP deposits and coatings. In order to avoid the accumulation of CaP precipitates on samples from supersaturated solutions, a vertical orientation was maintained throughout immersion. In this technique two different approaches were used when individual samples, or a set of a few samples were immersed into the SCPS solution.

Type II A. Samples were immersed vertically in individual 50 mL beakers containing approximately 50 mL of SCPS. The beakers were covered with a thin polyolefin film throughout immersion. The SCPS was poured from a single batch that underwent constant stirring between dispersions into each beaker to ensure solution homogeneity. Vertical immersion was achieved by implementing metal hangers attached to both the sample and to the parafilm covering the beaker. Distance of the samples from the sides of the glass beakers, which were being used as reaction cells, was held nearly constant throughout experimentation.

Type II B. This is an improved Type IIA immersion technique that eliminates difficulties associated with hanging samples at perfect vertical orientation. Painted metal clips, that underwent a cleaning regime similar to that used for the actual samples, were attached to the bottom of PS samples with minimal obstruction of surface area. The clips were shaped in a manner such that they allowed the samples to be perfectly vertical while in solution. Immersion took place in a 100 × 50 mm Pyrex dish containing approx. 250 mL of SCPS with even spacing between samples that exceeded 3 cm. The dish was covered with a glass lens during experimentation. From 4 to 8 samples could be placed in the same solution during experimentation, securing the same deposition conditions for each sample.

In each experiment, Type I and II, samples remained immersed in the SCPS for three days with daily solution refreshments. During solution refreshments, the old solution was carefully pipetted out of the beaker(s). New SCPS was carefully poured back in making sure not to disturb the samples. After three days of immersion, the samples were removed and gently washed with de-ionized water, placed in a covered Pyrex dish and dried in an L-C oven at 75 °C for 15 minutes.

Characterization of Calcium-Phosphate Coatings

Morphology and coverage of calcium phosphate coatings were examined under scanning electron microscopy, using a JEOL6400 SEM instrument. Approximately 12 mm wide sections were cut from the GS samples by a diamond scalpel. The size of the PS samples did not dictate that they be sectioned for observation. Samples were then mounted to aluminum pucks and sputter-coated with 2.5 nm of Au-Pd film to enhance resolution and reduce charging effects during SEM. SEM imaging was performed at an accelerating voltage of 20 kV and a working distance of 8 mm. Images were gathered via the digital image acquisition system in place with 1000 pixels per line and a dwell time of 0.1 msec for optimal image quality.

Field-emission scanning electron microscopy (FE-SEM, Hitachi S-4700, Hitachi Inc.) was employed to more closely examine structures of CaP films deposited via Type I and Type IIB immersion. An accelerating voltage of 5 kV and working distance of either 5.6 or 10.5 mm was used to resolve nanometer-scale features within the films.

The SEM and FE-SEM images were analyzed using Scion Image 4.02 (Scion Corporation) to gather information about the size of CaP particles/structures in the coatings.

Energy dispersive x-ray spectrometer (EDS) installed on the JEOL6400 SEM instrument was used to gather information about the calcium to phosphorus ratio of the deposited coatings. Carbon sputtering was used instead of Au-Pd sputtering for EDS analysis as the M_5-peak for gold (2.2057 keV) interfered with the K_α-peak of phosphorus (2.1455 keV). EDS analysis was performed at 20 kV of accelerating voltage with a working distance of 39 mm and a magnification of 100,000×. A live time of 100 seconds and dead time setting of approximately 22 % was used for all compositional characterization with the EDS. Measurements were made in five separate locations on each sample using the same settings and average value and standard deviation are presented in this paper. Information about the calcium to phosphorus ratio was obtained by the quantitative microanalysis tool inherent within the spectrum gathering software (SPIRIT).

A 31 × 41 transparent grid was super-imposed over SEM micrographs representative of the morphology of CaP films produced by Type I and Type IIA immersion techniques on the monolayers. "Void fraction" in the porous coating was determined as the ratio of grid points that fell on pores and the total number of grid points. These measurements are presented as percentage of surface area of the images.

Precipitates from Supersaturated Calcium Phosphate Solution

CaP particles were separated from one of the supersaturated solutions for evaluation of their size and morphology. An SCPS solution was prepared and poured through a glass micro-fiber filter (0.5 μm pore size, Whatman) under suction. The filter was then washed with acetone to remove water and placed in an L-C oven at 75°C for 48 hours prior to observation by FE-SEM. Imaging of particles was carried out with FE-SEM at accelerating voltage of 5 kV and working distance of 5.6 and 10.5 mm. The particle size was determined from collected images using Scion Image 4.02 software.

RESULTS AND DISCUSSION

Substrate Surface Characteristics

Figure 1 shows representative AFM images of GS and PS samples at different resolution. Both types of samples possessed nano-scale roughness such as shown in the left AFM image in Figure 1 for the GS sample. However, the nanoroughness was not analyzed in detail for PS samples due to the dominance of microscopic and sub-microscopic roughness for this type of sample as shown in the AFM image of Figure 1.

Grainy texture of the GS sample surface shown in Figure 1 is characteristic of glass slides coated with gold films by a sputtering technique and appearance of grainy structure results from the formation of nano-sized gold islands. From digital image analysis it was determined that the average diameter of the islands of gold was 56 ± 12.6 nm and this dimension is very similar to what we reported previously for substrates prepared according to the similar procedure.[14] Roughness values were also determined for the GS sample. The sample had a root-mean squared roughness of RMS = 3.7 nm, a geometrical roughness R_a= 3.0 nm, and a roughness parameter (representing the ratio between actual and projected surface area) r = 1.14.

Microscopic irregularities and characteristic "pitting" sites with a size of a few micrometers and variation in height of up to about 1 μm, characteristic of extruded amorphous-crystalline polymers, are evident on the surface of the PS samples (Figure 1, right image). Roughness values determined for the PS sample are R_{RMS} = 100 nm, R_a = 73 nm, and r = 1.13. The value of r is similar to that determined for the GS sample suggesting that nanoroughness had a major impact on the roughness parameter.

Contact angles measured for water drops on SAMs prepared in this study are shown in Table 1. Receding contact angle measurements were difficult to obtain on carboxyl and hydroxyl-terminated SAMs, as these values were near zero. The small profile of the

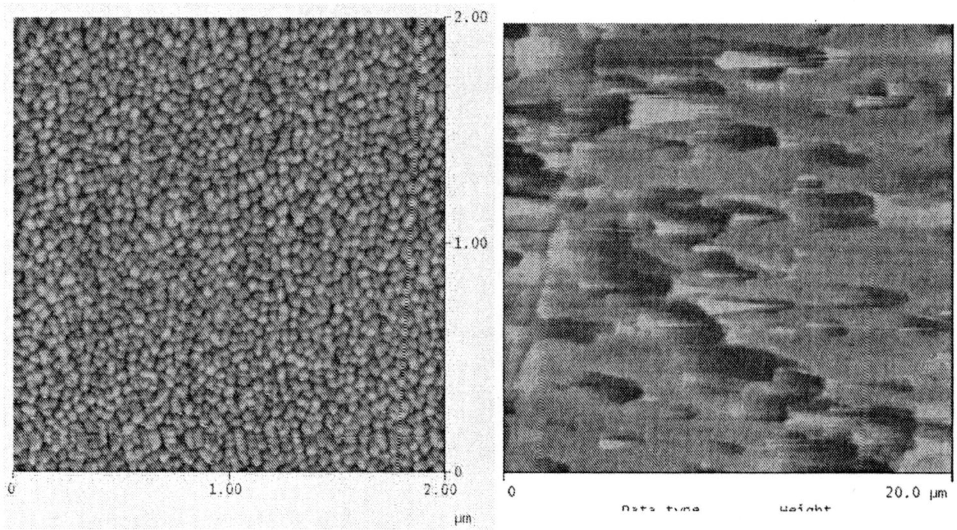

FIGURE 1 2 × 2 μm² (*left*) and 20 × 20 μm² (*right*) representative AFM images of topography of the GS and PS samples, respectively

TABLE 1 Advancing (Adv.) and receding (Rec.) contact angles for water drops on self-assembled monolayers. (NA = not available; GS = self-assembled monolayers prepared on gold-coated glass slides; PS = self-assembled monolayers prepared on gold-coated polystyrene samples; 1:1 etc. stands for the molar ratio of relevant thiols in solution used in preparation of SAMs and does not stand for the ratio of functionality formed on the substrate)

	Water Contact Angle [deg]			
	GS Sample		PS Sample	
Surface Functionality	Adv.	Rec.	Adv.	Rec.
COOH	47.5±0.4	14.5±0.9	23.6±1.7	<5
OH	46.4±0.4	26.2±2.0	38.2±1.6	<5
CH_3	111.4±2.7	88.0±2.0	95.7±0.8	83.7±2.6
OH+COOH (1:1)	39.8±1.1	NA	36.7±1.4	<5
OH+COOH (1:2)	NA	NA	37.6±1.5	15.9±1.2
OH+COOH (2:1)	NA	NA	55.9±2.3	23.6±4.5
OH+COOH (4:1)	NA	NA	56.0±1.3	24.0±1.0
OH+COOH (10:1)	NA	NA	59.2±2.7	19.0±2.4
CH_3+COOH (1:1)	98.9±5.3	61.6±2.6	84.2±2.5	50.8±4.1
OH+ CH_3 (1:1)	94.8±1.9	76.0±2.0	113.1±0.8	87.4±3.5
OH+COOH+CH_3 (1:1:1)	NA	NA	78.2±1.2	<5

receding drops challenged the ability of the drop shape analysis software to discern between the liquid and solid surface. Contact angles near zero are reported as smaller than 5 degrees.

The hydrophobic character of CH_3-terminated SAM is revealed by high advancing water contact angles, approximately 111 and 96 degrees for GS and PS samples, respectively. Typical advancing contact angle values measured for such SAMs are in the range of 110–112 degrees.[13] Contact angles several degrees smaller recorded for CH_3-terminated SAM formed on the PS sample suggests less dense packing of thiol molecules in the monolayer structure and/or submicroscopic imperfections of this particular sample. Receding water contact angles of 84–88 degrees were measured and they are also indicative of certain imperfections of the samples used in this study. Because both GS and PS samples covered with CH_3-terminated SAM had different roughness characteristics, but still showed similar receding contact angles, we expect that the samples had some degree of heterogeneity. In our previous research, we speculated that this heterogeneity might result from the intergranular area of the gold coating that is uncovered with thiols.[14]

COOH and OH-terminated SAMs served as hydrophilic surfaces in this study. Similar advancing water contact angles were recorded for both polar SAMs, COOH-terminated SAMs being more hydrophilic as revealed by smaller contact angles, particularly when prepared on gold-coated PS substrates. A value close to zero degrees for advancing water contact angles was reported for freshly prepared COOH and OH-terminated SAMs,[15] whereas as high as 84 and 65 degrees were reported in our previous paper.[14] Aging of the SAMs and its sensitivity to air-born contaminants are expected to be the major reasons for such a broad variation in measured contact angles for SAMs with COOH and OH functional groups. These relatively high-energy surfaces have a tendency to reduce their surface energy by adsorbing organics from laboratory air.

The low standard deviation for most of the measured contact angles and reasonable contact angle hysteresis values (difference between advancing and receding contact angles) are also indicative of a relatively good quality samples prepared in this study.

Any of the heterogeneous SAMs containing the methyl functionality produced large water contact angles suggesting that the thiols with methyl groups dominated the molecular structure of SAMs. The SAMs with mixed OH and COOH functionality demonstrated contact angles that are larger than contact angles measured on homogeneous COOH or OH-terminated SAMs. This result is of no surprise as 16-mercaptohexadecanoic acid is longer by four methylene units than 11-mercapto-1-undecanol and surfaces composed of these two thiols have triple surface functionality, two polar OH and COOH, and nonpolar CH_2.[16]

Morphology of Calcium Phosphate Coatings

Deposition characteristics of CaP on SAMs differed greatly depending on the conditions of immersion. Type I immersion experiments yielded CaP films with a spherical morphology, while Type IIa and IIb immersion methods produced films with a plate-like structure. The variance in morphology is indicative of a difference in the underlying deposition mechanism between the two immersion techniques and is discussed later in this paper.

Type I Immersion Experiment. The SEM images of CaP coatings obtained according to Type I SCPS immersion are shown in Figure 2. The coatings appear to be a network of many individual particles attached to each other that produce a porous structure. The clustered structure of Type I immersion CaP films may indicate a deposition mechanism based on solution-formed nuclei and subsequent deposition of particles.[17]

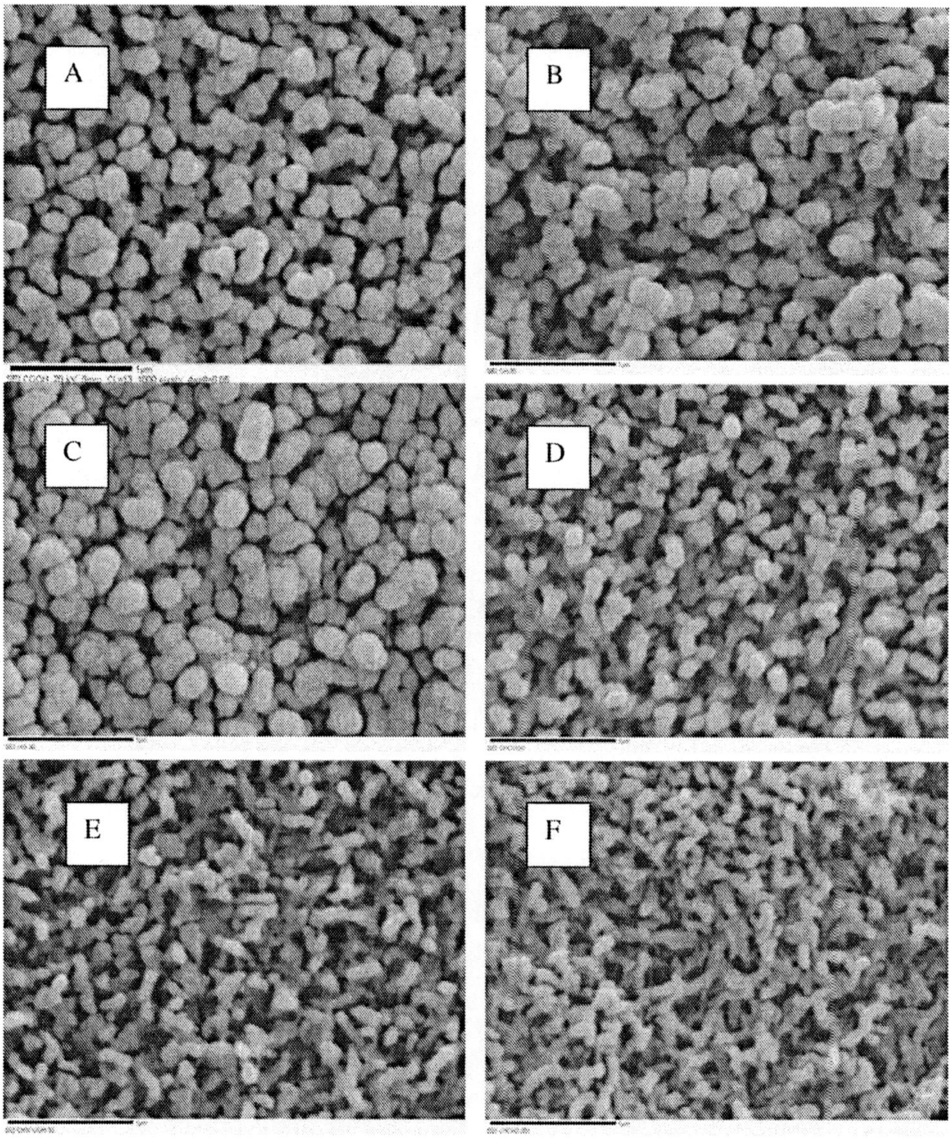

FIGURE 2 SEM micrograph of CaP films biomimetically deposited on COOH (A), OH (B), CH_3 (C), COOH+OH (D), COOH+CH_3 (E), OH+CH_3 (F) SAMs. CaP coatings were deposited on the SAMs in the same beaker (Type I immersion). A scale bar at the bottom of each image represents 1 μm.

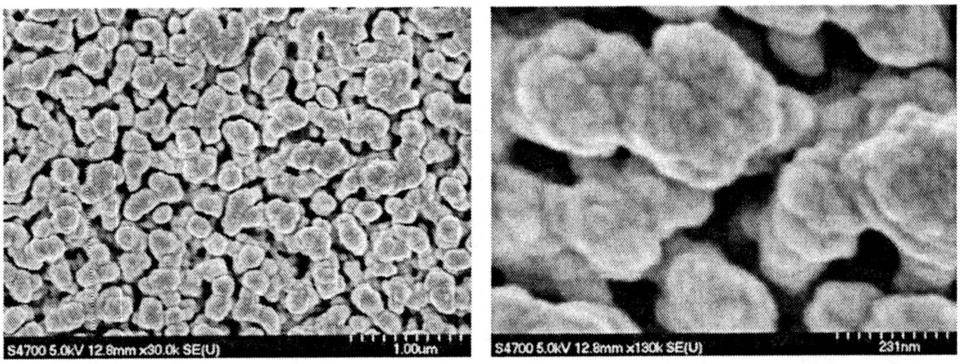

FIGURE 3 FE-SEM micrographs of CaP coating on COOH-terminated SAMs by Type I SCPS immersion

The particles on the homogeneous SAMs were quite spherical in nature and differed between functionalities in their densities and packing structures. CaP particles on heterogeneous functionality SAMs deviated from sphericity showing an elongated structure with the OH+CH_3 SAM having the largest aspect ratio. Type II immersion techniques failed to produce coatings of similar structures (see further part of this section).

CaP particle sizes were analyzed from the SEM images and found to be 230 ± 21 nm, 243 ± 19 nm, and 210 ± 14 nm for COOH, OH, and CH_3 terminated SAMs. The difference in aspect ratio between CaP particles deposited on heterogeneous functionality surfaces is noticeable. CaP particles were 152 ± 8 nm by 130 ± 11 nm on the OH+COOH surface, 260 ± 33 nm by 96 ± 6 nm for the CH_3+COOH surface, and 248 ± 37 nm by 88 ± 5 nm for the OH+CH_3 surface.

The reason for coatings of different morphologies formed on molecularly heterogeneous surfaces in this experiment is not clear to us. We are also unaware of any report that presents structures of CaP coatings similar to those shown in Figure 1. Possibilities for explanation could include a natural segregation of precipitated particles from solution on different surfaces due to different magnitude and nature of particle-surface interactions. Also, growth of particles could progress on homogeneous surfaces at the expense of dissolution of particles deposited on heterogeneous surfaces. Our experiments cannot support or reject either of these two mechanisms.

Further examination of CaP films on COOH-terminated substrates by FE-SEM (Figure 3) highlighted the details of coating morphology. The spherical particles deposited on the COOH monolayer by Type I immersion show an underlying structure that resembles the clustering of even smaller 10–20 nm particles. It was determined that this structure is similar to that for particles separated from the SCPS solution (Figure 4). However, the particles, or aggregates of nanoparticles, separated from solution seem to be smaller, 169 ± 5 nm. This difference may be a result of different batches of solutions used in coating deposition and preparation of precipitates, containing particles of different sizes. It can also suggest that indeed the aggregates of CaP coating grew to larger dimensions during the deposition experiment.

Type II Immersion Experiment. When Type II CaP deposition was used the CaP coating morphology was different (Figures 5 to 7). Coatings did not reveal particle-type

FIGURE 4 FE-SEM micrograph of CaP particles separated from SCPS solution in a glass microfiber filter

structures as in the case of coatings produced by Type I immersion method. The morphologies seem to be plate-like structures, but different from the "rose-like" morphology described by other researchers for CaP coatings deposited on metallic substrata.[1,18–20] For example, Figure 8 shows the SEM image of a CaP coating prepared from a similar supersaturated solution on a titanium alloy substrate in our previous studies.[1] The plates of the CaP film deposited on the titanium alloy are very discrete and well-defined. In contrast, the plate structures of coatings deposited on the monolayers seem to have greater depth in all dimensions and are less discrete. Each of the plates seem to intertwine with neighboring plates as exhibited in Figure 6.

It is hypothesized that the differences in coating morphology of CaP film formation between the immersion techniques are due mainly to differences in conditions during immersion of substrates in the SCPS. As discussed earlier, in the case of Type I SCPS immersion experiments, the spherical morphology of the particles (Figures 2 and 3) seemed to indicate precipitation of particles from solution and the subsequent build-up of CaP clusters as the dominant mechanism. Indeed, the similarity between particle morphologies on SAM surfaces deposited via Type I immersion and morphologies of particles separated from solution seem to support this hypothesis. In Type II immersion experiments the coating morphology (Figures 5 to 7) is indicative of a heterogeneous nucleation and subsequent crystal growth mechanism. The porous nature of this particular morphology appears to resemble that of human bone and would allow for cellular infiltration of the structure and may enhance osteoblast adhesion when compared to the spherical morphologies.[21,22]

Porosity of Calcium Phosphate Coatings

Void fraction of the coatings was examined in regions representative of particle morphologies. This parameter was determined from 2-dimensional SEM images and should not be mistaken with porosity, although is meant to characterize the overall morphology of the CaP films. It should also not be confused with surface area coverage that is described later. Void fraction of the coatings varied greatly between the surface functionalities. Evenly dispersed voids and 12.2% void fraction was noted on the carboxyl-terminated

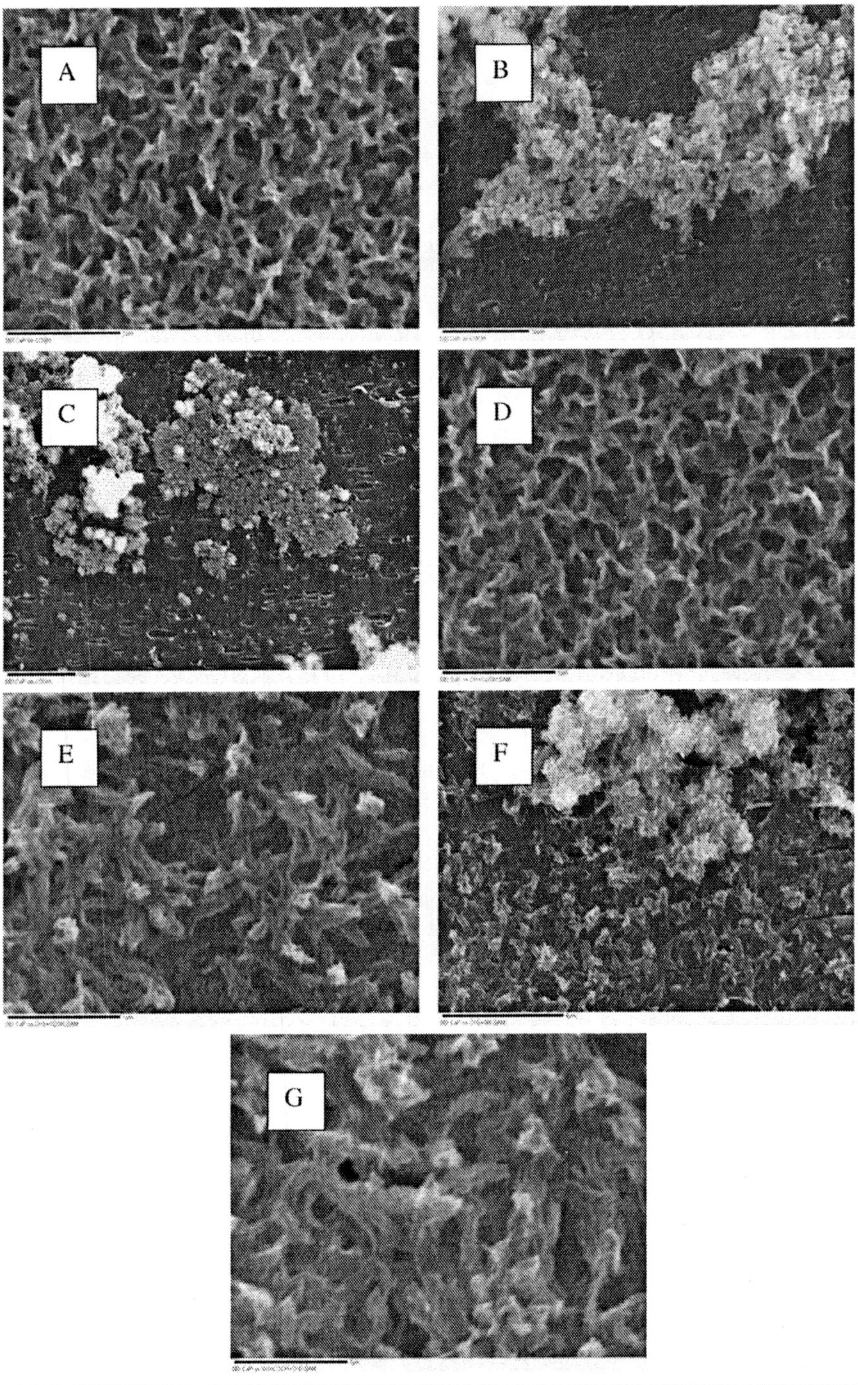

FIGURE 5 CaP coatings biomimetically deposited on COOH (A), OH (B), CH_3 (C), OH+COOH (D), CH_3+COOH (E), OH+CH_3 (F), OH+COOH+CH_3 (G) terminated SAMs by Type IIB SCPS immersion. A 1 μm scale bar is shown under each image.

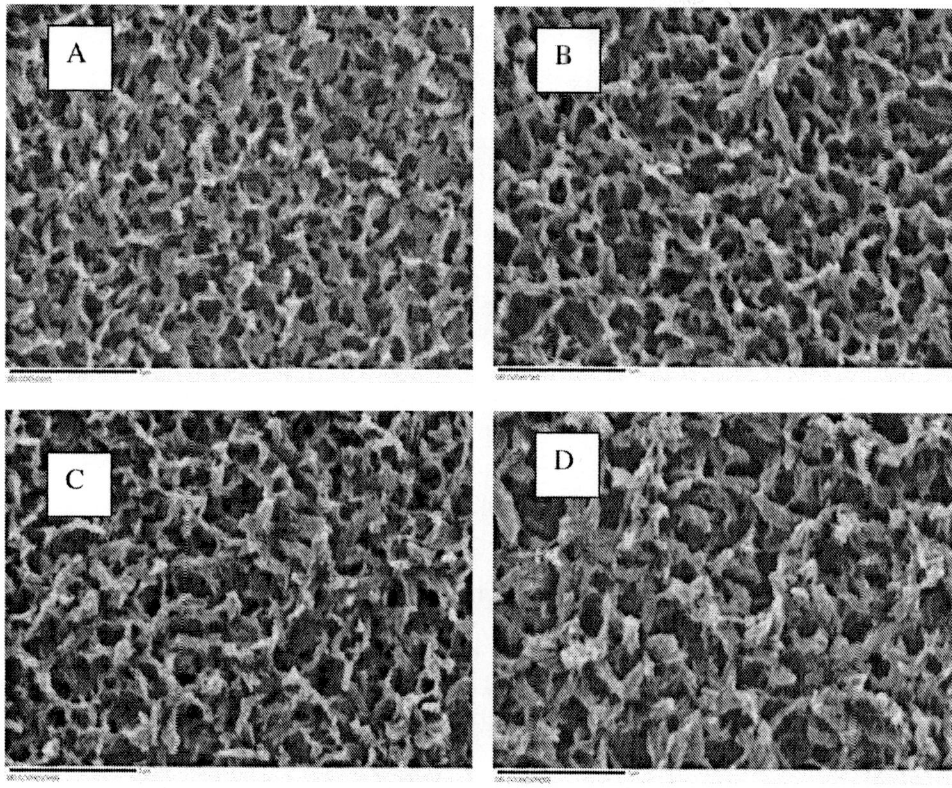

FIGURE 6 30,000× magnification SEM micrographs of coatings prepared on COOH(2):OH(1) (A), COOH(1):OH(2) (B), COOH(1):OH(4) (C), COOH(1):OH(10) (D) heterogeneous SAMs using Type IIA immersion methodology

FIGURE 7 FE-SEM micrographs of CaP coating on COOH-terminated SAMs by Type IIB SCPS immersion

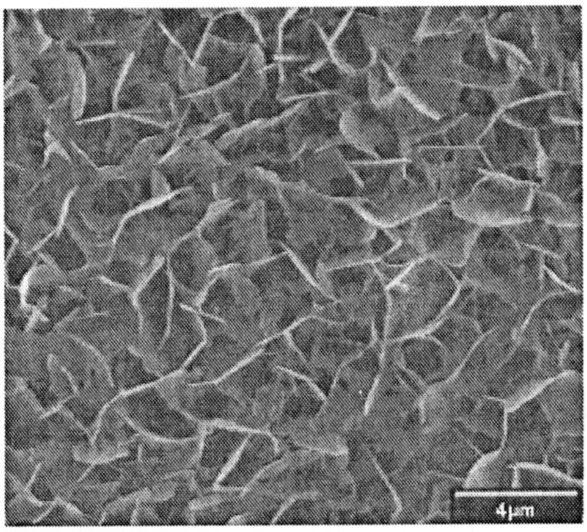

FIGURE 8 The scanning electron micrograph of calcium phosphate coating deposited on Ti-6Al-4V alloy from a supersaturated calcium-phosphate solution[1]

monolayer. Concentrated open spacing comprising 13.9% of surface area shown in the image was observed on methyl-terminated monolayers. Evenly dispersed openings and a limited 8.8% void fraction was calculated for the hydroxyl-terminated monolayer.

CaP films on heterogeneous functionality SAMs showed characteristics exhibited by each of the homogeneous functionality components used to produce the surface. For instance, dispersion of pores on the heterogeneous OH(1)+COOH(1) functional surface was less concentrated than those observed on homogeneous OH surfaces, but also less evenly distributed than those observed on homogeneous COOH surface. Open space was greatest on CH_3+COOH (12.9%) and OH+COOH (12.1%) surfaces. Limited void fraction (7.1%) was observed for the OH+CH_3 sample.

In implementing Type IIA immersion for heterogeneous OH+COOH monolayers with varying OH:COOH ratio, no clear relationship was found between increasing amounts of hydroxyl-terminated thiols in solution and porosity of the deposited CaP films. OH(4)+COOH(1) surfaces produced CaP films with void fraction accounting for 22.7% of the surface area shown in Figure 6C. OH(1)+COOH(2) surfaces elicited CaP films with void fraction making up 19.8% of the surface area shown in figure 6A. OH(2)+COOH(1) and OH(10)+COOH(1) surfaces produced films with the least amount of void fraction, 16.1% and 15.8%, respectively.

Coverage of Substrate by Calcium Phosphate Coating

A uniform coating was produced on COOH and OH(1)+COOH(1) (Figure 5) terminated SAMs using Type II immersion. OH+COOH+CH_3 terminated SAMs preliminarily showed excellent coverage when examined by optical microscopy. Closer observation with SEM showed a very thin CaP layer on the substrate, but with nearly all of the surface area covered. OH terminated SAMs lacked a continuous CaP film. Small agglomerations could be seen all over the surface. The agglomerations were also found on CH_3 terminated SAMs, but small areas of growth were discovered beneath these clusters as shown in Figure 5(C). These growth areas coalesced with growth areas of neighboring agglomerations in

TABLE 2 Calcium to phosphorus ratios of CaP coatings as a function of surface functionality and immersion technique

Surface Functionality	Ca:P Ratio [weight %]	Ca:P Ratio [atomic %]	Immersion Technique
COOH	2.00 ± 0.6	1.53 ± 0.5	I
	1.83 ± 0.1	1.42 ± 0.1	IIB
OH	2.94 ± 0.9	2.25 ± 0.7	I
	1.74 ± 0.1	1.34 ± 0.1	IIB
CH_3	2.19 ± 1.2	1.66 ± 0.9	I
	1.71 ± 0.1	1.32 ± 0.1	IIB
OH+COOH (1:1)	1.96 ± 0.1	1.51 ± 0.1	I
	1.80 ± 0.1	1.40 ± 0.1	IIB
OH+COOH (1:2)	1.85 ± 0.2	1.43 ± 0.2	IIA
OH+COOH (2:1)	2.37 ± 0.4	1.83 ± 0.3	IIA
OH+COOH (4:1)	2.22 ± 0.1	1.71 ± 0.1	IIA
OH+COOH (10:1)	3.45 ± 1.2	2.67 ± 0.9	IIA
CH_3+COOH (1:1)	2.03 ± 0.1	1.57 ± 0.1	I
	1.99 ± 0.2	1.54 ± 0.1	IIB
CH_3+OH (1:1)	3.42 ± 1.9	2.64 ± 1.5	I
	1.77 ± 0.1	1.37 ± 0.4	IIB
CH_3+COOH+OH (1:1:1)	1.76 ± 0.1	1.34 ± 0.1	IIA

some instances to begin the formation of a continuous CaP film. Heterogeneous functionality OH+CH_3 substrates exhibited even more coalescence of growth areas suggesting that chemical functionality may strongly influence the induction time of CaP formation.

Despite the differences in coating morphology, trends in substrate coverage were similar regardless of the immersion technique implemented. Substrate coverage of 100% was achieved on COOH, OH+COOH and OH+COOH+CH_3-terminated SAMs. As stated previously though, the coatings on the OH+COOH+CH_3 terminated SAM were extremely thin. CH_3+COOH terminated SAM substrates were approximately 80–85% covered with a CaP film. Heterogeneous functionality OH+CH_3 substrates elicited an approximate surface area coverage of 35–40%. Homogeneous CH_3 and OH substrates had the least amount of substrate coverage with 15% and 10%, respectively.

Varying the ratio of OH to COOH terminal sites on the SAMs had little effect on particle size and morphology as shown in Figure 7. Porosity of the coatings was slightly different between ratios. It was observed that an increasing amount of OH groups did reduce the substrate coverage and uniformity of the CaP films.

Calcium-to-Phosphorus Ratio in Coating

Calcium to phosphorus ratios of CaP films formed on self-assembled monolayers of differing functionalities varied for coatings produced in different experiments (Table 2). When Type I immersion was employed, the coatings were very calcium-rich. However, when Type IIB immersion was used, the films were calcium-deficient, with the overall Ca:P ratio lower than that of stoichiometric hydroxyapatite. This difference in coating compositions may highlight the difference in the underlying mechanism of deposition between the two immersion methods. The calcium-rich nature of the films, similarity in morphology to CaP particles separated from solution and the charge number of calcium ions (2+) compared to phosphate complexes (1−) imply that Type I immersion relies on an electrostatic interaction for the deposition of CaP films.

It is also worth noting that hydroxyl-dominated SAMs and SAMs with an equal ratio of hydroxyl and other groups elicited the most calcium-rich coatings when Type I and Type IIA immersion was employed. When Type IIB immersion was used, hydroxyl-containing SAMs had the lowest calcium to phosphorus ratios.

CONCLUSIONS

In this study, the self-assembled monolayers (SAMs) of alkanethiols served as model organic substrates in the biomimetic deposition of CaP coatings from supersaturated calcium phosphate solutions. It was found that homogeneous COOH-terminated SAMs and heterogeneous OH+COOH SAMs produced CaP coatings that covered 100% of the substrate surface area within three days of chemical immersion. SAMs containing a predominantly methyl, or hydroxyl character failed to elicit continuous CaP films covering all of the surface area. This trend of surface area coverage was observed regardless of the immersion method employed, which suggests that the deposition process is strongly dependent on surface functionality.

Both deposition methodology and substrate surface functionality had also influenced the chemical composition of the formed calcium phosphate coatings. Calcium to phosphorus ratios of CaP films varied from about 1.3 to 2.7 for coatings produced on SAMs of differing functionalities. OH-terminated SAMs induced formation of deposits with enriched calcium content whereas coatings with stoichiometry close to that for hydroxyapatite were preferentially formed on SAMs having COOH groups.

ACKNOWLEDGMENTS

Financial support from the MTU Department of Materials Science and Engineering is greatly appreciated. The authors would also like to thank: Dr. John Williams of the Chemistry Department for generously donating polystyrene materials and Mr. Owen Mills, Mrs. Ruth Schultz-Kramer and Mr. Ed Laitila for technical support with analytical equipment.

REFERENCES

[1] Baker KC, Anderson MA, Oehlke SA, Astashkina AI, Haikio DC, Donahue SW, Drelich JW. 2005. Growth, Characterization and Biocompatibility of Bone-Like Apatite Layers Biomimetically Deposited on Metallic Substrata. *Journal of Materials Science and Engineering Part C* (Article In Press).

[2] Liu Q, Ding J, Mante FK, Wunder SL, Baran GR. 2002. The Role of Surface Functional Groups in Calcium Phosphate Nucleation on Titanium Foil: A Self-Assembled Monolayer Technique. *Biomaterials.* Volume 23, 3103.

[3] Tanahashi M, Yao T, Kokubo T, Minoda M, Miyamota T, Nakamura T, Yamamuro T. Apatite Coated on Organic Polymers by Biomimetic Process: Improvement of Its Adhesion to Substrate by NaOH Treatment. 1994. *Journal of Biomaterials: Part B.* Volume 5, Number 4, 339.

[4] Rohanizade R, LeGeros RZ, Harsono M, Bendavid A. 2005. Adherent Apatite Coating on Titanium Substrate Using Chemical Deposition. *Journal of Biomedical Materials Research: Part A.* Volume 72, Issue 4, 428.

[5] Liu GJ, Miyaji F, Kokubo T, Takadama H, Nakamura T, Murakami A. 1998. Apatite-Organic Polymer Composites Prepared by a Biomimetic Process: Improvement of the Adhesion of the Apatite Layer to the Substrate by Ultraviolet Irradiation. *Journal of Materials Science: Materials in Medicine*. Volume 9, 285.

[6] Tanahashi M, Yao T, Kokubo T, Minoda M, Miyamota T, Nakamura T, Yamamuro T. 1995. Apatite Coated on Polymers by Biomimetic Process: Improvement In Its Adhesion to Substrate by Glow-Discharge Treatment. *Journal of Biomedical Materials Research*. Volume 29, 349.

[7] Liu Y, Hunziker EB, Van de Vaal C, de Groot K. 2003. Biomimetic Coatings vs. Collagen Sponges as a Carrier for BMP-2: A Comparison of the Osteogenic Responses Triggered In Vivo Using an Ectopic Rat Model. *Bioceramics: Key Engineering Materials*. Volume 16, 245.

[8] Barbanti SH, Santos AR, Zavaglia CA, Duek EA. 2004. Porous and Dense poly(L-lactic acid) and poly(D,L-lactic acid-co-glycolic acid) Scaffolds: In Vitro Degradation in Culture Medium and Osteoblasts Culture. *Journal of Materials Science: Materials in Medicine*. Volume 15, 1315.

[9] Kim HW, Knowles JC, Kim HE. 2005. Hydroxyapatite Porous Scaffold Engineered With Biological Polymer Hybrid Coating for Antibiotic Vancomycin Release. *Journal of Materials Science: Materials in Medicine*. Volume 16, 189.

[10] Wen HB, de Wijn JR, van Blitterswijk A, de Groot K. 1999. Incorporation of Bovine Serum Albumin in Calcium Phosphate Coating on Titanium. *Journal of Biomedical Materials Research: Part B*. Volume 46, Number 2, 245.

[11] Tanahashi M and Matsuda T. 1997. Surface Functional Group Dependence on Apatite Formation on Self-Assembled Monolayers in a Simulated Body Fluid, *Journal of Biomedical Materials Research* Volume 34, 305.

[12] Sato K, Kumagai Y, Tanaka J. 2000. Apatite Formation on Organic Monolayers in Simulated Body Environment. *Journal of Biomedical Materials Research* Volume 50, 16.

[13] Drelich J, Miller JD, and Good RJ. 1996. The Effect of Drop (Bubble) Size on Advancing and Receding Contact Angles for Heterogeneous and Rough Solid Surfaces as Observed with Sessile-Drop and Captive-Bubble Techniques. *Journal of Colloid Interface Science* Volume 179, 37.

[14] Tormoen GW, Drelich J, and Beach ER. 2004. Analysis of Atomic Force Microscope Pull-off Forces for Gold Surfaces Portraying Nanoscale Roughness and Specific Chemical Functionality. *Journal of Adhesion Science and Technology* Volume 18, Number 1, 1.

[15] Bain CD, Troughton EB, Tao Y-T, Evall J, Whitesides GM, and Nuzzo RG. 1989. Formation of Monolayer Films by the Spontaneous Assembly of Organic Thiols from Solution onto Gold. *J. Am. Chem. Soc.* 111, 321.

[16] Folkers JP, Laibinis PE, Whitesides GM. 1992. Self-Assembled Monolayers of Alkanethiols on Gold: Comparisons of Monolayers Containing Mixtures of Short and Long-Chain Constituents with CH_3 and CH_2OH Terminal Groups. *Langmuir* Volume 8, 1330.

[17] Tarasevich BJ, Chusuei CC, Allara DL. 2003. Nucleation and Growth of Calcium Phosphate from Physiological Solutions onto Self-Assembled Templates by a Solution-Formed Nucleus Mechanism. *Journal of Physical Chemistry: Part B*. Volume 107, 10367.

[18] Vatanatham M, Kimura S. 2000. Hydroxyapatite Coating on Thermally Oxidized Titanium Substrates. *Journal of American Ceramic Society* Volume 83, 2135.

[19] Habibovic P, Barrere F, van Blitterswijk C, de Groot K, Layrolle P. 2002. Biomimetic Hydroxyapatite Coating on Metal Implants. *Journal of American Ceramic Society* Volume 85, 517.

[20] Drummond JL, Steinberg AD, Krauss AR. 1997. X-ray Photoemission and Energy Dispersive Spectroscopy of Hydroxyapatite-Coated Titanium. *Journal of American Ceramic Society* Volume 80, 1868.

[21] Giavaresi G, Fini M, Cigada A, Chiesa R, Rondelli G, Rimondini L, Torricelli P, Aldini NN, Giardino R. 2003. Mechanical and Histomorphometric Evaluations of Titanium Implants with Different Surface Treatments Inserted in Sheep Cortical Bone. *Biomaterials* Volume 24, 1583.

[22] Fini M, Giavaresi G, Greggi T, Martini L, Aldini NN, Parisini P, Giardini R. 2003. Biological Assessment of the Bone-Screw Interface After Insertion of Uncoated and Hydroxyapatite-Coated Pedicular Screws in the Osteopenic Sheep. *Journal of Biomedical Materials Research: Part A* Volume 66, 176.

Measurements of Colloidal Forces Between Microscopic Probes and Substrates in Suspensions of Nanoparticles

J. Drelich,[*] J. Nalaskowski,[†] J. Long,[‡] Z. Xu,[‡] and J. Masliyah[‡]

The colloidal forces between atomic force microscopy probes and flat substrates in suspensions of nanoparticles were measured. Glass spheres with a diameter of 5 to 15 μm were used as the probes whereas mica, fluorite and silicon wafer were used as substrates. Aqueous suspensions were made of either 5 to 50 nm alumina or 10 nm silica particles. We report oscillatory force vs. distance curves, which indicate that the nanoparticles remain to be stratified in the liquid films separating the probes and the substrates during the force measurements. Such structural effects were observed particularly for systems featuring attractive and weak repulsive interactions of nanoparticle with probe and substrate. Oscillation of the structural forces shows a periodicity close to the size of nanoparticles used. When they are oppositely charged to the probes, the nanoparticles tend to coat probes and hinder probe-substrate contact during force measurements.

INTRODUCTION

Suspensions made of nanoparticles become increasingly important fluids in industrial activities. Systems such as nanofluids (this term is frequently used in the literature for suspensions containing dispersed nanoparticles) for cooling systems, dispersions for the formulation of nanocomposites, or nano-slurries for ultra-polishing and ultra-grinding of electronic materials have only recently been developed and utilized. Nanoparticles of

* Department of Materials Science and Engineering, Michigan Technological University, Houghton, Michigan

† Department of Metallurgical Engineering, University of Utah, Salt Lake City, Utah

‡ Department of Chemical and Materials Engineering, University of Alberta, Edmonton, Alberta, Canada

carbon black, titanium dioxide, calcite, and polystyrene are already used as pigments in printing inks and paints. Concentrated suspensions of silica and clay nanoparticles are currently used by the casting industry as refractory coatings and binders. Recently introduced nanofluids offer the potential for developing a new generation of miniature heat exchanger systems by employing flow of cooling media through microchannels.[1] In all these applications stability and rheology of the suspensions are important properties that require research attention. Understanding the nature and magnitude of colloidal forces, stability and behavior of the suspensions during their contact with solid surfaces is crucial to formulation of products with desired physical properties.

The mechanisms by which nanoparticles interact with surrounding fluids and neighbor nanoparticles are not understood at this moment and these mechanisms are expected to be more complex than those for larger particles. This is mainly caused by large interfacial area of such finely dispersed systems and the expectation that the surface properties of nanoparticles are different than the properties of macroscopic surfaces. Because of the small dimensions of nanoparticles, a significant fraction of the fluid phase in a suspension resides within a few nanometers of the particle surfaces and within narrow particle-particle interspaces. The nature of the particle-liquid and particle-particle interactions in nanofluid suspensions is, therefore, of fundamental importance to the eventual understanding of stability and properties of these systems.

Solid-liquid interactions, liquid ordering in the vicinity of the solid surface, and inter-particle forces must be understood if nanoparticles can be formed into stable systems with liquids and if the properties of such suspensions can be predicted and manipulated. In response to the demand in this research area, we study the interactions between nanoparticles mediated by a variety of polar and non-polar fluids and examine liquid structuring at nanoparticle surfaces. Previous research on thin aqueous films of concentrated suspensions made of microscopic particles suspended in air or on a solid surface has clearly shown that the particles tend to stratify into periodic colloidal structures, which cause the appearance of structural transitions during thin film thinning.[2–8] The thinning of the "suspension films" occurs in a stepwise manner, and the number of step transitions is directly proportional to the volume fraction of particles in the suspensions. Similar ordering of particles has been recorded in our atomic force microscopy (AFM) experiments using aqueous suspensions of nanoparticles. To our knowledge, little work has been reported on force measurements with colloidal probes in the suspensions of solid nanoparticles using the AFM technique. The only paper that we are aware of is that by McNamee et al.[9] in which they showed one set of force measurements between two silica surfaces in concentrated suspensions of 10 nm silica particles. However, their experiments were conducted with silica particles of a relatively high surface charge density. In such systems, as will be discussed in this paper, the force vs. separation curves showed only weak oscillations.

EXPERIMENTAL

The experiments were performed using a Nanoscope E AFM (Digital Instruments Inc.) in a fluid cell. Freshly cleaved mica, re-polished fluorite, or silicon wafer were used as substrates. Borosilicate glass spheres having a diameter of either 5 or 15 μm were glued onto triangular shaped cantilevers of a spring constant 0.12 N/m and 0.58 N/m, purchased from Digital Instruments. Glass probes were cleaned by UV irradiation prior to the experiments. The measurements of surface forces between glass probes and

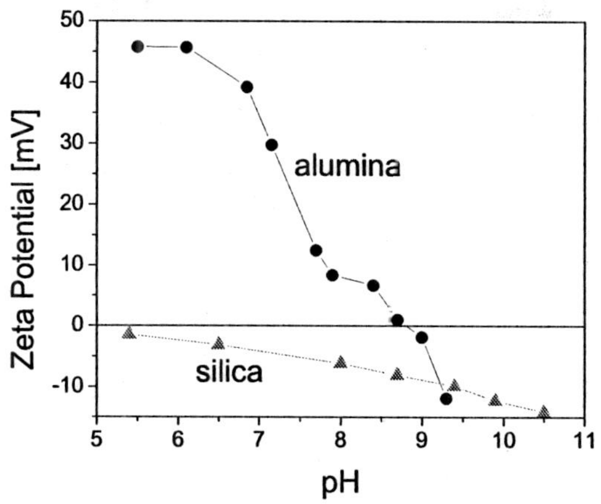

FIGURE 1 Zeta potential for alumina nanoparticles in 1 mM KCl and silica nanoparticles in Nyacol 830. The pH of suspensions was modified with either NaOH or HCl.

substrates were carried out in aqueous suspensions of nanoparticles. The force curves were analyzed with the SPIP software (Image Metrology, Lyngby, Denmark), which translates the cantilever deflection-distance data to force-separation profiles.

Alumina particles with a size from 5 to 50 nm were received from Argonne National Laboratory and suspended in deionized water or 1mM KCl solution (pH 5.8) at a concentration of about 5 wt%. The zeta potentials for these particles were calculated from electrophoretic mobility values measured with a Zetaphoremeter III (SEPHY/CAD) and are shown in Figure 1; see Ref 10 for details.

Suspension of 10 nm silica particles at a particle concentration of 30 wt%, pH 10.5, density 1.2 g/cm^3, and viscosity 8 cP was received from Eka Chemicals Inc. The commercial name for this product containing 0.55 wt% Na_2O is Nyacol 830. Its conductivity is 3.9 mS/cm and was measured using the Zetaphoremeter III instrument. The pH value of the silica suspension was reduced using a HCl solution before the AFM experiments. Zeta potential values of concentrated silica suspension were measured at varying pH using electroacoustic spectroscopy (DT-1200 Spectrometer, Dispersion Technology Inc.). The results are shown in Figure 1. The electroacoustic spectroscopy is described in details by Magual et al.[11]

RESULTS AND DISCUSSION

Structural Effects Caused by Silica Nanoparticles

Figure 2 shows an example of force vs. separation curves recorded for a 5 μm glass sphere approaching a surface of silicon wafer in a 30 wt% aqueous suspension of 10 nm silica particles. The forces recorded are repulsive (both glass and silicon wafer are negatively charged at pH>4 used in this study) but the force-separation curve is not smooth. A close examination of the force profile indicates a stepwise change as the probe approaches substrate. Four consecutive peaks at about 26 nm, 17 nm, 10 nm, and 5 nm of the sphere-to-substrate separation are of increasing magnitude, as shown in Figure 2.

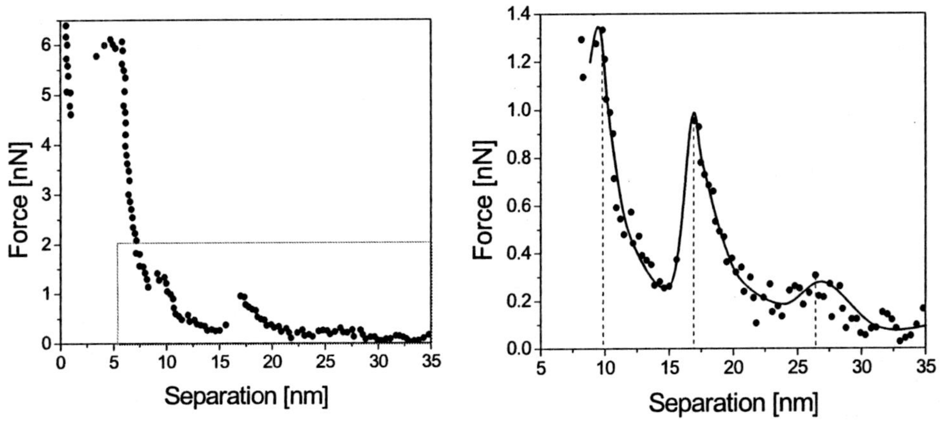

FIGURE 2 Force versus separation curves for a 5 μm glass sphere approaching a surface of silicon wafer in a 30 wt% aqueous suspension of 10 nm silica particles (pH 4.5). The graph on the right is the enlargement of the square area marked on the left graph. The line on the right graph was added to guide the readers.

Three steps of smaller magnitudes are also shown at a higher resolution force profile on the right graph of Figure 2.

The last step of the highest magnitude, at ~5 nm, shows attractive forces operating between the probe and substrate. Similar step was also detected during the force measurements with deionized water (pH 5.8), free of silica nanoparticles (not shown). Attractive forces of similar range were also reported for a silica probe interacting with mica[16] and silicon nitride probe interacting with mica[17] in aqueous solutions of electrolyte at pH 6. Bremmell et al.[16] speculated that the attraction of the silica to mica at 5 nm separation is caused by a hydrous silica gel layer surrounding silica surface. Such a gel layer is unlikely to be present in our glass-silicon system. Teschke et al.[17] on the other hand, simply attributed short-range attractions to van der Waals forces. Hamaker constant by one to two orders of magnitude larger $\{A = -6FD^2/R = [6 \times 6 \times 10^{-9} \times (5 \times 10^{-9})^2]/(2.5 \times 10^{-6}) = 3.6 \times 10^{-19}\ J\}$ than what we could expect for the glass-water-silicon ($\sim 4.7 \times 10^{-21}$ J by our estimate, and 1.2×10^{-20} J for silica-water-silica provided by McNamee et al.[9]) is needed to overcome the repulsive forces at 5 nm separation. We recognize that a certain degree of contamination of the probe and/or substrate of hydrophobic nature can also produce similar short-range attractive (hydrophobic) effects. Nevertheless, the conclusions on organized nanoparticle structures and their spacing remain the same regardless whether the probe in this experiment was indeed contaminated or not.

Figure 3 shows the effect of increasing pH of silica suspension on the force vs. separation curves for the same AFM probe and silicon substrate. Overall, as expected, repulsive forces between the glass sphere and silicon substrate in the concentrated suspension of silica nanoparticles increased with increasing pH. This is due to increasing surface charges for the nanoparticles, AFM probe and substrate.

As shown in Figure 3, weak-intensity oscillations superimposed on the long-range weak repulsive electrostatic force profiles were present for all the cases. Steps at 9–10 and 17–19 nm were evident on almost all the curves, and those at 25–28 nm were

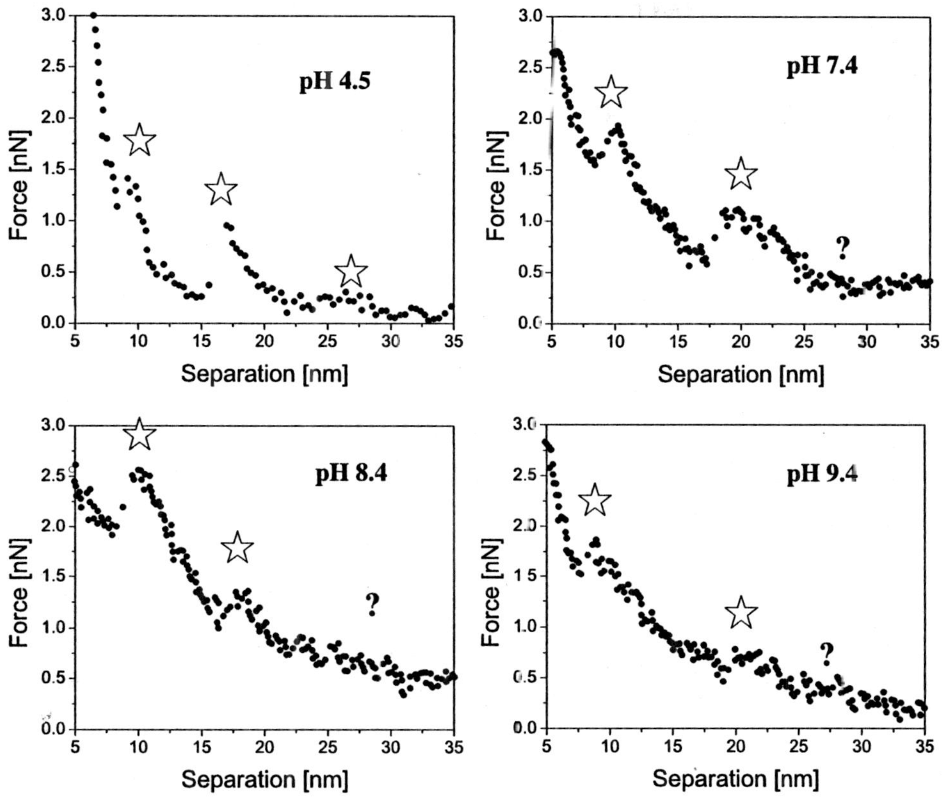

FIGURE 3 Force versus separation curves for a 5 μm glass sphere approaching a surface of silicon wafer in a 30 wt% aqueous suspension of 10 nm silica particles at pH 4.5, pH 7.4, pH 8.4, and pH 9.4. The stars on the graphs mark the identified positions for transition steps, not always sharp. The question marks on the graphs are used to indicate the regions of possible locations of steps but could not be confirmed due to significant scatter of the recorded points.

observed on only some of the curves. The intensity of the peaks however, diminished with increasing pH, just like their intensity reduces for peaks located at larger separations. As a result, the steps were more difficult to identify at curves produced in highly alkaline solutions than in acidic or close-to-neutral solutions. Some of the curves have a number of unexpected extra peaks of a smaller intensity. These "extra steps" are probably the result of false signals produced by an interference of the particles to the laser beam. It is interesting to note the gaps between consecutive peaks appear to increase with increasing solution pH. This increase is attributed to the increased repulsion between layers of organized nano particles as anticipated.

It is clear from the right graph of Figure 2 that the transition steps in the force vs. separation curve have an exponential character. Fitting their shape to the single exponential function ($y = y_0 + Ae^{-x/k}$) showed that the value of k varies from ~1.3 to 3.6 nm, but it remains consistent among curves recorded for silica suspensions with varying pH. Our results indicate that the scatter of the k value is the result of a limited number of experimental points for the step transition regions, which unfortunately were scattered

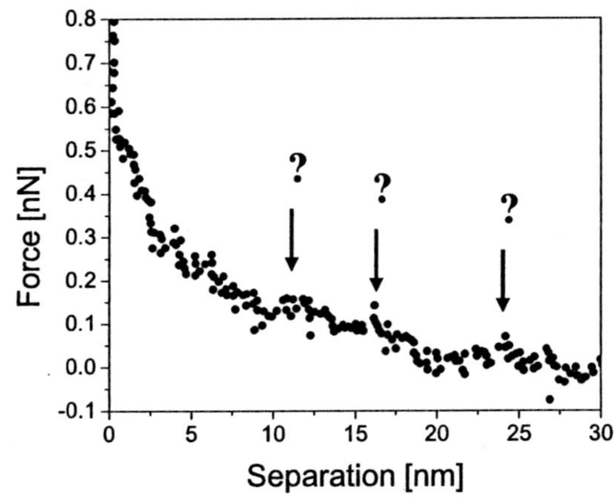

FIGURE 4 Force versus separation curve for a 5 μm glass sphere approaching a surface of mica in a 30 wt% aqueous suspension of 10 nm silica particles (pH 4.1)

significantly due to interference of suspended particles with the laser beam. Additionally, high ionic strength of silica suspensions used (conductivity = 3.9 mS/cm) and low values of zeta potential for silica nanoparticles (less than −14 mV) make the correlation between k and pH difficult to record. We are uncertain yet whether k value represents the Debye length value for the silica nanoparticles or represents compressibility of nanoparticle structures or both.

We expect that attractive forces, which were noticed between the probe and silicon wafer at distances less than ~5 nm, might also influence, to a smaller extent, the interactions between the probe and silica nanoparticles. If this hypothesis is correct, these forces enhance structuring of the silica nanoparticles in the thin film. In support of this hypothesis, we noticed that a stepwise character of the force vs. separation curves as shown in Figure 3 was difficult to reproduce for glass probes that lack attractive forces with mica substrates at a few nanometer distances. For example, Figure 4 shows the force vs. separation curve recorded for a hydrophilic glass sphere interacting with a freshly cleaved mica through the silica nanoparticle suspension at pH 4.1. The curve appears to have small steps at ~10, 16–17, and 24 nm, similar to those observed in Figures 2 and 3, but intensity of these steps is weak. These peaks, if present at all, are hidden in the noisy background of the curve.

The difficulties associated with recording force curves of stepwise character by the AFM colloidal probe technique, using hydrophilic probes and substrates of the same surface charge as dispersed particles, suggests that certain magnitude of attractive forces might be needed to keep the nanoparticles in a thin film of suspension formed between microscopic probes and substrates. If repulsive forces dominated interactions of silica nanoparticles with probes and/or substrates, especially if their magnitude was high, the oscillations were not observed in the force vs. separation curves, or at least they were extremely difficult to separate from the noise. This might indicate that the nanoparticles have a tendency to escape from the probe-liquid-substrate space if repulsive electrostatic forces are too strong. This observation is consistent with that reported by McNamee et al.[9] They also found that the oscillation in the AFM force curve is easier to reproduce if

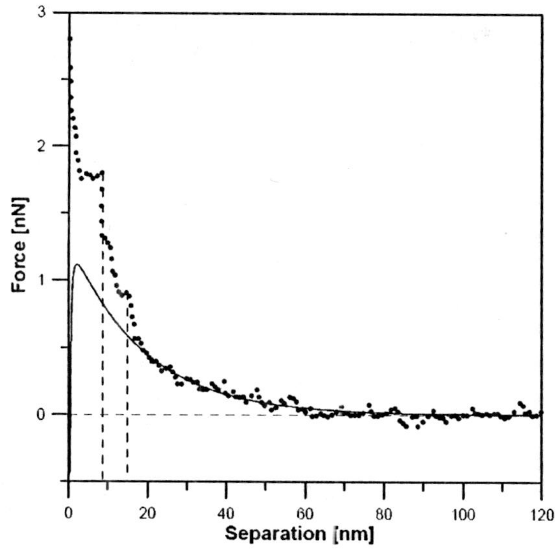

FIGURE 5 Force versus distance curves for the interaction of 15 μm glass sphere with a fluorite substrate in 30 wt% aqueous suspension of 10 nm silica. DLVO constant charge model fit parameters: Hamaker constant A=$4.7 \cdot 10^{-21}$ J (from combining relations), average surface charge = $1.03 \cdot 10^{-3}$ C·m^{-2}, average surface potential = 6.5 mV, Debye length = 11.7 nm.

the electrostatic repulsive forces in the colloidal systems are reduced or eliminated. However, this appears to be in contradiction to studies with foam films of larger areas (>100 μm in diameter) filled with silica nanoparticles.[7] According to Sethumadhavan et al.[7] increased repulsions between nanoparticles produce more homogeneous foam films of suspensions with nanoparticles ordered to more packed structures.

Structural effects are easy to monitor with the AFM colloidal probe technique when attractive electrostatic interactions operate between probes or substrates and nanoparticles. Figure 5 shows the results of surface force measurements between a 15 μm glass sphere and a fluorite surface in the same suspension of 10 nm silica particles. Repulsive interactions observed appeared to be unexpected as glass and fluorite surfaces are oppositely charged at pH 5.8, with glass carrying net negative[12] and fluorite net positive[13] surface charges. These repulsive forces clearly indicate that silica nanoparticles coated the fluorite surface, due to particle-fluorite electrostatic attractions, and reversed the surface charge of fluorite. Such scenario is supported by the research of Atkins et al.[14] They managed attractive electrostatic forces to fabricate coatings of 4–5 nm cerium oxide nanoparticles on mica, as well as coatings of 9–10 nm silica on top of the ceria-coated mica. The coated materials were used to measure colloidal forces in water of varying pH using surface force apparatus. Atkins et al.[14] found that the mica coated with nanoparticles reflects surface charge of nanoparticles rather than mica.

Lack of any strong attractive forces at distances less than the size of silica nanoparticle (10 nm) in Figure 5 also suggests the attached nanoparticles could remain on the fluorite surface even during a final contact of the glass probe with the "fluorite substrate."

The structuring of silica nanoparticles is evident from the results of Figures 2, 3, and 5. Periodicity of steps on the force vs. distance curves is quite regular for silica

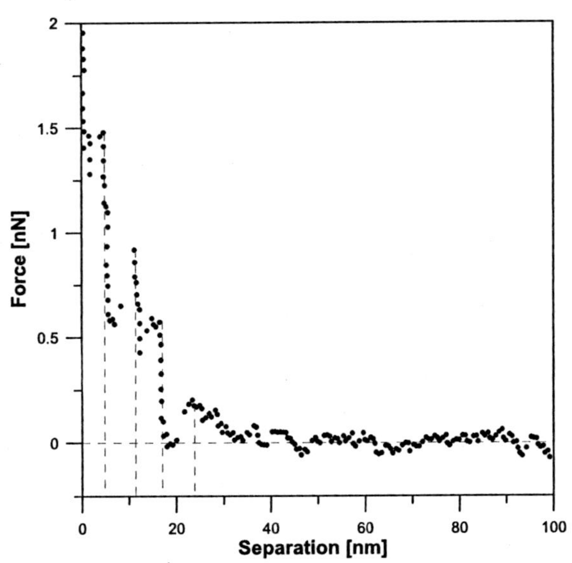

FIGURE 6 Force versus distance curve for the interaction of 15 μm glass sphere probe with freshly cleaved mica in a ~5 wt% water suspension of 5–50 nm alumina nanoparticles

nanoparticles, and varied by 8–9 nm for most of the steps, with the first step being observed at about 10 nm between the probe and substrate. Although formation of up to three layers of organized silica nanoparticles could be deduced from the results of Figures 2 and 3, the structuring of nanoparticles seems to extend beyond 24–28 nm in the experiments with fluorite substrates (Figure 5) with the first two steps being more clear than others. The typical value of 8–9 nm that we measured as a distance between steps suggests close packing of silica spheres in the ordered layers. Close packing predicts that the transition between layers of organized spheres should happen every $a = \sqrt{3}r_p$ distance, which is 8.7 nm for 10 nm silica particles. These calculations however, do not take into account the solvation layer thickness (≤1 nm) surrounding silica particles, which increases the apparent size of nanoparticles and is probably not compressed during the force measurements and the collapse of the ordered layers.

Structural Effects Caused by Alumina Nanoparticles

A number of surface force measurements were also performed in this study with alumina nanoparticles. Although these particles were not uniform and their diameter varied from about 5 to 50 nm, they were available in a dried form and therefore, could be used to prepare suspensions of controlled electrolyte concentrations.

In general, similar oscillatory character of force vs. separation curves was observed during the measurements of colloidal forces between glass probes and mica substrates in suspensions of alumina nanoparticles at pH less than pH<9.5 down to pH~5. Figures 6 and 7 show examples of the typical force profiles.

Strong repulsive interactions as well as the stepwise character of forces are evident on both graphs. For alumina nanoparticles of varying sizes (5 to 50 nm) the oscillation on the force profiles was less regular than that for more uniform silica nanoparticles. The

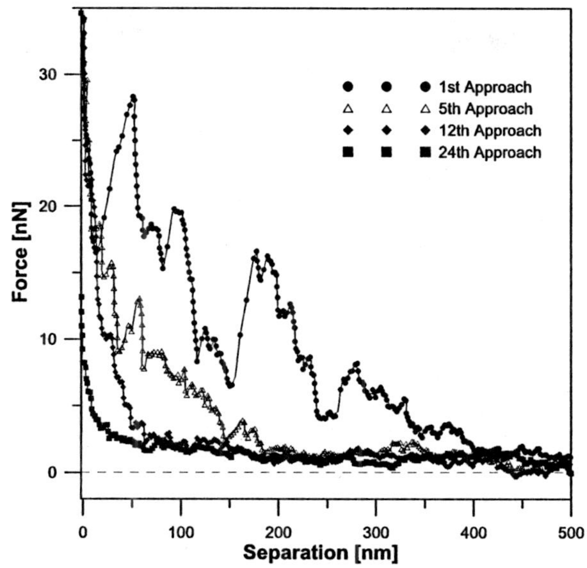

FIGURE 7 Effect of subsequent approaches on the interaction force between mica surface and a 15 μm glass spherical AFM probe in a suspension of alumina nanoparticles. The probe and substrate remained in contact with suspension for several minutes before measurements were performed. Connecting lines are added for clarity.

step changed from about 5 to 50 nm, reflecting a broad variation in alumina particle sizes used. The results in Figures 6 and 7 were obtained with suspensions at pH 5.8. At this pH, both glass[12] and mica[15] were negatively charged, whereas the alumina particles were positively charged (Figure 1). As a result of electrostatic interactions, positively charged alumina nanoparticles attached to both the AFM probes and substrates. An increase in pH of the alumina-in-water suspension up to about 9.2, had little effect on the shape of the force vs. separation curves and the characteristic steps were recorded, although their intensity appeared to be weakened (not shown). The steps on the force profiles nearly disappeared for suspensions with pH 9.5 and could not be located on the force vs. separation curves recorded with suspensions at pH 9.8 (not shown). This observation supports our observation made with silica nanoparticles. A certain value of attractive forces between suspended nanoparticles and one of the macroscopic surfaces, either the AFM probe or substrate, is beneficial for detecting structural effects caused by hydrophilic nanoparticles with the colloidal probe technique.

In selected experiments, we also investigated the dynamic effects during the formation of nanoparticle structures. It was found that the velocity of the probe at which it approaches the substrate through the suspension had little effect on the force vs. separation curves. Similar curves were recorded for scan rate varying from about 0.5 Hz to 14 Hz. However, the oscillation of forces was sensitive to the time elapsing between measurements as well as frequency at which the cantilever probe is brought into contact with the substrate. For example, when the colloidal force measurements were conducted after an incubation time of several minutes or longer, the range of oscillatory forces increased, from 30–40 nm in Figure 6 to 200–500 nm in Figure 7 for alumina suspensions. This finding suggests that the alignment of nanoparticles into structures is a continuous process, at least at the time scale of several minutes of our experiments. In fact, this process might be quite dynamic and takes place over a very long period for alumina suspensions where particles of different sizes compete for lowest-energy positions in multi-layered structures.

Additionally, when the probe was brought into contact with the substrate surface frequently, the oscillatory character of the force vs. separation curve vanished (Figure 7). Decay and finally disappearance of oscillation in measured forces suggest that the configuration of aligned nanoparticles can be destroyed by applied loads and/or turbulences of nanofluids during the AFM measurements. Whether the effect shown in Figure 7 is associated with i) a gradual removal of attached nanoparticles from the substrate and/or probe or ii) destruction of a network of nanoparticles forming more-or-less periodic structure between interacting macroscopic surfaces remains to be determined. We also found that destroyed structures start to rebuild themselves soon after the probe was immobilized and left in the suspension at a large distance from the substrate (not shown).

Effect of Dispersed Nanoparticles on Oscillatory Character of Force-Separation Curves

We limit our interpretation of stepwise changes on force profiles measured in concentrated particle suspensions to a triple layer model of uniform nanoparticles. Such a simplified model is shown in Figure 8 and is expected to be conceptually valid in explanation of three characteristic steps of Figures 2 and 3, one at about 10 nm, second at 17–19 nm and third at 25–28 nm.

At a large separation of the AFM probe from the substrate (Figure 8-A), nanoparticles are probably less ordered. When the probe is at a short distance from the substrate, less than 28–30 nm in the experiments discussed in this section (but it is expected to reach the distances of several nanoparticle diameters in some of the systems according to Wasan et al.[5]), the thermal movement of nanoparticles remaining in the thin film between the probe and the substrate is space restricted. These particles tend to form more organized layers with smaller inter-particle distances than those nanoparticles dispersed in the bulk liquid.

Theoretical analysis on layering of particles in thin liquid films suggests that the structure of layers depends on the concentration of particles and position of a layer in the thin film.[2] Indeed, we found that such structuring of silica nanoparticles used in this study is only detectable by the AFM colloidal probe technique for high concentration suspensions. The results like those in Figures 2 and 3 were recorded for ~30 wt% silica suspensions and we could not reproduce them for suspensions having 6 wt% or less. Previous studies also showed that the oscillation character of force curves could not be recorded for volume fractions of silica nanoparticles less than 0.3.[9] When the concentration of particles in suspensions is low, the particles are dispersed randomly and the film is liquid-like.[5] The situation is different when nanoparticles are charged oppositely to AFM probe and/or substrate (such cases are, for example, represented by results in Figures 5–7). Due to strong electrostatic attractive interactions, nanoparticles tend to coat larger surfaces, even if concentration of nanoparticles is much less than several weight percent (surface inducing structure).

The layered structure of organized particles is represented as a triple layer in Figure 8-B. It is interesting to notice that Wasan and Nikolov[6] also observed layered structures with a thickness of three diameters of particles in a wedge aqueous films formed in air

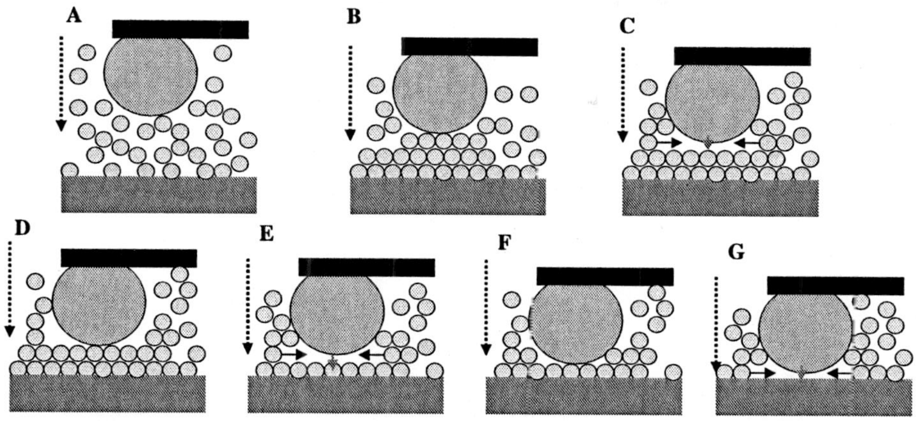

FIGURE 8 Schematic of possible scenarios that take place during the measurements of colloidal forces between a glass sphere and a silicon substrate in a suspension of 10 nm silica particles. Each drawing shows arrangement for nanoparticles in a space between the probe and substrate at different stage of the probe approaching the substrate surface. Probe is far away from the surface of substrate (A); probe-substrate distance is equivalent to about thickness of the three (B), two (D) or one (F) layer of nanoparticles; distance is less than three but larger than two layers of nanoparticles (C), less than two but larger than one layer of nanoparticles (E), less than monolayer of nanoparticles (G).

bubble-sphere latex suspension-glass substrate system. In their study, the latex particles of 1 μm diameter were used in preparation of suspensions. Wasan et al.[8] additionally demonstrated that the first and second layer of ordered particles are colloid crystal-like and have predominantly hexagonal packing structure whereas the third layer is already less ordered. Such two layer crystal-like plus one layer liquid-like model for organized nanoparticles could explain our AFM results with uniform silica nanoparticles. Third layer of nanoparticles seems to be easily displaced by the AFM probe whereas two layers closest to the surfaces of probe and substrate show more resistance to displacement. Obviously, at this stage of our research, this interpretation and our simplified drawing in Figure 8 should be considered as a crude analysis of the AFM results.

Organized structure of nanoparticles as shown schematically in Figure 8 resists thinning of the film due to so-called structural forces and the first steps on the force-separation curves are formed (at about 25–28 nm in Figures 2 and 3). The probe must penetrate a more condense phase and overcome the geometrical barrier of aligned nanoparticles with their solvation layers. At some point, the pressure that builds in the thin film of the organized suspension by the AFM probe pressing onto a substrate, would exceed the structural energy barriers holding together the adjacent two nanoparticle layers. At this point, one of the three nanoparticle layers collapses (Figure 8-C). Although the drawing in Figure 8-C shows that the layer adjacent to the probe is removed, practically the layer with weakest attractive forces will collapse first. At this point we cannot speculate which one.

After the collapse of one of the three organized layers, probe-substrate repulsive forces are reduced. This reduction is the result of both attractive depletion force that results from an osmotic pressure difference and probable weaker electrostatic forces operating over a larger distance between the probe surface and the next layer of the nanoparticles (if the layer of nanoparticles adjacent to the probe surface collapses). As a result, the probe jumps to the next layer of nanoparticles (Figure 8-D). Now, the probe presses the second layer of nanoparticles, which according to results in Figure 2, were held more strongly on the substrate. The sequence repeats (Figure 8-E to Figure 8-G) until all nanoparticles are removed from the space between the probe apex and substrate. As a result, the force profiles for a microscopic probe interacting with a flat substrate in a suspension with much smaller particles become oscillatory. In some cases, it is also possible that the monolayer of silica nanoparticles remain in the probe-substrate space to the end of surface force measurements. Such cases however, were not clearly distinguishable in our measurements and they require the nanoparticles to be located at the axis of the probe apex.

CONCLUSIONS

We have demonstrated that oscillatory structural forces can be studied for suspensions of nanoparticles using atomic force microscopy. Oscillation of forces was recorded for 10 nm silica and 5–50 nm alumina particles suspended in aqueous phases with a periodicity close to the size of nanoparticles used. The periodicity of characteristic steps on the force vs. separation curves recorded for suspensions of 10 nm silica nanoparticles suggest close packing of the particles in the thin nano-scaled films.

This preliminary study suggests that attractive forces between suspended nanoparticles and either probes or substrates or both enhance the appearance of the steps on the force curves. The attractive forces help the ordered nanoparticles sustain certain level of pressure from the probes approaching the substrates. The experiments conducted in this research cannot rule out the possibility of hydrophilic nanoparticles to form ordered structures in thin films between hydrophilic probes and substrates with dominantly repulsive forces, but it shows that these structures, if present, are difficult to study with the colloidal probe technique.

ACKNOWLEDGMENTS

The authors would like to express appreciation to David Schaefer from Eka Chemicals Inc. and Dr. Jeffrey Eastman from Argonne National Laboratory for providing samples of nanoparticles that were used in this study. Laboratory training and assistance with zeta potential measurements provided by Jim Skwarok and Riley Beauchamp are also appreciated. Financial support provided to JD by the University of Alberta is gratefully acknowledged.

REFERENCES

[1] Keblinski, P.; Phillpot, S.R.; Choi, S.U.S.; Eastman, J.A. 2002. *Intern. J. Heat Mass Transf.* **45**, 855.

[2] Nikolov, A.D.; Wasan D.T. 1992. *Langmuir* **8**, 1985.

[3] Marinova, K.G.; Gurkov, T.D.; Dimitrova, T.D.; Algarova, R.G.; Smith, D. 1998. *Langmuir* **148**, 2011.

[4] Sethumadhavan, G.N.; Nikolov, A.D.; Wasan, D.T. 2001. *J. Colloid Interface Sci.* **240**, 105.

[5] Wasan, D.; Nikolov, A.; Moudgil, B. 2005. *Powder Technol.* **153**, 135.

[6] Wasan, D.T.; Nikolov, A.D. 2003. *Nature* **423**, 156.

[7] Sethumadhavan, G.; Bindal, S.; Nikolov, A.; Wasan, D. 2002. *Colloids Surf. A: Physicochem. Eng. Aspects* **204**, 51.
[8] Wasan, D.; Nikolov, A.; Henderson, D. 2003. *AIChE Journal* **49**, 550.
[9] McNamee, C.E.; Tsuji, Y.; Ohshima, H.; Matsumoto, M. 2004. *Langmuir* **20**, 1953.
[10] Liu, J.; Zhou, Z.; Xu, Z.; Masliyah, J. 2002. *J. Colloid Interface Sci.* **252**, 409.
[11] Magual, A.; Horvath-Szabo, G.; Masliyah, J. 2005. *Langmuir* **21**, 8649.
[12] Jacobasch, H.-J.; Grundke, K.; Mader, E.; Freitag, K.-H.; Panzer, U. 1992. *J. Adhesion Sci. Technol.* **6**, 1381.
[13] Haas, S.R.; Nascimenmto, F.R.; Schneider, I.A.H.; Gayloude, Ch. 1999. *Rev. Microbiol.* **30**, 225.
[14] Atkins, D.; Kekicheff, P.; Spalla, O. 1997. *J. Colloid Interface Sci.* **188**, 234.
[15] Scales, P.J.; Grieser, F.; Healy, T.W. 1990. *Langmuir* **6**, 582.
[16] Bremmell, K.E.; Jameson, G.J.; Biggs, S. 1998. *Colloids Surf. A: Physicochem. Eng. Aspects* **139,** 199.
[17] Teschke, O.; Ceotto, G.; de Souza, E.F. 2001. *Phys. Rev. E* **64**, 011605.

PART 5

Applications of Nanomaterials and Nanominerals

Particulate-Based Additives for Enhancing the Efficacy of Sunscreens

Ashoke SenGupta* **and Ralph Spindler***

INTRODUCTION

In recent years, the use of sun protection products has grown considerably due to the increased public awareness of the UV radiation damage of the human skin. The SPF (sun protection factor) rating system has been developed to help consumers select the appropriate sunscreen product for any given outdoor activity involving exposure to the sun. The higher the SPF, the longer is the duration of protection by a properly applied sunscreen.

Formulating high-SPF sun protection products poses a considerable challenge. The number of UV absorbing materials that are approved for use in drug products–sunscreens, is limited. Also, these UV-filters can be used only up to a specified maximum dosage which varies from one UV-filter to another. In addition, these sunscreen actives are considerably expensive, and are mostly organic, oil-like substances that tend to cause skin irritation. Maximizing the SPF for a given dosage of sunscreen actives, therefore, is a strategy widely sought by cosmetic chemists.

Polargel® UV, a nanodispersion-based additive available from AMCOL, enables meeting the aforementioned formulation challenge, being capable of boosting the SPF of sunscreen emulsions by as much as 100% or higher. At its typical use level, it has no significant, inherent SPF-contribution but leads to a more effective utilization of the UV-absorbers, thus resulting in an SPF boost.

* AMCOL International Corporation, Arlington Heights, Illinois

TECHNOLOGY

The Polargel® UV products represent a family of unique, multifunctional additive for cosmetic and personal care products, designed to offer multiple functional benefits that include:

- SPF boost of organic sunscreens; plus
- Thickening of the water phase in oil-in-water (O/W) emulsions
- Dry after-feel
- Antioxidant property
- Polymeric stabilization of O/W emulsions
- Deflocculation or dispersion of pigments in color cosmetics

Manufactured through proprietary processes, these additives are aqueous nanodispersions of particulate materials selected from titanium dioxide (TiO_2), fumed alumina, zinc oxide (ZnO), and sodium bentonite clay. A phenolic, natural polymer (spruce extract polymer) is used as the primary surface-modifier for the particulate materials. Once adsorbed on the particle surface, the surface-modifier sets in particle-to-particle repulsion which in turn acts to deflocculate the particles. The primary particles are thus released from their aggregates, with concomitant reduction in particle size. The finely divided state of the particles coupled with the interfacial properties of the spruce extract polymer provides for a mechanism of producing a more uniform film of the UV-filters on the skin. The outcome is an increased coverage of the skin by the sunscreen actives, leading to the SPF boost.

The spruce extract polymer and particulate network-forming materials such as sodium bentonite clay and fumed alumina render the additive multifunctional. The spruce extract polymer-coated particulate materials present phenolic groups known for having antioxidant properties. The interfacial-activity of the polymer allows the additive to function as an emulsification aid for oil-in-water (O/W) emulsions. On the other hand, the dry after-feel, and the thickening of the water phase in O/W emulsions are attributed to the bentonite clay or the fumed alumina component of the additive.

FUNCTIONAL FEATURES

SPF Boost

In emulsion-based formulations, the Polargel® UV products have been found to increase the SPF by as much as 100% or higher, when used at a dosage level of about 3.5–8% by weight of sunscreen formulations. The result is the exciting possibility of formulating high-SPF products with lower levels of oily, organic sunscreens that tend to be expensive and potentially skin-irritating. Typical SPF boosts (3-subject in-vivo SPF measured as per United States Federal Drug Administration protocol) due to the additive are shown below in Figures 1 and 2, while the reduction in sunscreen dosage achievable with the use of the additive is depicted in Figure 3.

Antioxidant Property

Being a phenolic polymer, the spruce extract can quench free radicals, a basic requirement for the antioxidant property. Based on tests involving the quenching of a relatively

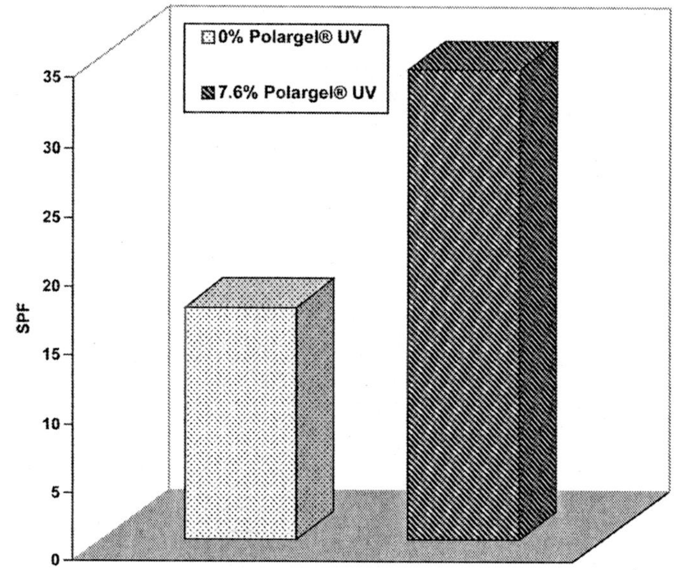

FIGURE 1 SPF boost in SPF 15+ sunscreen (O/W emulsion) (organic sunscreens: 7.5% octyl methoxycinnamate, 5% octyl salicylate, 4% oxybenzone)

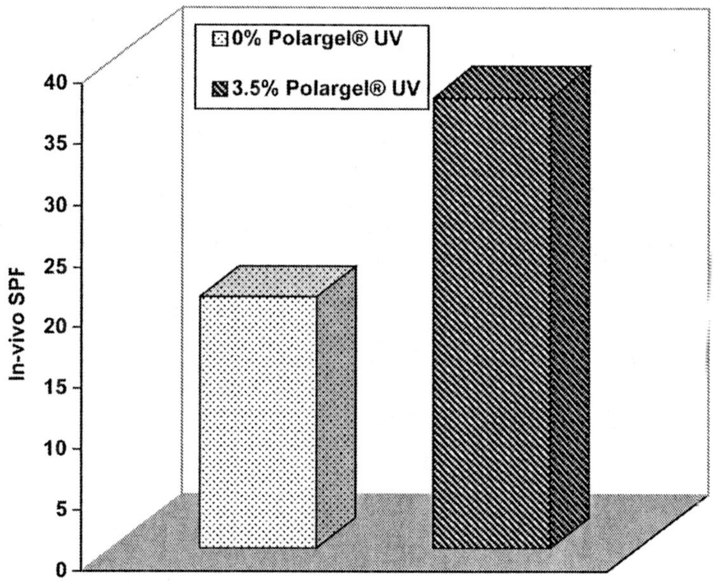

FIGURE 2 SPF boost in SPF 20 sunscreen (W/O emulsion) (organic sunscreen: 7.5% octyl methoxycinnamate, 5% octocrylene)

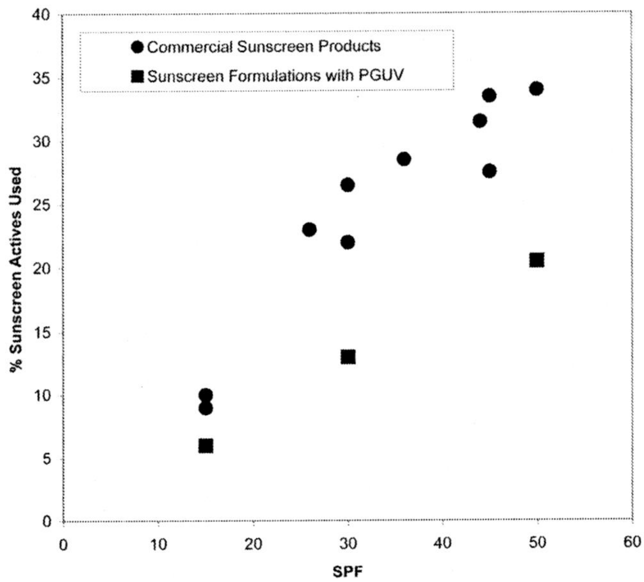

FIGURE 3 Sunscreen active dosage—commercial sunscreen products versus Polargel® UV-containing sunscreen formulations

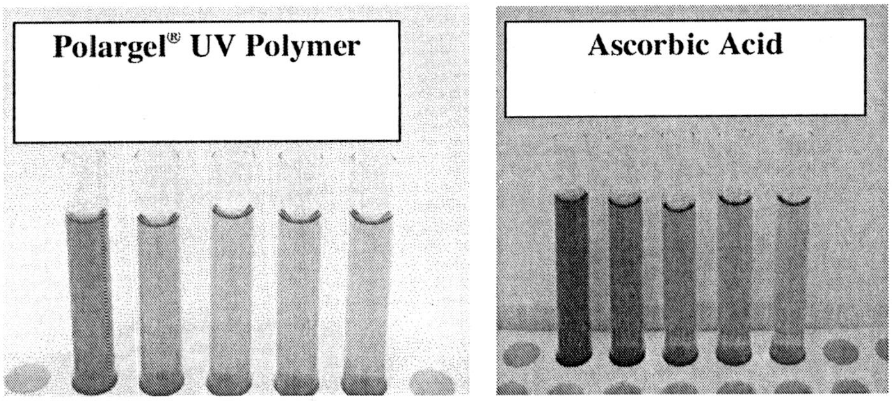

FIGURE 4 DPPH quenching by ascorbic acid and a Polargel® UV polymer

stable free radical diphenylpicryl hydrazyl (DPPH), it was found that adding 7.5% of Polargel® UV is equivalent to adding about 0.01–0.06% (depending on the phenolic group-content of the spruce extract polymer) by weight of ascorbic acid, a potent antioxidant. The DPPH quenching, as indicated by the vanishing of the characteristic purple color of the ethanol-water solution of the free radical, is shown in Figure 4 for both ascorbic acid and a Polargel® UV polymer.

Pigment Dispersion

This feature is particularly beneficial for liquid makeup formulations. The medium-to-high molecular weight anionic, spruce extract polymers can adsorb on a variety of particulate materials. Once adsorbed on the particle surface, they can stabilize the suspended particles against flocculation by providing for electro-steric repulsion between the particles. The polymers, therefore, potentially could work as a dispersant for a variety of particulate materials including the pigments used in color cosmetics.

Extensive deflocculation of iron oxide particles can be achieved with the use of Polargel® UV polymers. An outcome of such deflocculation is the reduction in suspension-viscosity upon addition of a spruce extract polymer to concentrated suspensions of iron oxide (obtained from two different commercial sources), as depicted in Figure 5. The particle size data (Coulter LS 230 particle size analyzer) presented in Table 1, provide a further confirmation of a high level of deflocculation.

Emulsification and Stabilization

The spruce extract polymers' ability to adsorb at an oil-water interface, while providing for interdroplet repulsion, allows them to act as an emulsifier. High-shear equipment such as the rotor-stator mixer, colloid mill, and pressure-homogenizer may be required to produce good emulsions, if no other surfactant is added. In the presence of an added surfactant, the spruce extract polymers may serve as a co-emulsifier. The following demonstrates the emulsifying ability of a Polargel® UV product.

Emulsion Formulation:

Phase	Ingredient	% By Weight
A	Deionized Water	63.89
A	Propylene Glycol	5.0
A	Xanthan Gum (2%)	7.5
A	Polargel® UV	7.6
A	Phenonip (Preservative)	1.0
B	DC 245 Fluid/Finsolv TN	15

Procedure:

- Combine all Phase A ingredients in a Waring blender.
- Mix well until uniform.
- Add Phase B (either 245 fluid or Finsolv TN) slowly to the batch and homogenize the mixture in the Waring blender @ 22,000 rpm.

Emulsion Stability Testing:

- The emulsions remained stable after heating to 60 °C, followed by centrifuging @ 3,000 rpm for 30 minutes.

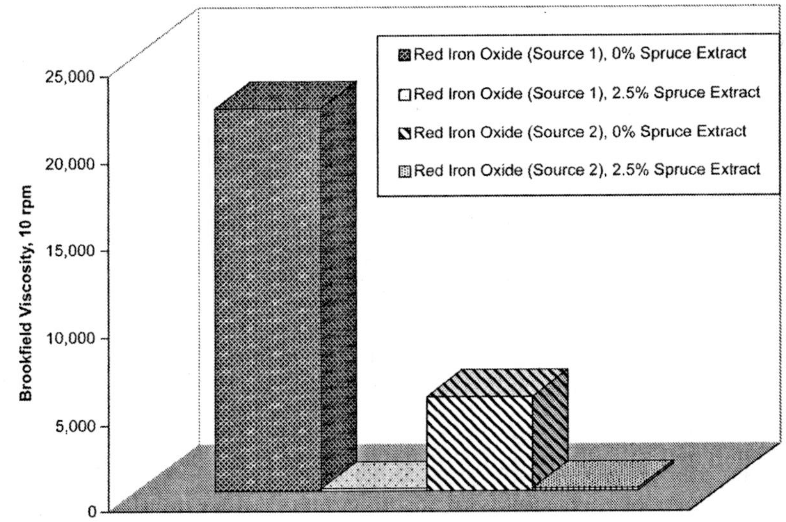

FIGURE 5 Suspension-viscosity in 50% (w/w) iron oxide dispersions, with and without a Polargel® UV (PGUV) polymer

TABLE 1 Particle size of iron oxide in suspensions, with and without a Polargel® UV (PGUV) polymer

Suspension ID (Polymer Dosage is Based on the Weight of Iron Oxide)	Weight Percent of Pigment, Based on the Total Weight of Pigment and Water	Mean Particle Diameter, micron	Number % of Particles with Diameter Smaller than a Given Diameter
Iron Oxide (Source 1), 0% PGUV Polymer	7.8%	3.22	90% < 5.096 micron
Iron Oxide (Source 1), 2.5% PGUV Polymer	7.8%	0.529	90% < 0.674 micron
Iron Oxide (Source 1), 2.5% PGUV Polymer	50%	0.549	90% < 0.738 micron
Iron Oxide (Source 2), 0% PGUV Polymer	20%	2.999	90% < 4.746 micron
Iron Oxide (Source 2), 2.5% PGUV Polymer	20%	0.619	90% < 0.898 micron

TYPICAL FORMULATIONS

Sunscreen Formulation with 7.5% OMC, 5% Octyl Salicylate, 4% Oxybenzone and 7.6% Polargel® UV, In-vivo SPF: 34

Phase	Ingredient	% By Weight
A	Deionized Water	26.00
A	Polargel® UV (48.6% solids)	7.60

(continued)

Phase	Ingredient	% By Weight
B	Sodium Caproyl Lactylate (12.5% aqueous solution)	20.0
B	Propylene Glycol	2.50
B	Xanthan Gum (2%)	7.50
B	Sodium Hydroxide (15%)	0.40
B	Phenonip (preservative)	1.00
C	Octyl Methoxycinnamate	7.50
C	Octyl Salicylate	5.00
C	Oxybenzone	4.00
C	Finsolv TN (C_{12-15} alkyl benzoate)	6.00
C	DC 245 Fluid (Cyclomethicone)	5.00
C	Isopropyl Myristate	7.50

Procedure:

- Combine the Phase A ingredients and mix well until uniform or lump-free in a Silverson rotor-stator homogenizer.
- Add the Phase B ingredients to the batch and mix well.
- Combine the Phase C ingredients and mix well until the solid components dissolve completely.
- Add the combined Phase C to the batch. Mix well until uniform, at a Silverson speed of at least 5,000 rpm.

Polargel® UV, although a water-based additive, does not impede formulating very water resistant (VWR) sunscreens. Given below are examples of VWR SPF 15+ and 50+ (determined clinically, i.e., in-vivo) formulations that contain Polargel® UV.

VWR SPF 15+ Formulation			VWR SPF 50+ Formulation		
Phase	Material	% By Wt.	Phase	Material	% By Wt.
A	DI Water	73.16	A	DI Water	63.515
	Polargel® UV	3.5		Polargel® UV	7.6
	Polargel NF (Montmorillonite)	0.6		Disodium EDTA	0.1
	50% NaOH	0.48		50% NaOH	0.3
	Glycerin	1		Glycerin	1
	Phenonip (Preservative)	0.9		Phenonip (Preservative)	1
	Keltrol T (Xanthan Gum)	0.3		Keltrol T (Xanthan Gum)	0.4
	Ultrez 21 (Carbopol)	0.2		Carbopol 940	0.1
B	Isopropyl Myristate	7	B	Isopropyl Myristate	3
	Cetyl phosphate	0.625		Lanolin Oil	0.5
	Cetyl alcohol	0.625		Cetyl phosphate	0.625
	Octyl methoxycinnamate	5		Octyl methoxycinnamate	7.5
	Avobenzone (Parsol 1789, organic sunscreen)	2		Oxybenzone	4
				Octyl Salicylate	5
	Polyethylene AC 1702 (water-resistance enhancer)	2.5		Avobenzone (Parsol 1789)	2
				Polyethylene AC 1702	1.25

(continued)

VWR SPF 15+ Formulation			VWR SPF 50+ Formulation		
Phase	Material	% By Wt.	Phase	Material	% By Wt.
C	DI Water	2	C	DI Water	2
	Allantioin	0.1		Allantioin	0.1
	Ascorbic Acid	0.01		Ascorbic Acid	0.01
	Total	100.0		Total	100.0

SUMMARY OF BENEFITS

- High SPF with reduced amounts of sunscreens
- Elegant formulations–less oily and skin-irritating
- Natural ingredients including a polyphenol
- Secondary benefits
- Significant cost savings

ACKNOWLEDGMENTS

The authors thank Ms. Ilona Lin and Mr. David McGregor for carrying out most of the developmental work leading to the technology presented herein. We also thank Mr. Ro Oteri for compiling the data presented in Figure 3. Permission from AMCOL Research and Development to publish this work is also gratefully acknowledged.

Synthesis of Mesoporous Silica Coatings on Magnetic Particles for Mercury Removal Applications

Jie Dong* and Zhenghe Xu*

Magnetic particles with mesoporous coatings of high specific surface area and controlled pore sizes were synthesized by optimizing synthesis conditions. After functionalization with mercapto-propyl-trimethoxy-silane, the particles were tested as a novel and effective sorbent for mercury removal from industrial waste effluents. SEM, TEM, BET surface area measurement and atomic adsorption spectroscopy were used to characterize the sorbents and their affinity for mercury. Loading capacities of mercury as high as 2.6 mg/g and 3.6 mg/g were obtained at aqueous pH values of 2 and 7, respectively. The nanocomposites exhibit a strong magnetism which allows the subsequent magnetic separation of the sorbents from a complex multiphase system.

INTRODUCTION

The extreme toxicity of mercury poses a serious threat to the environment. An estimate of a total annual global input of mercury to the atmosphere from all sources including natural, anthropogenic, and oceanic emissions is 5500 tons (U.S. Environmental Protection Agency 2003). Discharge of mercury-containing effluents from relevant industries has been a concern of environmental authorities. Considerable efforts have been devoted to the study of effective removal of heavy metals, including mercury from waste water.

The discovery of mesoporous materials has been attracting worldwide attention due to their potentials in a wide range of applications, ranging from emission control, detoxification of industrial effluents, removal of heavy metals, recovery of precious metals from industrial processes and production of ultra-clean water or solvents to being used as

* Department of Chemical and Materials Engineering, University of Alberta, Edmonton, Alberta, Canada

catalysis, reactive and specialty sorbents, nanoreactors, templates of nano-synthesis, biological sensors and electronic devices. In 1992, scientists at Mobil Oil Corporation announced the direct synthesis of the first broad family of mesoporous silicates known as the Mobil composite of matter, or MCM for short. The formation of mesopores is based on a liquid-crystal templating (LCM) mechanism (Kresge et al. 1992 (a, b)). Following this method a large number of highly porous materials with regularly arranged and uniform pores in 2 nm to 5 nm diameter range has been prepared. A unique feature of this type of materials is their large surface area desirable for many designated applications.

Taking heavy metal removal or precious metal recovery from an industrial effluent as an example, it is highly desirable if not necessary to achieve a strong and specific affinity for target metal ions on mesoporous silica of large specific surface area. The specific affinity of silica-based mesoporous materials is accomplished by surface functionalization. For environmental cleanup applications, silanation has been identified as a most promising avenue of introducing specific surface functional groups on mesoporous silica surfaces (Feng et al. 1997; Mercier and Pinnavaia 1998 (a, b); Lim et al. 1998; Brown et al. 1999, 2000; Antochshuk and Jaroniec 2002). The comparatively large (nanometer-scale) pore structure of open frame work provides an uninhibited access of the metal ions to every ligand site in the structure, which is considered as the most important advantage of these types of adsorbents. However, the functionalized mesoporous silica itself has limitations in industrial applications. One challenge is to separate the loaded fine powder silica from a complex, multiphase system for safe disposal or recovery of the adsorbed metal ions and recycle of the sorbents.

The magnetic separation could be a viable method to separate the spent sorbents from multiphase systems if these mesoporous particles are of magnetic property. There are three major preparative methods to embed magnetic materials into a non-magnetic matrix. The first is to entrap colloidal iron oxides (such as Fe_2O_3) or ions inside the pores of a silica matrix and then reduce them to magnetic iron oxides (Fe_3O_4) or metallic iron under H_2 gas environment at high temperatures. The drawback of this method is a relatively low level of iron content (< 5%) achievable in the silica matrix, which makes it necessary to use extreme experimental conditions for magnetic phase formation (Nourlinos et al. 2001). The second approach is to use zeolites as porous materials. In this case, the desired metal ions (such as Pb, Cu, and Cd ions) are exchanged with cations in zeolites from dilute solutions, together with magnetite-$Fe(OH)_3$ coprecipitation, followed by magnetic filtration (Feng et al. 2000). The third method involves coating of tiny inexpensive magnetic particles with functional ligands to fabricate inorganic and organic hybrid particles (Caruso et al. 1999; Caruso 2000; Ohmori and Matijevic 1993; Xu and Liu 1997). The third method is considered to be the most versatile and the easiest to be adopted. The sorbents prepared by this method can provide sufficient magnetization by choosing the proper size of core magnetic particles for the subsequent magnetic separation.

To obtain a uniform and well covered mesoporous silica layer on magnetic particles of high specific surface areas, the control of synthesis conditions is extremely important. Many synthesis conditions, such as reaction time, reaction temperature, reactant concentration, type and amount of catalyst, quality of solvent, could affect the properties of the resultant silica coatings. A classical approach to optimize these reaction parameters is to use so called "one-variable-at-a-time" strategy (Groupy 1993). In this approach, only one variable is varied and optimized each time. Although frequently applied, this classical methodology may not be efficient in determining the true optimal conditions, because it neglects synergetic effect of the variables. To minimize the number of synthesis while considering variable interactions, factorial design of statistical nature is often used to study major trends and determine a promising direction for further experimentation (optimization). The basic idea of the factorial design approach is to devise a small set of experiments, in which all pertinent factors are changed systematically using standard factorial design tables. Subsequent analysis of the experimental results allows factors of major influence and synergetic effect to be identified. Applying the information derived above to a statistical model, one can effectively search for optimal synthesis conditions. The main applications of this methodology have been to the development of new products or processes and to optimization of existing product and manufacturing processes in industrial practices (Steinberg and Hunter 1984).

The main objective of this paper is to optimize synthesis conditions by factorial design and analysis of synergetic effects among different factors, for producing mesoporous silica coatings of high specific surface area on magnetic particles for mercury removal applications.

EXPERIMENTAL

Materials

Magnetite particles of <5 μm sizes (98% Fe_3O_4) from Aldrich Canada were used as received. Sodium silicate ($Na_2SiO_3 \cdot 5H_2O$), cetyl-trimethyl-ammonium chloride (CTAC, 25 wt% in water), tetraethyl orthosilicate (TEOS, 98%) and mercapto-propyl-trimethoxy-silane (MPTS, 95%) from Aldrich were used in dense liquid silica coating, molecular templating, sol-gel reaction and functionalization, respectively. Hydrochloric acid (37 wt%) and ammonium hydroxide (29.4 wt%) were purchased from Fisher Scientific and used as catalysts. Absolute ethanol, Millipore de-ionized water (Millipore, Canada) and HPLC grade toluene (>99%, Aldrich) were used whenever needed. Mercury standard solutions (1000 ppm in 10% HNO_3) from Fisher Scientific were diluted to different concentrations as needed for adsorption tests.

Synthesis Procedures

A typical procedure for the preparation of magnetic sorbents is described as follows. Dense liquid silica coating was applied first to protect the magnetic cores from being leached under severe industrial application conditions (Liu et al. 1998) and to prepare a surface amenable for molecular-template assembly (Wu and Xu 2004). In molecular templating, a CTAC aqueous solution (25 wt%) was added to obtain a final surfactant concentration of 75 mM, which was well above the critical micelle concentration of the surfactant. The dense silica coated particles after thorough-washing with de-ionized water without drying were placed in the above CTAC solutions which were gently agitated by a mechanical stir for 30 minutes. A desired amount of TEOS solution was then added, followed by drop-wise addition of catalysts. The synthesis under mechanical agitation was allowed for a given period. At the end of synthesis, the magnetic particles were collected by a hand magnet and the collected particles were washed repeatedly by ethanol and then dried in a vacuum oven at room temperature for overnight.

Since the sol-gel reaction was identified to be affected by many factors, with the amount of catalyst and TEOS, and reaction time being the major process variables, these

TABLE 1 Conditions for 2^3 factorial design

	Catalyst ($NH_3 \cdot H_2O$)/mL (A)	Sol-gel reaction time/h (B)	TEOS/mL (C)
+	13.5	15	0.9
–	9	10	0.45

TABLE 2 Complete 2^3 factorial design: factors and levels

No.	Catalyst (A)	Time (B)	TEOS (C)
1	–	–	–
2	+	–	–
3	–	+	–
4	+	+	–
5	–	–	+
6	+	–	+
7	–	+	+
8	+	+	+

three factors were selected as variables to be optimized in synthesis. A standard 2^3 (two levels three variables) factorial design was applied to study relative significance of these parameters in the synthesis of mesoporous silica coatings. The levels for each variables and designs of these experiments are given in Tables 1 and 2, respectively. The specific surface area of the synthesized composite particles was used as the response variable. The composition of base solutions in volume ratio of absolute ethanol: DI water: 25% CTAC was set at 180: 13.5: 18. The addition of catalyst and TEOS was adjusted to reflect various levels in Table 1.

Functionalization by Thiol Ligand

A given amount (0.5 gram in this study) of mesoporous silica-coated magnetite particles was steamed over boiling de-ionized water for 2 hours. The hydrolyzed sample was dried in a vacuum oven over night at room temperature. The dried sample was suspended by vigorous stir in 100mL of toluene. The suspension was contained in a three-neck flask with a reflux condensing tube. The refluxing was performed under a N_2 atmosphere to prevent oxidation of the thiol groups. After dispersion, 1 mL of MPTS was added to the flask with the suspension under continuous agitation. The reaction was allowed to proceed for 24 h under a reducing reflux atmosphere. The resulting product was separated magnetically, washed with toluene and dried overnight under vacuum at room temperature.

Characterization

The BET surface areas of magnetite particles after different treatments were analyzed by measuring N_2 adsorption-desorption characteristics using an AutoChem II 2920 analyzer. Scanning electron microscopy (SEM, Hitachi S-2700) and X-ray spectrometer (Princeton Gamma-Tech, PRISM IG (Intrinsic Germanium) detector) were used for the

TABLE 3 Results of the 2^3 factorial design experiments

No.	Mean	Catalyst (A)	Time (B)	TEOS (C)	AB	AC	BC	ABC	β_j (m^2/g)	β_j' (m^2/g)	β_j'' (m^2/g)	S_j^2
1	+	−	−	−	+	+	+	−	91.6	92.0	91.2	0.32
2	+	+	−	−	−	−	+	+	65.0	65.6	64.4	0.72
3	+	−	+	−	−	+	−	+	72.2	71.2	73.2	2.00
4	+	+	+	-	+	−	−	−	70.1	70.3	70	0.05
5	+	−	−	+	+	−	−	+	76.1	76.6	75.6	0.50
6	+	+	−	+	−	+	−	−	96.5	96.7	96.3	0.08
7	+	−	+	+	−	−	+	−	150.9	151.1	150.7	0.08
8	+	+	+	+	+	+	+	+	68.1	67.0	69.1	2.20
Divisor	8	4	4	4	4	4	4	4				
Effect	86.3	−22.8	8.0	23.2	−19.7	−8.5	15.2	−31.9				

cross-section analysis. High resolution TEM (JEOL-2010) was used to examine the morphology of the synthesized mesoporous silica layer and to measure pore sizes. Magnetic properties of the particles before and after treatment were determined with a Quantum Design (San Diego, CA, USA) 9T-PPMS dc magnetometer. Atomic Absorption Spectrophotometer (VARIAN 220FS (Fast Sequential) AAS) was used to determine mercury concentrations in solution before and after adsorption.

RESULTS AND DISCUSSION

Factorial Design

The results of the 2^3 factorial design are shown in Table 3. The specific surface area was measured twice for each sample (β_j', β_j'') and β_j is the average of the two values. One of the most important outcomes using factorial design is the analysis of relative importance of variables, individually and collectively. The details of the analysis are given in Appendix and the results are summarized in Table 4. As shown in Table 4, the main effects of the amount of catalyst and TEOS are more significant than that of the reaction time, reflected by larger surface area changes in response to their variations. Although the results indicate the main effect to be significant, the interactive effects such as A×B and B×C (−19.7 and 15.2) can not be neglected. However, the overall interactive effect of A×B×C (−31.9) turns out to be the most dominant effect for this system.

The existence of a significant interaction among the variables for the sol-gel process suggests that the main effects must not be interpreted individually when optimizing synthesis conditions. The interactive effects can be visualized by the diagram shown in Figure 1. The shadowed face represents the optimum results for specific surface areas. When interpreting this diagram, the three-factor effect is verified to be dominating. The above analysis clearly shows that it would be impossible to carry out a complete analysis of the situation by a univariate study. The factorial design method would suggest optimal conditions for the sol-gel reaction at low catalyst concentration, long reaction time and high TEOS dosage. When the time is fixed at high level (+), reducing the amount of catalyst can increase specific surface area, especially at high level of TEOS.

TABLE 4 Summary of effects and standard errors for the 2^3 factorial design

Effect	Estimate ± Standard Error
Average	86.3±0.22
Main effect	
Catalyst (A)	−22.8±0.43
Time (B)	8.0±0.43
TEOS (C)	23.2±0.43
Two-factor interaction	
A×B	−19.7±0.43
A×C	−8.5±0.43
B×C	15.2±0.43
Three-factor interaction	
A×B×C	−31.9±0.43

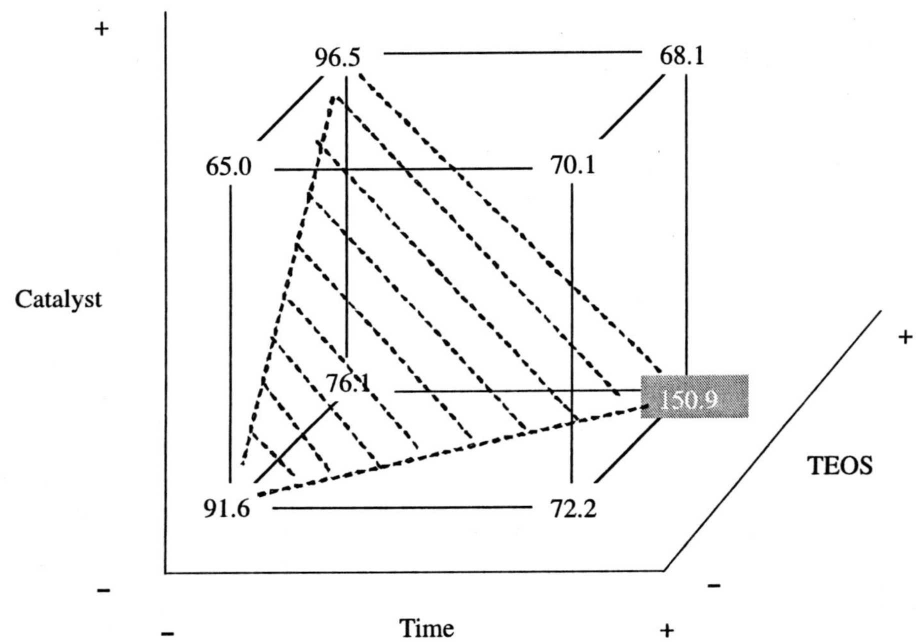

FIGURE 1 Factorial design diagram for the surface area of the sorbents

Optimization

From the 2^3 factorial design experiments, it can be concluded that the interaction effects are dominant. Compared with the other two variables, the effect of reaction time is relatively minor and will be kept constant (15 hours) in the following experiments. The other reaction conditions with the corresponding results are given in Table 5. The surface areas for the tests coded as No.11 (149.2 m^2/g) and No. 7 (150.9 m^2/g) are similar.

TABLE 5 Optimization of the experiment

No.	Catalyst (mL)	Time (h)	TEOS (mL)	BET (m^2/g)
(7)*	9	15	0.9	150.9
(9)	6.8	15	1.125	104.3
(10)	5.6	15	1.350	121.6
(11)	3.4	15	1.575	149.2
(12)	2.2	15	1.800	123.4

* (7) is the No.7 sample of 2^3 factorial design

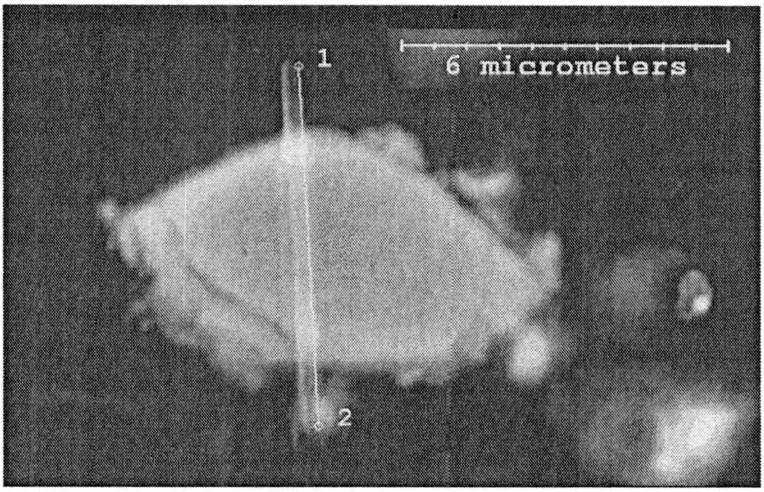

FIGURE 2 SEM cross-section analysis

It appears that a specific surface area of about 150 m^2/g is the optimum for coating mesoporous silica layers on micron-size magnetite particles.

Gravimetric analysis showed a weight increase of about 10%–15% by silica coating. With this mass increase as coated silica, the specific surface area of the mesoporous silica layer is estimated to be about 1,500–1,000 m^2/g, which is among the best results of bulk mesoporous silica materials reported in literature.

SEM Cross-section Analysis

The cross-section SEM image of a polished magnetite after mesoporous silica coating is shown in Figure 2. X-ray microanalysis was applied across the sample from point 1 to point 2. The corresponding X-ray intensity of silicon is shown in Figure 3 from left to right. Two obvious silicon peaks at the left and right edges in Figure 3 were observed. The X-ray intensity from silicon diminished towards the center of the particle where iron oxide was exposed by polishing. The result confirms a silica layer coated around the exterior of the magnetite particles.

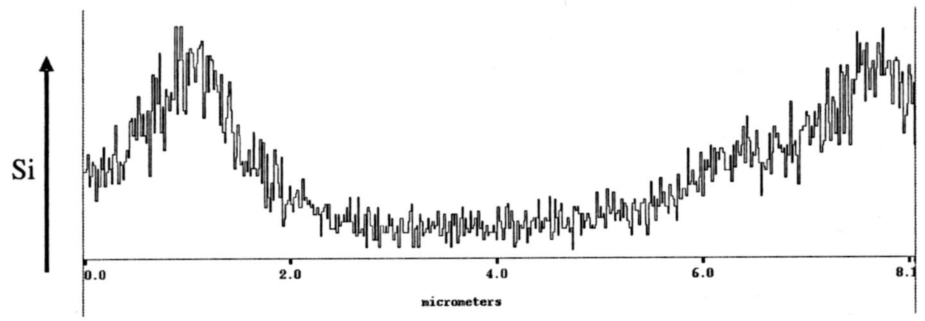

FIGURE 3 X-ray microanalysis result of the particle shown in Figure 2

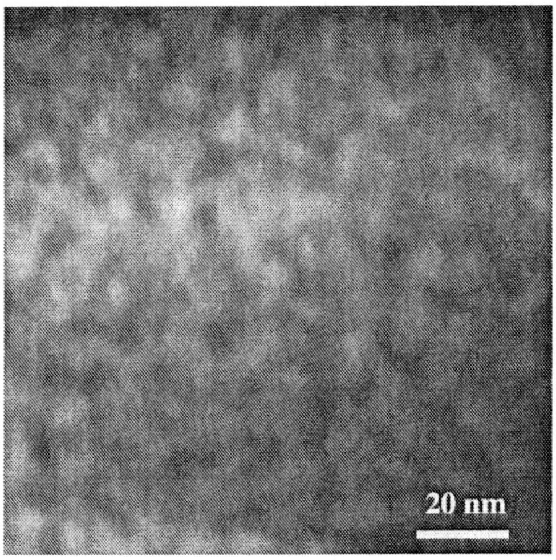

FIGURE 4 High resolution TEM of the silica layer

TEM Results

High resolution TEM study of the silica layer revealed the honeycomb structure of mesoporous silica coatings (Figure 4), as anticipated. The pore size is estimated to be around 7–8 nm. This value is in excellent agreement with the size of the templates assembled on silica wafer, determined using in-situ atomic force microscopy (AFM) (image is not shown here).

Magnetization Measurement

The room temperature magnetization curves for bare and mesoporous silica-coated magnetite particles are shown in Figure 5. A saturation magnetization of 85.5 emu/g was obtained for magnetite particles without coating. The saturation magnetization of particles coated with diamagnetic silica decreased to 73.0 emu/g. Since the saturation magnetization can be considered to be proportional to the mass fraction of magnetic

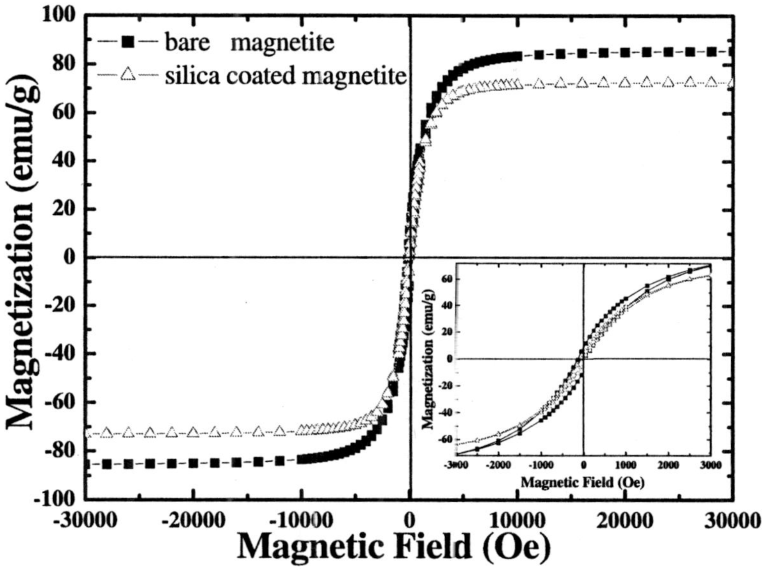

FIGURE 5 **Room temperature magnetization measurements**

material in the particles, the observed decrease of 15% in saturation magnetization by coatings of 15% diamagnetic silica is anticipated. Since a particle with a saturation magnetization of 73.0 emu/g remains strongly magnetic, such a reduction in saturation magnetization does not interfere with the subsequent magnetic separation after the resultant magnetic particles are loaded with mercury. More importantly, magnetite particles coated with mesoporous silica can be considered to remain super-paramagnetic as shown by minimal coercivity and hysteresis on the magnetization curves (insert in Figure 5). This magnetization characteristic ensures that the magnetite particles do not become permanently magnetized after exposure to an external magnetic field, which in turn permits the particles to be readily re-dispersed when the magnetization field is removed. It is clear that the magnetic sorbents prepared as such can be isolated in practice by magnetic separation for reuse or recycle.

Mercury Adsorption Isotherms

The amount of mercury adsorbed per gram of sorbents, Γ, was determined by Eq. (1).

$$\Gamma = \frac{(C_i - C_e) \times V}{1{,}000 \times M} \quad \textbf{(EQ 1)}$$

where Γ is in mg/g, V is the volume of mercury-containing aqueous solution (in milliliters), M is the mass of sorbents (in grams), C_i is the initial mercury ion molar concentration (before contact) and C_e is the equilibrium mercury ion concentration (after contact). In the current study, mercury adsorption on mesoporous silica-coated magnetite as prepared was found to be negligible. After functionalization, the silica-coated mesoporous magnetite showed a strong affinity for mercury in aqueous solutions, indicating the importance of surface functionalization in sorbent applications. The mercury adsorption

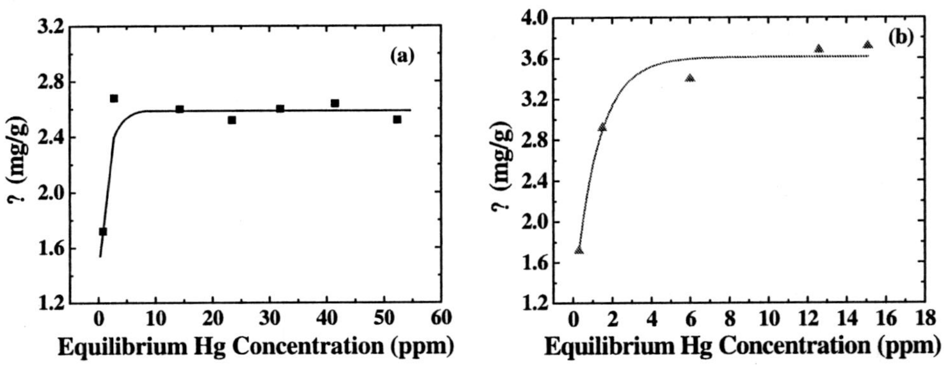

FIGURE 6 Mercury adsorption at (a) pH 2 and (b) pH 7 by functionalized magnetic sorbents

on the functionalized magnetic sorbent exhibited the Langmuir type of isotherms as shown in Figure 6, suggesting a monolayer adsorption. The mononlayer adsorption indicates a chemisorption mechanism of mercury on functionalized mesoporous magnetite surfaces. The maximum loading capacity of mercury increases from 2.6 mg/g at pH 2 to 3.6 mg/g at pH 7 as shown in Figure 6, indicating a competitive adsorption of mercury with protons in acid medium for surface thiol groups.

CONCLUSIONS

Using factorial design method, the concentration of catalyst, reaction time and TEOS dosage were found to affect collectively the coating of mesoporous silica on magnetite particles. The optimal synthesis conditions for increasing surface area of mesoporous silica layer on magnetite particles were identified to use a small amount of catalyst with large dosage of TEOS over an extended reaction period. Under the optimum synthesis condition, the specific surface area of the total material at 10%–15% silica coating was as high as 150 m^2/g, which translates to 1,000 m^2/g–1,500 m^2/g per gram of coated silica. The presence of silica layers on magnetite was confirmed with SEM imaging and X-ray microanalysis. The pore size of honeycomb silica coatings were determined from high resolution TEM to be about 7–8 nm, similar to the size of templates measured by in situ atomic force microscope. The resultant composite particles were determined to be strongly magnetic with a specific saturation magnetization of 73 emu/g. Such particles could be easily separated by conventional magnetic separation for regeneration and recycle. The functionalized mesoporous silica exhibited a strong adsorption of mercury from aqueous solutions. The adsorption was found to be pH dependent: the saturation loading capacity about 3.6 mg/g at pH 7 was higher than that at pH 2 (2.6 mg/g).

ACKNOWLEDGMENTS

The financial support of NSERC/EPCOR/AERI is gratefully acknowledged. We also would like to thank Feng Wang (Physics Department at University of Alberta) for his great help in obtaining TEM micrographs.

APPENDIX: EFFECTS ANALYSIS OF FACTOR DESIGN

A quicker method to analyze the effects is to use a table of contrast coefficients (Table 3). For example, the estimate of the mean is calculated from the first column,

$$\frac{+91.6 + 65.0 + 72.2 + 70.1 + 76.1 + 96.5 + 150.9 + 68.1}{8} = 86.3$$

and the A × C interaction from the sixth column,

$$\frac{+91.6 - 65.0 + 72.2 - 70.1 - 76.1 + 96.5 - 150.9 + 68.1}{8} = 8.5$$

The remaining effects are obtained similarly to yield the estimates already displayed in Table 3. To compare and emphasize the main effects of the synthesis process, it was necessary to calculate the standard error of the effects. Since the adopted experimental design was constituted by eight experiments, the error (σ) was calculated as a linear combination of the eight average values obtained in the respective duplicates (see Eq. (2)

$$\sigma = \left[\left(\frac{1}{8} + \frac{1}{8}\right) \times \left(\sum \frac{s_j^2}{8}\right)\right]^{\frac{1}{2}} \quad \textbf{(EQ 2)}$$

in which s_j is the j th experimental variance).

REFERENCES

Antochshuk, V. and M. Jaroniec, 1-Allyl-3-propylthiourea Modified Mesoporous Silica for Mercury Removal, Chem. Commun., **3**, 258–259 (2002).

Brown, J., L. Mercier and T.J. Pinnavaia, Selective Adsorption of Hg^{2+} by Thiol-functionalized Nanoporous Silica, Chem. Commun., **1**, 69–70 (1999).

Brown, J., R. Richer and L. Mercier, One-step Synthesis of High Capacity Mesoporous Hg^{2+} Adsorbents by Non-ionic Surfactant Assembly," Microporous Mesoporous Mater., **37**, 41–48 (2000).

Caruso, F., Nanoengineering of Particle Surfaces, Adv. Mater., **13**, 11–22 (2000).

Caruso, F., A.S. Susha, M. Giersig and H. Mohwald, Magnetic Core-shell Particles: Preparation of Magnetite Multilayers on Polymer Latex Microspheres, Adv. Mater., **11**, 950–953 (1999).

Feng, D., C. Aldrich and H. Tan, Removal of Heavy Metal Ions by Carrier Magnetic Separation of Adsorptive Particles, Hydrometallugy, **56**, 359–368 (2000).

Feng, X., G.E. Fryxell, L.-Q. Wang, A.Y. Kim, J. Liu and K.M. Kemner, Functionalized Monolayers on Ordered Mesoporous Supports, Science, **276**, 923–926 (1997).

Goupy, J.L., Methods for Experimental Design-principles and applications for physicists and chemists, Data Handling in Science and Technology, Vol. 12, pp 1–8, Elsevier, Amsterdam, 1993.

Kresge, C.T., M.E. Leonowicz, W.J. Roth, J.C. Vartuli, J.S. Beck, Ordered Mesoporous Molecular Sieves Synthesized by a Liquid-crystal Template Mechanism, Nature, **359**, 710–712 (1992 a).

Kresge, C.T., M.E. Leonowicz, W.J. Roth, J.C. Vartuli, Synthetic Mesoporous Crystaline Material, US Patent 5098684, 1992 b.

Lim, M.H., C.F. Blanford and A. Stein, Synthesis of Ordered Microporous Silicated with Organosulfur Surface Groups and Their Applications as Solid Acid Catalysts, Chem. Mater., **10**, 467–470 (1998).

Liu, Q., Z. Xu, J.A. Finch and R. Egerton, A Novel Two-Step Silica-Coating Process for Engineering Magnetic Nanocomposites, Chem. Mater., **10**, 3936–3940 (1998).

Mercier, L. and T.J. Pinnavaia, A Functionalized Porous Clay Heterostructure for Heavy Metal Ion (Hg^{2+}) trapping, Microporous Mesoporous Mater., **20,** 101–106 (1998 a).

Mercier, L. and T.J. Pinnavaia, Heavy Metal Ion Adsorbents Formed by the Grafting of a Thiol Funcitonality to Mesoporous Silica Molecular Sieves: Factors Affecting Hg (II) Uptake, Environ. Sci. Technol., **32**, 2749–2754 (1998 b).

Nourlinos, A., A. Simopoulos, D. Petridis, H. Okumura and G. Hadjipanayis, Silica-maghemite nanocomposites, Adv. Mater., **13,** 289–291 (2001).

Ohmori, M. and E. Matijevic, Preparation and Properties of /uniform Coated Colloidal Particle: Silica on Iron, J. Colloid Interface Sci., **160,** 288–292 (1993).

Steinberg, D.M. and W.G. Hunter, Experimental Design: Review and Comment, Technometrics, **26,** 71–97 (1984).

U.S. Environmental Protection Agency, Mercury Study Report to Congress: Overview. http://www.epa.gov/airprogm/oar/mercover.html (accessed June 2003).

Wu, P., J. Zhu and Z. Xu, Template-Assisted Synthesis of Mesoporous Magnetic Nanocomposite Particles, Adv. Funct. Mater., **14,** 345–351 (2004).

Xu, Z. and Q. Liu, Silanation and stability of 3-Aminopropyl Triethoxy Silane on Nanosized Superparamagnetic Particles: I. Direct Silanation, Appl. Surf. Sci., **120**, 269–278 (1997).

Synthesis of Silicon Carbide by Reaction Milling in a Dual-drive Planetary Mill

Debasis Chaira,* B.K. Mishra,* and S. Sangal*

Reaction milling has become a very effective means of synthesizing compounds from the corresponding elemental constituents. In this work, silicon carbide is synthesized from elemental silicon and graphite powder by high-energy reaction milling in a specially built dual drive planetary mill. Reaction milling in this type of mill is most affected by the size of the grinding media. The phase evolutions, particle size distribution, and morphology of particles during milling are studied during a 40 hour grinding period. Particle size distribution shows that 70% of particles is below 10 micron. X-ray diffraction study indicates complete conversion of silicon and graphite to silicon carbide. The crystallite size varies from 150 nm from the start to 10 nm after 40 hours of milling. The lattice strain increases with milling up to about 20 hours when SiC forms and subsequently with the formation of SiC the lattice strain is reduced. The SEM pictures clearly reveal the progress of size reduction and the extent of agglomeration of particles. Finally, dispersion of SiC in aluminum matrix is studied at different temperatures. While excellent compatibility between SiC and Al is evident from the SEM micrographs, the problem of contamination and agglomeration remains.

INTRODUCTION

Silicon carbide is a promising filler material due to its good thermal and chemical stability. This is valid both during synthesis and under severe service conditions. The SiC particulates can be used with aluminium for synthesis of Al-SiC metal matrix composites. Aluminium based metal matrix composites are becoming very popular due to their lightweight, high strength to weight ratio, stiffness and wear resistance properties. The specific applications of these composites include engine blocks, pistons, brake-system components, seals, solid lubricants, wear and abrasion resistant structures, electromechanical contacts and

* Indian Institute of Technology, Kanpur, India

chassis components. Owing to its high temperature electrical properties, high breakdown voltage, and high electron mobility silicon carbide is used in electronic devices that are particularly suitable for high temperature applications. It is also used in various other types of high temperature applications such as heating elements and refractory material for furnaces due to its high fracture strength, excellent wear and creep resistance, high resistance to corrosion, good thermal conductivity and relatively low coefficient of thermal expansion.

Silicon carbide can be produced by many methods (Vogt et al., 1985; Canon et al., 1982). The traditional method which is to a large extent commercialized is known as the Acheson process. Here a solid-state reaction between sand and petroleum cokes at very high temperature (2500°C) in an electric arc furnace using two large graphite electrodes leads to the formation of silicon carbide. The commercial product has a large grain size and is invariably contaminated with oxygen. Other fabrication processes for obtaining SiC include polymer pyrolysis, chemical vapor deposition (CVD) and even hot pressing. Liquid phase reactions have also been used to synthesize silicon carbide, as reported by Ritter (1995). Most of the above mentioned approaches require temperatures higher than 1500°C. However, silicon carbide can also be synthesized without heating, such as high energy milling of elemental powders at ambient temperature, which is termed as reaction milling.

The mechano-chemical synthesis (reaction milling) of metal carbides was proposed by Le Caer et al. (1990). They prepared carbides of Ti, V, Cr, Mn, Fe, Co, Ni, Zr, Nb, Mo, Ta, W, Re, Al, and Si by milling the above elements with graphite powder in either a SPEX 8000 shaker mill or a Fritsch P-7 planetary mill for a fixed amount of time (Matteazzi and Caer, 1991). El-Eskandarany et al. (1995) prepared stoichiometric SiC powders by solid-state reaction of elemental silicon and carbon powders via the room temperature mechanical alloying process. Complete fcc–SiC alloy powders were obtained after 300 hours of milling. The mechanical alloying process was performed in a high-energy ball mill. Yang and Shaw (1996) also produced SiC from silicon and graphite by attrition milling. Based on the x-ray line broadening they concluded that product compound was β-SiC where the average crystallite size was of the order of 70 nm.

REACTION MILLING

Reaction milling is a process where simultaneous milling and chemical reaction takes place in a highly energetic environment. It can be practiced in planetary mills where the force field could be made to vary from one to two orders of magnitude compared to equivalent size ball mills. Reaction milling uses mechanical processing to induce chemical reactions. The major areas where mechano-chemical processing could be applied include ultrafine powder production, mineral and waste processing, metals refining, combustion reactions, production of a fine dispersion of second phase particles, extension of solubility limits, refinement of the matrix microstructure/nanograin formation, formation of amorphous phases. An important feature of mechanochemical process is the refinement of microstructure (i.e., grain size and particle size) associated with simultaneous particle deformation, fracture, and welding processes that accompany ball/powder collision events. The energy transmitted to crystalline powders during milling may result in dislocation cell structure that develops into random nanostructured grains with increasing milling time. It should be realized that even though nanometer grain sizes are realized during mechano-chemical processing, the particle size typically decreases only to micrometer level.

Planetary milling can induce chemical reactions in a variety of powder mixtures. In fact it has been shown that mechanical activation substantially increases the kinetics of solid-state chemical reactions (Schwarz and Petrich, 1988; Schwarz et al., 1985). In the systems of interest here, reaction does not occur at the beginning of the milling process. However, it becomes possible after a certain activation time, as milling reduces the particle size, thoroughly mixes the components, and increases the number of chemically active defect sites. The activated states or the reaction zones not only increase as particle and grain sizes decrease, but are also regenerated through the repeated particle fracture and welding events. As a consequence, reactions that require high temperatures to occur will occur at low temperatures in a planetary mill without any need for external heating. Thus, a wide range of chemical reactions can be mechanically initiated.

PLANETARY MILLING

The dual drive planetary mill consists of a gyratory shaft and two cylindrical steel jars, both are rotated simultaneously and separately at high speed. Such high-speed rotation of both jars and the shaft makes the balls to move strongly and violently, leading to large impact energy of balls that improves the grinding performance. This type of mill has been used not only for grinding but also for mechanical alloying (MA) and mechanochemical (MC) operations (McCormick and Froes, 1998).

The dual drive planetary mill developed specifically for the synthesis of nanocarbides has a rotating shaft that sweeps a circle of diameter 750 mm. The two steel jars of 13.5 mm diameter (2500 ml each) rotate about their own axes around the common axis of the main shaft. The planetary mill is powered by two motors. A 5 HP motor works on the main rotating shaft and a 3 HP motor drives the jars. The rotating speed of both motors can be varied independently and continuously by a frequency controller. This mill design follows closely the patented design of Rajamani et al. (2000) with several minor modifications to improve its performance and suit to the local conditions. Figure 1 shows a configuration of one half of the planetary mill where a jar rotates about a primary axis O. Here G is the diameter of rotation of the axis of the jar and D is the diameter of the jar itself. Once these parameters are fixed, it turns out, that the planetary mill could be simply characterized by the ratio of the diameter of rotation of the mill axis G to the diameter of the mill D. Let R be the ratio of speed of rotation of the mill axis to the speed of rotation of the mill about its center. Then, if the shaft rotates at an angular velocity of ω, the mill rotates at $R\omega$ on the shaft. A negative value of R indicates that the mill and the shaft rotate in opposite directions. During operation, four different kinds of forces act on a ball inside any one of the jars as shown in Figure 1. These are (i) centrifugal force about the fixed gyrating axes, (ii) centrifugal force about the mill axis, (iii) Coriolis force due to the planetary motion (rotating coordinate system). These forces produce a special acceleration field inside the mill which is difficult to visualize.

The dynamics of the planetary mill dictates that

$$R = -1 \pm \sqrt{\frac{G}{D}} \quad \textbf{(EQ 1)}$$

which allows computation of the critical speed. The percent critical speed is

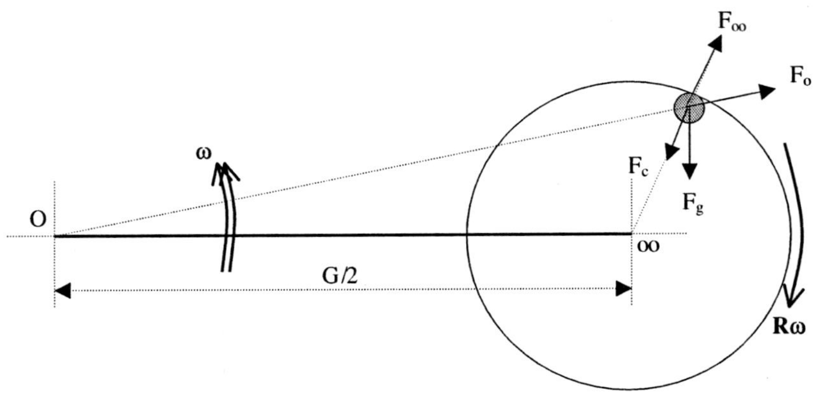

FIGURE 1 Schematic of the acceleration field in a planetary mill

$$\%CS = \frac{N_2}{R \times N_1} \times 100$$

where N_1 is the gyrating speed and N_2 is the jar speed. In our case, the mill was operated at 63% critical speed. This was arrived at by doing the following calculation:

$$R = -1 \pm \sqrt{\frac{750}{135}} = -1 \pm 2.36$$

Taking the minus sign for opposite direction of motion between the jar and the main shaft we get,

$$R = -3.36$$

The percent critical speed is calculated by considering the shaft speed and jar speed as 225 and 475 rpm. Therefore

$$\%CS = \frac{475}{3.36 \times 225} \times 100 = 62.8$$

The mechanical description of the milling process is best understood by considering the overall motion of the balls inside the milling chamber and the mechanics of an individual collision including the elastic and plastic deformation of the powder. Much research along these lines was performed in connection with modeling the mechanical alloying process and a summary of this research is available in the literature (Courtney, 1995). It is clear that the force field that exists inside the mill is quite complex and for this reason predicting the motion of the grinding media is not a trivial task. Nonetheless, using the discrete element method (DEM) this complex motion could be visualized (see Mishra, 1995; Cleary and Hoyer, 2000). DEM not only allows visualization of the dynamic charge mass but also provides information on the frequency of collisions with specific relative velocities that can be used for the modeling purpose.

MATERIALS AND METHODS

Starting materials used for milling were silicon and graphite with 99.5 and 99% purity respectively. Steel balls of diameter 6 and 12 mm were used for milling. When setting up a milling experiment, one has to select the number and size of the milling balls and the total mass of the powder. Many investigators choose a ball-to-powder ratio of about 10:1 or a somewhat larger value for a planetary mill. Such conditions work well in most cases. Here, silicon and graphite powders were mixed in 1:1 atomic ratio. Milling was carried out for 40 hours in two jars each containing 125 gm of powder and 2.5 kg steel balls.

Interruption and room temperature aging is known to significantly alter the kinetics of some mechanochemical reactions. However owing to practical convenience, all the experiments were carried out for 5 hours on a continuous basis and then the jars were allowed to cool. To prevent agglomeration a dispersant known as 1% by weight of the powder Licowax was added. Once the jar temperature attained the room temperature, it was opened to collect the powder. A small sample was carefully retained for size and X-ray analysis by using the pulversite sample splitter. The remaining powder was put back into the jars and the milling was continued with fresh addition of Licowax. The aging effect at room temperature could not be studied.

The powder samples were analyzed at different intervals of milling time to study the progress of reaction. Powder particles were characterized by X-ray diffraction (CuK_α), laser particle size analyzer (Malvern Master sizer 2000), and scanning electron microscope (FEI Quanta).

Preparation of Al-SiC Composites

By froth flotation technique the unreacted graphite particles were separated from the bulk powder that was obtained after 40 hours of milling. The concentrate and the tailing fractions were allowed to dry in an oven. The tailing that mostly contained silicon carbide particles was subjected to further separation using a heavy liquid (density 2.6 gm/cc). The float and sink fractions contained silicon and silicon carbide particles respectively. The overall yield after the separation processes was calculated to be 66%. The silicon carbide powder was mixed with aluminium powder for the synthesis of Al-SiC metal matrix composite. SiC powder at 10 and 20 vol.% were mixed with Al powder and pellets were prepared by cold compaction. The cold compacted pellets were then sintered at 600°C for 1 hour to study the mechanical property. The scanning electron microscope (SEM) with the built in energy dispersive spectroscopy (EDS) were used to study the morphology of particles.

RESULTS AND DISCUSSION

XRD Analysis

In order to determine the phase change of the particle mixtures during the reaction milling process, a small sample of milled product was repeatedly picked up at regular intervals for X-ray diffraction analysis. Figure 2 shows the XRD graph of powder milled for 40 hours. The two different X-ray patterns correspond to milled powder prepared by using 6 and 12 mm diameter steel balls. In either spectra the peaks corresponding to Si_5C_3, SiC and βSiC are present. However, the intensity of silicon carbide peaks is higher corresponding to the powder that was processed with 12 mm diameter steel balls as

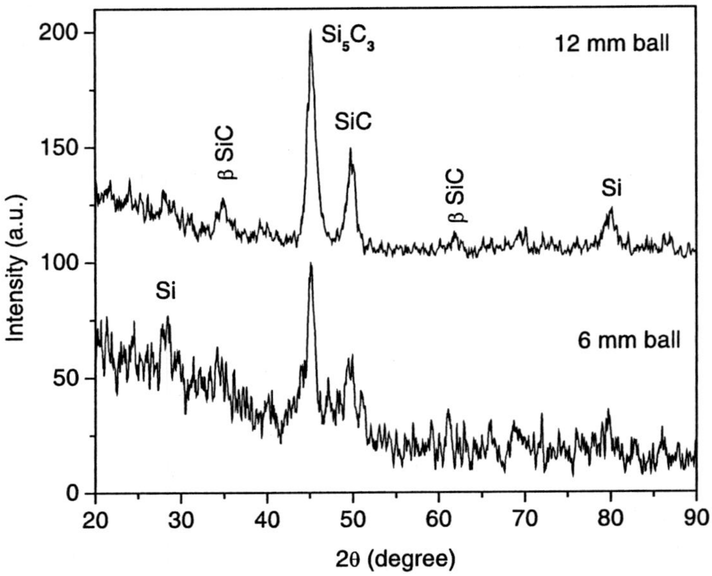

FIGURE 2 **XRD patterns of milled powder at 40 hours of milling times for two different balls**

compared to the powder processed by 6 mm balls under identical conditions. At the end of the experiment, it was also observed that the 6 mm balls were invariably found coated with the powder whereas no such coating was observed on the 12 mm balls. This shows that the intensity of impact in case of 12 mm diameter ball is sufficient to dislodge any coating that forms on the surface of the ball.

The XRD patterns of the end product (as milled SiC powder) are shown in Figure 3 for different intervals of milling time. Several interesting observations are made. First, the graphite peak disappeared after about 10 hours of milling. This is a general observation in this study, which may be due to amorphization of graphite during milling. However, at this stage the disappearance of the graphite peaks was not accompanied by a shift in the silicon peaks, which indicate that there was no formation of Si(C) solid solutions. In fact, the shifts in the peaks would have suggested that a transitional bonded state between Si and C formed, although no trace of SiC was observed. Second, with increased milling time, silicon peaks also disappeared and silicon carbide peaks were observed. The silicon peak disappeared around 20 hours of milling. Finally, as evident from Figure 3, the Bragg peaks for the milled product (after about 10 hours of milling) are broad, suggesting accumulation of lattice strain and reduction in crystallite size. These two individual contributions to broadening have been analyzed in the next section.

Crystal Size and Lattice Strain

Some features of the activation process, particularly the crystal size, can be investigated by analyzing the X-ray diffraction pattern. For this purpose, as received and milled powders were analyzed using X-ray diffraction (XRD) methods with CuK_α radiation. The XRD peak broadening was used to measure the particle size and internal strain. The broadening due to small crystal size was evaluated through Scherrer formula

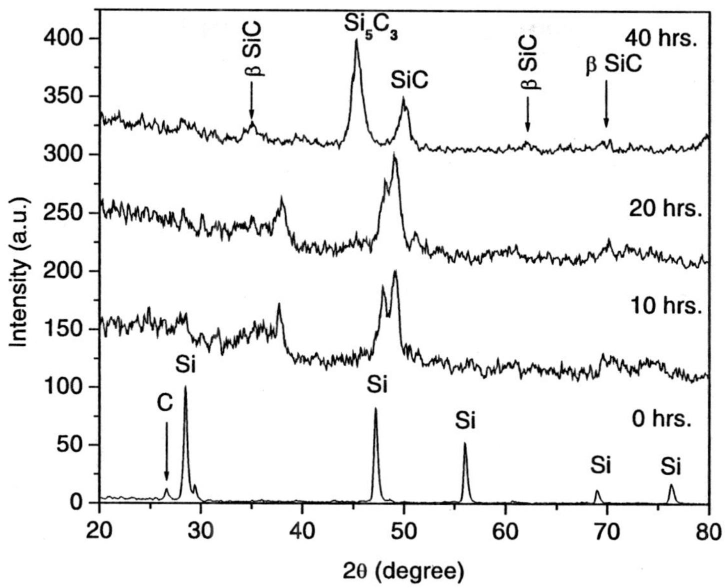

FIGURE 3 XRD patterns of milled powder at selected milling times

$$B_p(2\theta) = \frac{0.9\lambda}{D\cos\theta}$$

where D is the average crystal size, B_p *(2θ)* is the broadening of the diffraction line measured at full width half maximum intensity (FWHM), λ is the wave length of the x-ray radiation and θ is the Bragg angle. The strain broadening can be calculated as

$$B_s(2\theta) = 4\eta\frac{\sin\theta}{\cos\theta}$$

where B_s *(2θ)* is the broadening due to internal strains and η is the effective internal strains. The total broadening, B_t *(2θ)*, due to both the crystal size and internal strains was assumed to be the linear addition of the two contributions. From the slope and the intercept of the plot of $B_t\cos\theta$ versus $\sin\theta/\lambda$, the crystallite size and internal strain were calculated. Correction for instrumental broadening, which comes due to the superposition of peaks due to $K_{\alpha1}$ and $K_{\alpha2}$, when peaks are not resolved, was taken into account in the measurement of the peak broadening. This was done by running XRD of pre-milled powder under the same conditions as the milled powder. The breadth at half maximum intensity (FWHM) of the desired curve B_t (2θ), which could be obtained if there were no instrumental broadening, was calculated by linear subtraction of breadth of before milled powder from total breadth.

The decrease of the grain size and lattice strain to characterize the activation process has been determined from the X-ray diffraction patterns. Although the accumulation of lattice strain is a measure of defect formation, determining the defect structure was found to be more difficult. The crystallite size and the lattice strain of the powder measured from the XRD peak broadening is shown as a function of milling time in

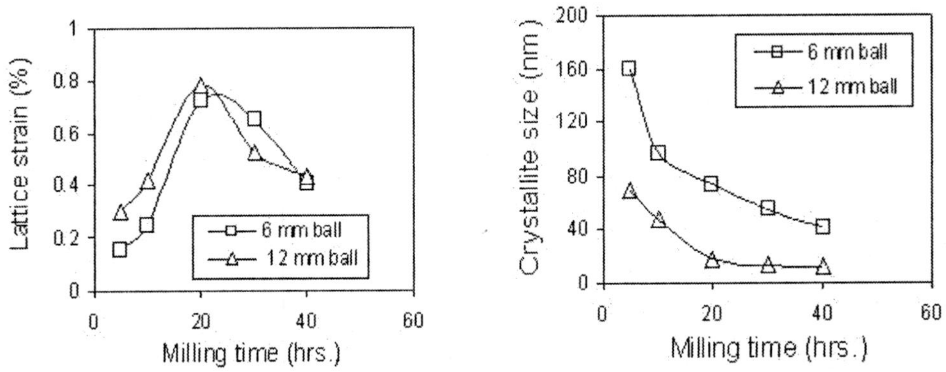

FIGURE 4 Effect of ball size on lattice strain and the crystallite size of the milled powder calculated from X-ray diffraction patterns; left: lattice strain; right: crystal size

Figure 4. It can be seen that the crystallite size decreases and internal strain increases rapidly with milling time up to about 20 hours. With further milling the crystallite size remains almost constant but internal strain appears to decrease. This could be the onset of phase change (i.e., nucleation of SiC). Finally, from Figure 4 it is noticed that the different mass of the balls results in different crystal size and strain accumulation rates.

Particle Size

During the early stage of milling, the powder particles of Si and C are mixed together and then mechanically activated such that individual particles of Si and C could form composite particles. At a later stage, after about 20 hours of milling, the solid-state reaction starts and considerable amounts of the product (SiC) are formed. Figure 5 shows the particle size distribution of powder at different intervals of milling time. While the d_{25} (25% passing size) size continuously decreased from 10 to 1 micron, the d_{75} size decreased from 25 to 15 micron. Similar experiments under identical conditions with 12 mm balls produced a coarser particle size distribution after 40 hours of grinding. However, the rate of SiC formation during milling was better with 12 mm ball.

There was no evidence of welding, as particle size did not increase during the initial stages of milling. It should be mentioned that the size spectra corresponding to individually milled product of Si and C did not show any evidence of growth in particle size. In contrast, in TiC system, El-Eskandarany (1996) analyzed the polished and etched particles representing the early stage of grinding. They showed that the individual particle contains many thick layers of the diffusion couples of constituent powders. They believed that a solid-state reaction occurred at the clean-interfaces of these layers, and a new single phase was formed after further milling. When the new phase formed the reactant layers disappeared. Therefore, in such systems the only way the particles could grow is by cold welding. However, similar mechanism of phase formation was not observed in the SiC system.

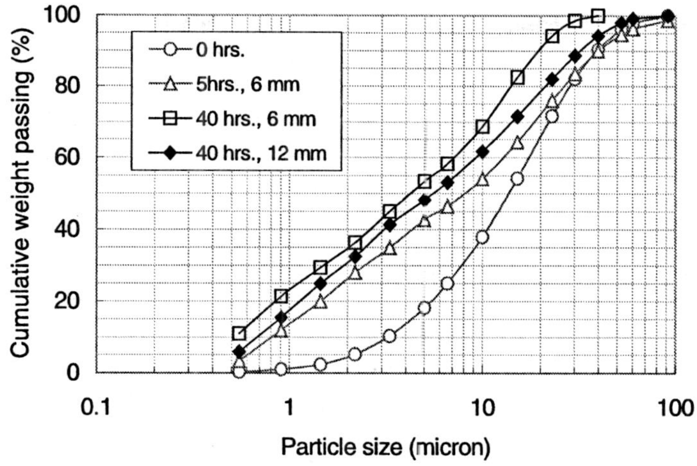

FIGURE 5 Particle size distribution of the milled product at different intervals of time

Temperature

The temperature of outer surface of the jars during milling was measured by a high precision laser pyrometer. The temperature measurements were carried for a total period of 5 hours. During this period milling was carried out continuously. It was observed that after about 3 hours of grinding a steady state condition was attained when the temperature remained constant. The variation in the measured temperature of the outer surface of the jar during milling is shown in Figure 6. Here the temperature profiles correspond to milling of pure graphite and silicon carbide (silicon and graphite). It is clear from the figure that the outer surface temperatures reach 90° and 75°C during SiC and graphite milling respectively. Under the steady state condition, the temperature of the inner surface of the jar was calculated by a simple heat balance. The measured temperature of the outer surface of the jar and calculated temperature of the inner surface of the jar for different materials are presented in Table 1.

The actual processing of the powder takes place during the collisions between two balls or a ball and the wall of the mill. From the point of view of reaction milling, a particularly important question is the temperature increase during collision. The possible approaches were reviewed by Koch (2003) and recently worked out in some detail by Murty et al. (2004). Depending on the model, rather different estimates of the temperature increase were obtained. For example, when Fe powder was impacted by an 8-mm steel ball, the inner wall temperature was estimated to be 150°C. These temperatures appear to be sufficient for chemical reaction given the state of mechanical activation achieved in the planetary mill.

Scanning Electron Microscopy (SEM)

The SEM technique was used to follow the changes in shape and size of the milled powders during the different stages of the reaction milling process. Figure 7 shows the SEM micrographs of the milled powders after different intervals of milling time. In all the micrographs the marker is set at 50 micron. At the initial stage of milling, the powders of

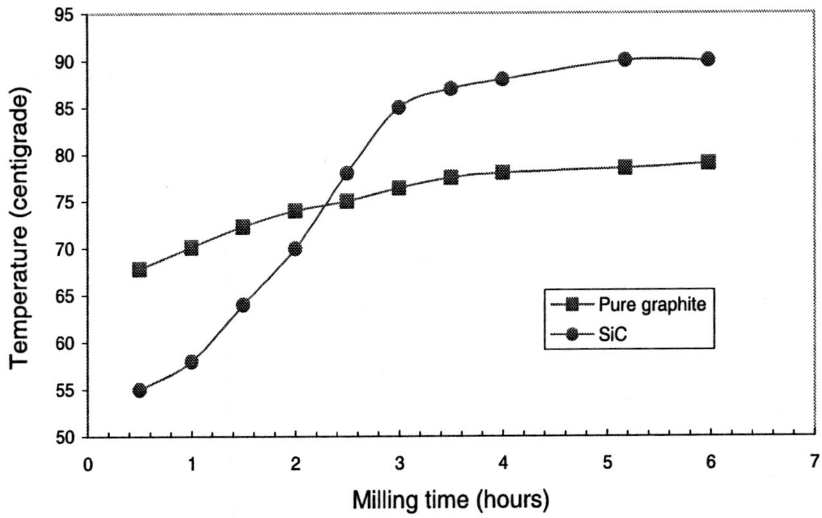

FIGURE 6 Rise in temperature of the outer surface of the jar

TABLE 1 Inside and outside vial surface temperature for different materials

Materials	Measured Outside Temperature (°C)	Calculated Inside Temperature (°C)
Pure iron	116	150
Pure graphite	79	113
Iron + graphite (C: Fe = 1.5)	93	127
Iron + graphite (C: Fe = 2.0)	83	117
SiC	90	124
Limestone	85	119

the reactant materials are bulky, with random shape and size. As the milling progressed the powders became spherical as evident from Figure 7(d), which correspond to 40 hours of milling. As regards particle size. It is also evident from the SEM images that it decreases gradually with increasing milling time. However, no evidence of particle coarsening could be obtained through SEM.

Characterization of Al-SiC Composite

To study the morphology and distribution of silicon carbide particles in the Al matrix, SEM was used. Figure 8 shows the SEM picture and the EDS spectra of Al–10 vol.% SiC composite that was sintered at 600°C in an inert atmosphere for 1 hour. The micrograph shows that SiC particles are agglomerated and non-uniformly distributed in the Al matrix. The interface between the metal and the particle plays a critical role in determining the properties of metal matrix composites. The SEM micrograph shows bonding between Al matrix and SiC particles. The EDS of embedded particle shows Si and Fe while the matrix shows only Al as major constituents. This indicates that the SiC powder

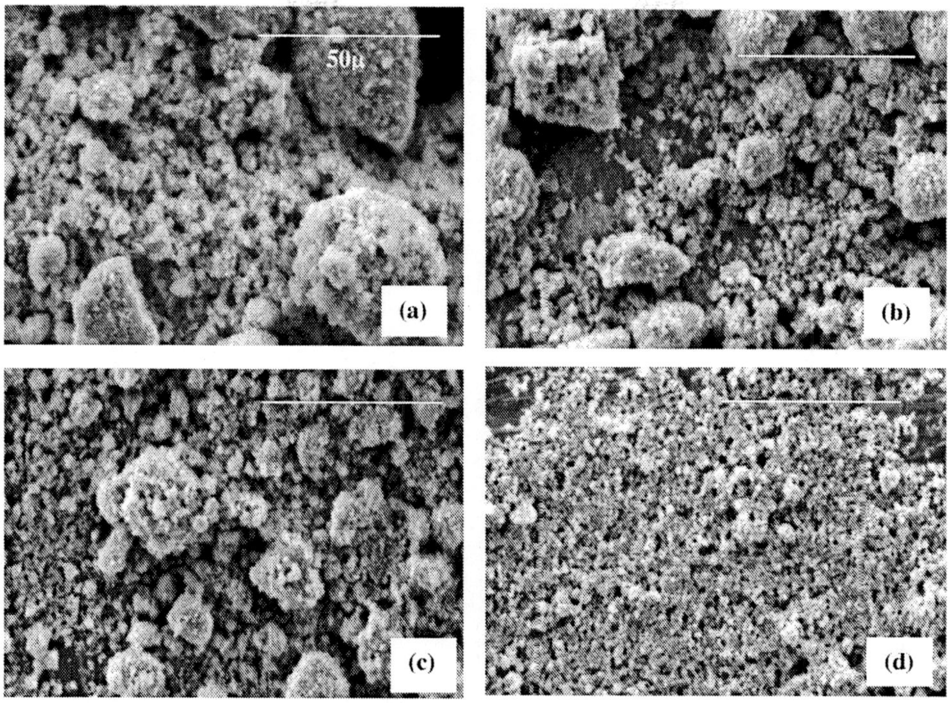

FIGURE 7 SEM photograph of the powder after (a) 10 (b) 20 (c) 30 (d) 40 hours of milling

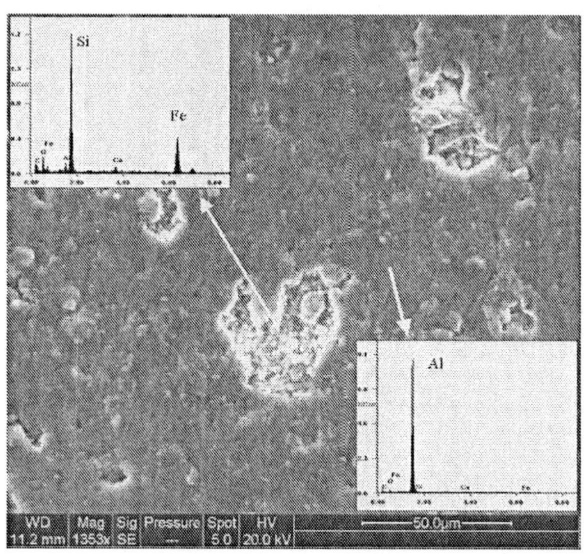

FIGURE 8 SEM picture and EDAX spectra of the Al-10 vol.% SiC composite sintered at 600°C in an inert atmosphere for 1 hour

is contaminated with Fe that was generated during milling as a product of wear of steel balls and liner. Interestingly, the contamination due to Fe was not evident from the X-ray diffraction data.

CONCLUSIONS

Silicon carbide has been synthesized in a specially built dual drive planetary mill using the concept of reaction milling. The mill was operated at 62% critical speed. Two different sized balls, 6 and 12 mm diameter, were tested. It was found that 12 mm balls enhanced intensity of impact which was evident from the quality of powder and lattice strain rates attained during processing. However, particle size distribution was found to be coarser when the milling was carried out using 12 mm balls.

At room temperature, under dry conditions, using 12 mm balls, elemental silicon and graphite powders were milled for 40 hours. As the milling progressed, powder particles became finer and assumed to contain more defects. These powders in their activated state formed the basis for the synthesis of crystalline SiC at ambient temperature. At the end of 40 hours of milling, SiC of crystallite size of 10 nm was formed and the powder contained 20% less than one micron. Excessive amount of agglomeration of particle was evident. Unlike the commercially available planetary mills where it takes 300 hrs to obtain SiC, the dual drive mill produces the same in about 40 hours. The composites made out of SiC and Al show acceptable compatibility at the interface. Several steps are being taken to improve the quality of dispersion by eliminating agglomeration and contamination.

Finally, we believe among different mechano-chemical synthesizing techniques, reaction milling process–which combines mechanical activation and chemical reaction–appears to be the most interesting technique for its capability to obtain nanocrystalline powders that can be used as filler materials. A host of materials being synthesized using the concept of reaction milling to investigate the mechanisms of the process at a very fundamental level. Several ingenious methods are being explored to control the mechano-chemical reaction for controlling the size of the particles. There are several other issues that need to be tackled; one of the major issues is powder contamination, which is not discussed here (see Koch, 2003).

REFERENCES

Cannon, W.R., Danforth, S.C., Flint, J.H., Haggerty, J.S., and Marra, R.A., 1982, "Sinterable ceramic powders from laser-driven reactions - 1. Process description and modeling." *J. Am. Ceramic Soc.*, 65, 324–330.

Cannon, W.R., Danforth, S.C., Haggerty, J.S., and Marra, R.A., 1982, "Sinterable ceramic powders from laser-driven reactions - 2. Powder characteristics and process variables," *J. Am. Ceramic Soc.*, 65, 330–335.

Courtney, T.H., 1995, "Process modeling of mechanical alloying (overview)," *Mater. Trans. JIM*, 36, 110.

Cleary, P.W., Hoyer, D., 2000, "Centrifugal mill charge motion and power draw: comparison of DEM predictions with experiments," *International Journal of Mineral Processing*, 59, 131–148.

El-Eskandarany, M.S., Sumiyama, K., Suzuki, K., 1995, "Mechanical solid state reaction for synthesis of -SiC powders," *Journal of Materials Research*, 10: (3), 659–667.

El-Eskandarany, M.S., 1996, "Synthesis of nanocrystalline titanium carbide alloy powders by mechanical solid state reaction," *Metall. Trans.* A, 27, 2374–2382.

Joardar, J., Pabi, S.K., and Murty, B.S., 2004, "Estimation of entrapped powder temperature during mechanical alloying" *Scripta Materialia*, 50(9), 1199–1202.

Koch, C.C., 2003, Top-down synthesis of nanostructured materials: Mechanical and thermal processing methods, *Rev. Adv. Mater. Sci.*, 5, 91–99.

Le Caer, G., Bauer-Grosse, E., Pianelli, A., Bouzy, E., and Matteazzi, P., 1990, "Mechanically driven syntheses of carbides and silicides," *J. Mater. Sci.*, 25, 4726–4731.

Matteazzi, P., and Caer, G.L., 1991, "Room temperature mechano-synthesis of carbides by grinding of elemental powder," *Journal of American Ceramic Society*, 74 (6), 1382–1390.

McCormick, P.G., and Froes, F.H., 1998, "The fundamentals of mechanochemical processing." 1998, Journal of Metals, 50 (11), 61–65.

Mishra, B.K., "Ball charge dynamics in a planetary mill." 1995, *KONA* No. 13, 151–158.

Murthy, G.S., Johnston, D., and Singh, V. 2004, "Comparison of dry and wet milling degerm and defiber processes for ethanol production," *Proceedings of the Corn Utilization and Technology Conference*. U.S. Dept. of Agriculture, Agricultural Research Service.

Rajamani, R.K., Milin, L., Howell, G., 2000, "Dual drive planetary mill" United States Patent no. 6,086,242.

Ritter, J.J., 1995, in Better Ceramics through Chemistry II, Materials Research Society Symp. Proc., Vol. 73, eds. Brinker, C.J., Clark, D.E., and Ulrich, D.R, *Materials Research Society*, Pittsburgh, PA, p. 367.

Schwarz, R.B., and Petrich, R.R., 1988, "Calorimetry study of the synthesis of amorphous Ni-Ti alloys by mechanical alloying," *Journal of Less-common Metals*, 140, 171–184.

Schwarz, R.B., Petrich, R.R., and Saw, C.K., 1985, "The synthesis of amorphous Ni–Ti alloy powders by mechanical alloying," *Journal of Non-Crystalline Solids*, 76 (2–3), 281–302.

Vogt, G.J., Vigil, R.S., Newkirk, L.R., and Tukula, M., 1985, Proc. of the 7th International Symp. on plasma Chemistry, Vol. 2, ed. Timmermans, C.J, Eindhoven, H, *International Union of Pure and Applied Chemistry*, p. 668.

Yang, Z.G., and Shaw, L.L., 1996, "Synthesis of nanocrystalline SiC at ambient temperature through high energy reaction milling." *Nanostructured Materials*, Vol. 7, No. 8, 873–886.

Nanocrystalline Coatings for Composite Copper Electrowinning Anodes

Michael S. Moats,[*] Carl W. Brown Jr.,[†] and Kenneth L. Hardee[†]

Lead anodes cause product contamination and increased power consumption in electrowinning systems. Several attempts have been made to develop a replacement, but they have failed principally due to lead's low cost and robustness. Recently, a composite anode has been developed that reduces power consumption and improves product quality. This composite structure, known as a Mesh on Lead[‡] Anode, combines a lead alloy plate with the electrocatalytic properties of a nanocrystalline precious metal oxide coating on titanium mesh. The performance of these anodes during extensive field testing, including a three year demonstration at a commercial copper electrowinning tankhouse, will be discussed.

BACKGROUND

Lead anodes have been used in metal electrowinning systems since their early development. The strength, robustness and corrosion resistance of lead anodes have been improved over time through alloying, cold rolling and occasionally surface pretreatment. Even so, lead based anodes cause product contamination and high power consumption. Cathode contamination occurs through mechanical inclusion of anode corrosion products and/or electrodeposition of soluble lead ionic species. High power consumption results from oxygen evolution occurring primarily on PbO_2 (exact oxide species is highly complex) which has a high overpotential. Lead anodes when placed in sulfuric acid based electrolytes corrode and quickly form a surface film of $PbSO_4$. When a sufficiently high potential is applied, $PbSO_4$ converts to PbO_2, which will then support

* University of Utah, Salt Lake City, Utah

† ELTECH Systems Corporation, Fairport Harbor, Ohio

‡ Mesh on Lead is a trademark of ELTECH Systems Corporation

the anodic reaction of water decomposition/oxygen evolution. The overpotential for oxygen evolution on PbO_2 is quite high (ca. 400–600 mV), resulting in excessive power consumption for electrowinning systems with lead anodes. The $Pb/PbSO_4/PbO_2$ interface is also not stable. With time there is a thinning of the lead anode that increases the power consumption due to a larger electrode gap, while the lost material accumulates as sludge in the bottom of the tank. Flakes of material from the anode lead become attached to the cathode causing dendrite formation, which can cause short circuiting and a loss of current efficiency.

Because of these disadvantages, numerous attempts have been made to replace lead anodes with other materials. Specifically in copper electrowinning, attempts were made in the 1970s and 1980s to implement various forms of dimensionally stable anodes based on titanium substrates. None of these were successful due to high costs, short lifetimes and/or the robustness of the design.

In the late 1990s, the concept of a composite lead/titanium/catalytic coating anode was introduced (Hardee and Brown 1999). This design became known as a Mesh on Lead (MOL™) anode. It incorporates the current carrying capability, the robustness and low cost of a lead substrate, the electrocatalytic properties of a precious metal oxide coating and the dimensional stability of titanium for coating attachment. These anodes have been commercially demonstrated in numerous tankhouses. Their performance has been reported (Moats, Hardee and Brown 2003a; Moats, Hardee and Brown 2003b; Hardee and Moats 2000) as achieving a reduced power consumption of 11–17%, the production of LME Grade copper in cobalt-free electrolyte, an improved current efficiency and minimal lead sludge generation.

This paper delves into three important characteristics of these anodes that have yet to be discussed. These are the nanocrystallinity of the electrocatalytic coating, the mesh attachment and long term life.

ELECTROCATALYTIC COATING

Coating Composition

The development of an economic, power saving electrocatalyst that can operate in a commercial electrowinning system was the largest problem in the MOL design. It has been previously reported that to maximize the economic benefits, the coating needed to significantly reduce the oxygen evolution overpotential compared with Pb/PbO_2. This was achieved by the development of a different class of precious metal oxide coatings (Hardee, Ernes and Brown 2002; Hardee, Ernes and Brown 2004). Typically RuO_2 catalysts (RuO_2-TiO_2) are used for chlorine evolution and IrO_2 catalysts (IrO_2-Ta_2O_5) are used for oxygen evolution. Unfortunately, IrO_2 catalysts are more expensive and do not reduce the oxygen evolution overpotential as much as RuO_2. The historic problem with using RuO_2 during oxygen evolution is its dissolution (electrochemical wear). It was discovered that RuO_2 could be stabilized by the introduction of another precious metal such as Pd and Rh without the addition of Ti (the typical stabilizing element). Thus, RuO_2, stabilized by Pd and/or Rh, was utilized for MOL anodes and resulted in significant power savings. Further improvements were made to this type of coating resulting in a coating composition used for MOL anodes that maximizes lifetime on a per cost basis, with an adequate power savings. The precise composition is considered to be proprietary information by ELTECH.

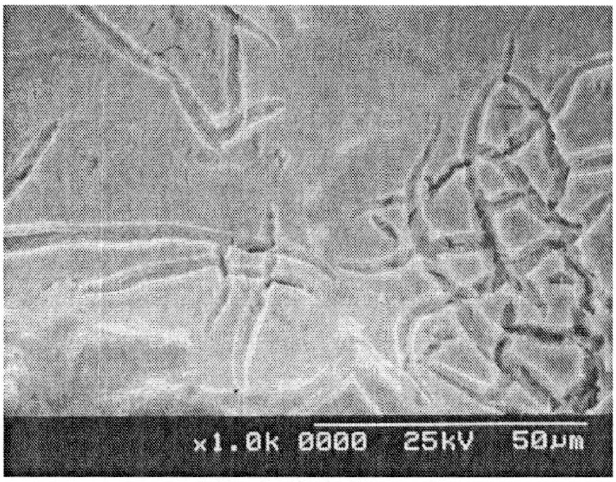

FIGURE 1 SEM surface view of coating at 1000×

Nanocrystallinity

During the development of the PGM (platinum group metal) stabilized RuO_2, physical characterization of the coatings was performed. This included using techniques such as scanning electron microscopy (SEM) and x-ray diffraction (XRD). Figure 1 shows an image (taken on a Hitachi S2300 SEM) of the surface of the catalyst at a relatively low magnification typically used for examining electrocatalytic coatings. The surface reveals some areas of mud-cracking typical of RuO_2-type coatings along with other areas that appear smooth. The catalyst surface was also examined using a High Resolution SEM (HRSEM-Hitachi S4500) and is shown in Figure 2. At very high magnifications, the previously smooth appearing areas actually have a morphology composed of nanocrystallites and nano-pores. The crystallites appear to range from about 10–30 nm in size while the pores appear to be 2 to 3 times larger. The crystallite size and the porosity can affect the electrochemical characteristics of the coating including the oxygen overpotential and the coating stability.

To further examine the observed nanostructure, x-ray diffraction patterns were taken on a coated sample with a grazing incidence beam using a Scintag X1 250 diffractometer. The XRD pattern, shown in Figure 3, revealed relatively broad RuO_2 peaks along with the peaks for Ti from the underlying substrate. The average crystallite size was calculated for the RuO_2 phase and determined to be ~15 nm. This adequately correlates with the observed nanocrystallite size in HRSEM.

ANODE IMPLEMENTATION

Since the early design concept of the mesh on lead anode, full scale demonstrations have been a large part of the development effort to determine the product's capabilities and shortcomings in industrial settings.

From the very first experiment, MOL anodes exhibited the expected power savings and cathode quality anticipated by reducing the oxygen evolution overpotential and stabilizing the lead substrate. Later, a test was conducted in an industrial research facility

FIGURE 2 High resolution SEM at 100K magnification revealing nanocrystalline particles of coating

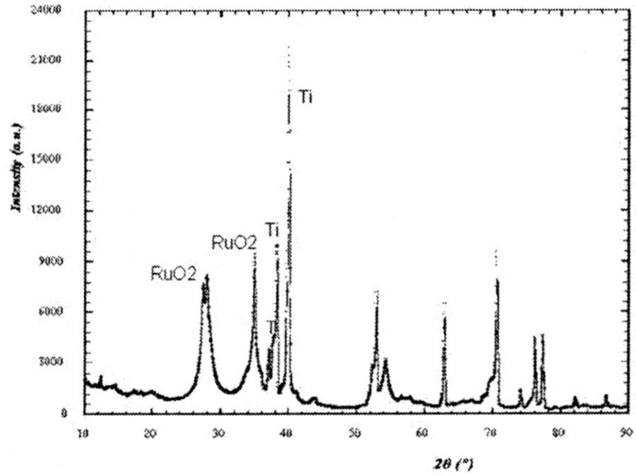

FIGURE 3 XRD pattern of nanocrystalline coating

that proved the ability of the MOL anodes to operate in cobalt-free electrolyte. (Cobalt is a costly addition to the electrolyte for stabilizing lead anodes.) The demonstrations also revealed that early versions of the product needed improvements to have an adequate lifetime in service. The most significant improvements that were needed were in the welds attaching the titanium mesh to the lead substrate and in the ability of the coating to resist current reversals thus extending its overall lifetime. Each of the initial shortcomings and the improvements made to overcome them will be discussed in more detail in the following sections.

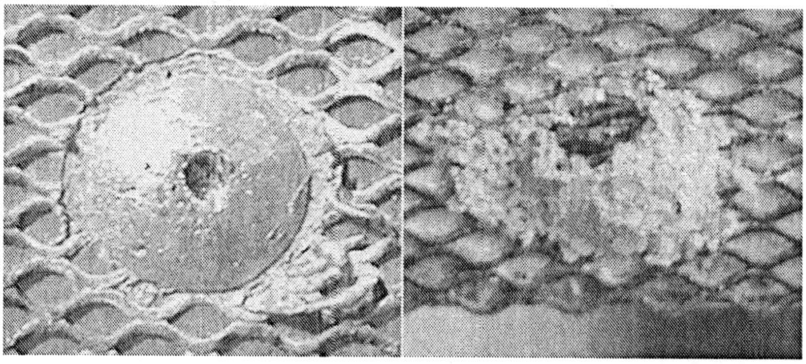

FIGURE 4 Initial pinhole in lead weld nugget (left) and resulting corrosion (right)

Mesh Attachment

The Mesh on Lead design utilizes the attachment of a coated titanium mesh to a lead substrate (Brown et al. 2000; Brown et al. 2002). This attachment could be achieved by different techniques including brazing, soldering and/or welding. The attachment material needs to be corrosion resistant, conductive and mechanically stronger than the titanium mesh. After a round of laboratory and industrial testing, it was decided that a technique that melts a nugget of lead material around the coated titanium mesh and penetrates into the lead substrate would be the most feasible method of attachment.

Following this decision, hundreds of full-scale MOL anodes were installed in commercial plants to demonstrate the technology. From these tests it was discovered that corrosion of the weld material occurred at an elevated rate, with some degrading substantially. However, due to the abundance of attachment points, there was never an instance of de-attachment of the titanium mesh from the lead substrate. It was felt, though, that improvements were needed to insure a long anode life.

Microscopic examination of the weld nuggets indicated that the corrosion appeared to be initiating from residual "pin-holes" left after the attachment process. An example of the "pin-holes" and the resulting corrosion are shown in Figure 4. Improvement was thus focused on reducing/eliminating the formation of these "pin-holes." To achieve this improvement, a simple design of experiments was conducted in which the weld material and three other process parameters were evaluated as factors. The results were very conclusive and changes to the weld material and process parameters were made. These changes resulted in the near elimination of "pin-holes" and during a recent three-year test of MOL anodes in commercial cells, less than 0.5% of the welds (<30 weld locations) have experienced degradation in service out of over 6,000 weld points utilized for the full compliment of anodes within a cell.

Current Reversal

During the initial demonstrations of MOL anodes, it was necessary to prove that the lead substrate was stable at the lower operating potential imposed by the electrocatalytic coating. This was shown by taking the cells offline every three to four months to drain the cell and check for accumulated lead sludge. Simultaneous cleaning of the MOL cells and adjacent cells containing standard Pb-Ca-Sn anodes provided a direct comparison on the amount of lead sludge generated. In all cases, the amount of lead sludge found at the bottom of the MOL cells was less than approximately one kilogram compared with 20–50 kilograms of material from the lead anode cells after only 3–4 months of operation.

An unfortunate and unintended consequence of these cell cleanings, in almost all cases, was a reversal of current when the cells were shorted from the system to allow operator access. The current reversal is not the result of having MOL anodes in the electrowinning circuit, but from the lead anodes in the adjacent cells still in electrical contact with the MOL anodes. To short the cells to be cleaned, a jumper frame is placed over the cells. Typically the frame bridges multiple cells resulting in spontaneous electrochemical reactions, i.e., a battery effect. In the case of lead anodes, this allows lead oxide to convert back to lead sulfate and copper to dissolve from the cathodes, which is analogous to a lead-acid battery. During a simulated acid cleaning event in the laboratory, an instantaneous reverse current flow of up to 350 A/m^2 was measured. This is equivalent to 30–40,000 amps of reverse current in a full size electrowinning cell.

Unlike lead anodes, the MOL anodes do not have PbO_2 to convert back to $PbSO_4$ to support the current reversal. Other reactions such as copper plating on the electrocatalytic mesh of the anode structure and reduction reactions within the electrocatalytic coating are driven by the current coming from the adjacent lead anodes in the jumpered cells. The reduction reactions within the electrocatalytic coating are detrimental to the long term stability of the coating. Hence with each current reversal the lifetime of the electrocatalyst was shortened. To minimize the effect of the current reversal, protocols were established for attaching the jumper switch (Hardee et al. 2003) and a coating was developed that is more tolerant to current reversal.

Lifetime

With the improvements mentioned in the previous sections, the latest version of MOL anodes was installed in a commercial copper electrowinning cell during October of 2002. These anodes have been in continuous commercial operation for approximately three years at the time this paper was written. Over this time, the anodes have demonstrated an average cell voltage savings of 208 mV compared to seven cells with lead anodes used as reference. This translates to an average power savings of 10.4 %.

SUMMARY

Mesh on Lead anodes consist of standard Pb-Sn-Ca anodes coupled with an electrocatalytic coating on a titanium mesh. The development of a corrosion resistant attachment method and a nanocrystalline coating tolerant of current reversals produced MOL anodes that have been demonstrated to be robust and capable of long term service in a copper electrowinning environment. These anodes have been in service for approximately 3 years in a commercial copper electrowinning facility. The stabilized ruthenium oxide coating has resulted in an average voltage savings of >200 mV compared with standard lead alloy anodes.

REFERENCES

Brown, C.W., J.I. Bishara, L.M. Ernes, A.W. Getsy, K.L. Hardee, B.L. Martin, G.R. Pohto, T.J. Schue and T.R. Turk. 2000. US patent 6,139,705 (31 October 2000).

Brown, C.W., J.I. Bishara, L.M. Ernes, A.W. Getsy, K.L. Hardee, B.L. Martin, G.R. Pohto, T.J. Schue and T.R. Turk. 2002. US patent 6,352,622 (5 March 2002).

Hardee, K.L. and C.W. Brown. 1999. Electrocatalytic titanium mesh surfaces combined with standard lead substrates for process improvements and power savings in copper electrowinning. *Proceedings of the Copper 99–Cobre 99 International Conference–Volume III*, Eds. *J. E. Dutrizac, J. Ji, V. Ramachandran*, October 1999, 575–584.

Hardee K.L., L.M. Ernes, C.W. Brown. 2002. US patent 6,368,489 (9 April 2002).

Hardee K.L., L.M. Ernes, C.W. Brown. 2004. US patent 6,368,489 (12 October 2004).

Hardee, K.L. and M.S. Moats. 2000. Application of Titanium Mesh-on-Lead Technology to Metal Electrowinning Systems. *Electrochemistry in Mineral and Metal Processing V*. The Electrochemical Society. May 2000. 292–302.

Hardee, K.L., M.S. Moats, C.W. Brown, R.L. Wilhelm, E.M. Halko, Z.A. Wade. 2003. US patent application 20030066759 (April 10, 2003).

Moats, M.S., K.L. Hardee and C.W. Brown. 2003a. Copper electrowinning in cobalt-free electrolyte using mesh-on-lead anodes. *Proceedings of the Copper 2003–Cobre 2003–Volume V*, Eds. J.E. Dutrizac, C.G. Clement. 543–551.

Moats, M., K. Hardee and C. Brown. 2003b. Mesh-on-Lead Anodes for Copper Electrowinning. *JOM*. Vol. 55 No. 7. 46–48.